Yeast Protocols

Methods in Molecular Biology™
John M. Walker, SERIES EDITOR

60. **Protein NMR Protocols**, edited by *David G. Reid, 1996*
59. **Protein Purification Protocols**, edited by *Shawn Doonan, 1996*
58. **Basic DNA and RNA Protocols**, edited by *Adrian J. Harwood, 1996*
57. **In Vitro Mutagenesis Protocols**, edited by *Michael K. Trower, 1996*
56. **Crystallographic Methods and Protocols**, edited by *Christopher Jones, Barbara Mulloy, and Mark Sanderson, 1996*
55. **Plant Cell Electroporation and Electrofusion Protocols**, edited by *Jac A. Nickoloff, 1995*
54. **YAC Protocols**, edited by *David Markie, 1995*
53. **Yeast Protocols**: *Methods in Cell and Molecular Biology,* edited by *Ivor H. Evans, 1996*
52. **Capillary Electrophoresis**: *Principles, Instrumentation, and Applications,* edited by *Kevin D. Altria, 1996*
51. **Antibody Engineering Protocols**, edited by *Sudhir Paul, 1995*
50. **Species Diagnostics Protocols**: *PCR and Other Nucleic Acid Methods,* edited by *Justin P. Clapp, 1996*
49. **Plant Gene Transfer and Expression Protocols**, edited by *Heddwyn Jones, 1995*
48. **Animal Cell Electroporation and Electrofusion Protocols**, edited by *Jac A. Nickoloff, 1995*
47. **Electroporation Protocols for Microorganisms**, edited by *Jac A. Nickoloff, 1995*
46. **Diagnostic Bacteriology Protocols**, edited by *Jenny Howard and David M. Whitcombe, 1995*
45. **Monoclonal Antibody Protocols**, edited by *William C. Davis, 1995*
44. ***Agrobacterium* Protocols**, edited by *Kevan M. A. Gartland and Michael R. Davey, 1995*
43. **In Vitro Toxicity Testing Protocols**, edited by *Sheila O'Hare and Chris K. Atterwill, 1995*
42. **ELISA**: *Theory and Practice,* by *John R. Crowther, 1995*
41. **Signal Transduction Protocols**, edited by *David A. Kendall and Stephen J. Hill, 1995*
40. **Protein Stability and Folding**: *Theory and Practice,* edited by *Bret A. Shirley, 1995*
39. **Baculovirus Expression Protocols**, edited by *Christopher D. Richardson, 1995*
38. **Cryopreservation and Freeze-Drying Protocols**, edited by *John G. Day and Mark R. McLellan, 1995*
37. **In Vitro Transcription and Translation Protocols**, edited by *Martin J. Tymms, 1995*
36. **Peptide Analysis Protocols**, edited by *Ben M. Dunn and Michael W. Pennington, 1994*
35. **Peptide Synthesis Protocols**, edited by *Michael W. Pennington and Ben M. Dunn, 1994*
34. **Immunocytochemical Methods and Protocols**, edited by *Lorette C. Javois, 1994*
33. **In Situ Hybridization Protocols**, edited by *K. H. Andy Choo, 1994*
32. **Basic Protein and Peptide Protocols**, edited by *John M. Walker, 1994*
31. **Protocols for Gene Analysis**, edited by *Adrian J. Harwood, 1994*
30. **DNA–Protein Interactions**, edited by *G. Geoff Kneale, 1994*
29. **Chromosome Analysis Protocols**, edited by *John R. Gosden, 1994*
28. **Protocols for Nucleic Acid Analysis by Nonradioactive Probes**, edited by *Peter G. Isaac, 1994*
27. **Biomembrane Protocols**: *II. Architecture and Function,* edited by *John M. Graham and Joan A. Higgins, 1994*
26. **Protocols for Oligonucleotide Conjugates**: *Synthesis and Analytical Techniques,* edited by *Sudhir Agrawal, 1994*
25. **Computer Analysis of Sequence Data**: *Part II,* edited by *Annette M. Griffin and Hugh G. Griffin, 1994*
24. **Computer Analysis of Sequence Data**: *Part I,* edited by *Annette M. Griffin and Hugh G. Griffin, 1994*
23. **DNA Sequencing Protocols**, edited by *Hugh G. Griffin and Annette M. Griffin, 1993*
22. **Microscopy, Optical Spectroscopy, and Macroscopic Techniques**, edited by *Christopher Jones, Barbara Mulloy, and Adrian H. Thomas, 1993*
21. **Protocols in Molecular Parasitology**, edited by *John E. Hyde, 1993*
20. **Protocols for Oligonucleotides and Analogs**: *Synthesis and Properties,* edited by *Sudhir Agrawal, 1993*
19. **Biomembrane Protocols**: *I. Isolation and Analysis,* edited by *John M. Graham and Joan A. Higgins, 1993*
18. **Transgenesis Techniques**: *Principles and Protocols,* edited by *David Murphy and David A. Carter, 1993*
17. **Spectroscopic Methods and Analyses**: *NMR, Mass Spectrometry, and Metalloprotein Techniques,* edited by *Christopher Jones, Barbara Mulloy, and Adrian H. Thomas, 1993*
16. **Enzymes of Molecular Biology**, edited by *Michael M. Burrell, 1993*
15. **PCR Protocols**: *Current Methods and Applications,* edited by *Bruce A. White, 1993*
14. **Glycoprotein Analysis in Biomedicine**, edited by *Elizabeth F. Hounsell, 1993*
13. **Protocols in Molecular Neurobiology**, edited by *Alan Longstaff and Patricia Revest, 1992*
12. **Pulsed-Field Gel Electrophoresis**: *Protocols, Methods, and Theories,* edited by *Margit Burmeister and Levy Ulanovsky, 1992*
11. **Practical Protein Chromatography**, edited by *Andrew Kenney and Susan Fowell, 1992*
10. **Immunochemical Protocols**, edited by *Margaret M. Manson, 1992*
9. **Protocols in Human Molecular Genetics**, edited by *Christopher G. Mathew, 1991*
8. **Practical Molecular Virology**: *Viral Vectors for Gene Expression,* edited by *Mary K. L. Collins, 1991*
7. **Gene Transfer and Expression Protocols**, edited by *Edward J. Murray, 1991*
6. **Plant Cell and Tissue Culture**, edited by *Jeffrey W. Pollard and John M. Walker, 1990*
5. **Animal Cell Culture**, edited by *Jeffrey W. Pollard and John M. Walker, 1990*

Methods in Molecular Biology™ • 53

Yeast Protocols

Methods in Cell and Molecular Biology

Edited by

Ivor H. Evans

*School of Chemical and Life Sciences,
University of Greenwich, London, UK*

Humana Press Totowa, New Jersey

© 1996 Humana Press Inc.
999 Riverview Drive, Suite 208
Totowa, New Jersey 07512

For additional copies, pricing for bulk purchases, and/or other information about other Humana titles, contact Humana at the above address or at any of the following numbers: Tel.: 201-256-1699; Fax: 201-256-8341; E-mail: humana@interramp.com

All rights reserved. No part of this book may be reproduced, stored in a retrieval system, or transmitted in any form or by any means, electronic, mechanical, photocopying, microfilming, recording, or otherwise without written permission from the Publisher. Methods in molecular biology™ is a trademark of the Humana Press Inc.

All authored papers, comments, opinions, conclusions, or recommendations are those of the author(s), and do not necessarily reflect the views of the publisher.

This publication is printed on acid-free paper. ∞
ANSI Z39.48-1984 (American National Standards Institute) Permanence of Paper for Printed Library Materials.

Cover illustration: From Figs. 1 and 2 in Chapter 39, "Fluorescence Microscopy Methods," by Jirí Hasek and Eva Streiblová.

Photocopy Authorization Policy:
Authorization to photocopy items for internal or personal use, or the internal or personal use of specific clients, is granted by Humana Press Inc., **provided** that the base fee of US $5.00 per copy, plus US $00.25 per page, is paid directly to the Copyright Clearance Center at 222 Rosewood Drive, Danvers, MA 01923. For those organizations that have been granted a photocopy license from the CCC, a separate system of payment has been arranged and is acceptable to Humana Press Inc. The fee code for users of the Transactional Reporting Service is: [0-89603-319-8/96 $5.00 + $00.25].

Printed in the United States of America. 10 9 8 7 6 5 4 3 2 1

Library of Congress Cataloging in Publication Data

Main entry under title:
Methods in molecular biology™.

Yeast protocols ; methods in cell and molecular biology / edited by
  Ivor H. Evans.
    p. cm. — (Methods in molecular biology™ ; 53)
  Includes index.
  ISBN 0-89603-319-8 (alk. paper)
    1. Yeast—Biotechnology. 2. Yeast fungi—Research—Methodology.
  I. Evans, Ivor H. (Ivor Howell) II. Series: Methods in molecular
  biology™ (Totowa, NJ) ; 53.
    [DNLM: 1. Yeasts —physiology. 2. Yeasts—isolation and
  purification. 3. Yeasts—cytology. W1 ME9616J v. 53 1996 / QW
  180.5.Y3 Y415 1996]
  TP248.27.Y43Y416 1996
  589.2'330487—dc20
  DNLM/DLC
  for Library of Congress 95-43900
                                                    CIP

Preface

Yeast Protocols intends to offer a selection of well-proven protocols in cell and molecular biology, applicable to yeasts including, but certainly not exclusively aimed at, *Saccharomyces cerevisiae*. *Saccharomyces cerevisiae* and its very distant cousin, *Schizosaccharomyces pombe,* are of course now foremost model eukaryotic organisms, and the focus of wide-ranging experimental studies, especially those using molecular genetic techniques. Many of the latter, such as DNA sequencing and in vitro mutagenesis, are general DNA techniques, well covered in other volumes of Humana's *Methods in Molecular Biology* series, and elsewhere. The inclusion of a number of non-DNA techniques in this book is meant to reflect the resurgent interest in yeast cell biology sparked by the development of gene manipulation methods—for example, cellular localization of cloned gene products using microscopical techniques.

The presentation of protocols follows the successful *Methods in Molecular Biology* series format, with a clear sequence of steps and extensive troubleshooting notes. It is our hope that these protocols will be useful not only to established members of the full-time research community, but also to the less experienced—first degree level and masters students undertaking project work, as well as PhD students starting their experimental programs; I am well aware that these young apprentice scientists are not always able to receive the supervision time that they, and indeed their supervisors, would like.

Yeast Protocols would also be succeeding in a particularly valuable objective if it persuaded yeast workers to try a novel experimental approach or line of inquiry they might not otherwise have considered—perhaps because it had seemed just too difficult or remote from their previous experience. Another of the book's objectives is to encourage the application of techniques—many initially fine-tuned for *Saccharomyces cerevisiae*—to other interesting representatives of the yeast world.

I would like to extend hearty thanks to all the authors who have generously made the time available for, and had the commitment to contribute to, *Yeast Protocols*. The pressures of university and research life are now such that many scientists are, understandably, increasingly reluctant to involve themselves in the writing of books; thus, our contributors' efforts are all the more appreciated. As editor, I would also like to thank the series editor, John Walker, and the staff of Humana Press for their support and encouragement.

Finally, I would like to express my great sorrow on hearing of the recent death of Dorothy Spencer. Dorothy died earlier this year after a typically courageous struggle against breast cancer, which had metastasized to the liver. Many in the yeast world will recall with affection the humorous and lively presences of Dorothy and her husband, Frank, at yeast meetings great and small, as well as their numerous contributions to yeast biology: It is very sad to record the untimely breaking of that unique partnership.

Ivor H. Evans

Contents

Preface v

Contributors xi

Ch. 1. Isolation and Identification of Yeasts from Natural Habitats,
  John F. T. Spencer and Dorothy M. Spencer 1
Ch. 2. Maintenance and Culture of Yeasts,
  John F. T. Spencer and Dorothy M. Spencer 5
Ch. 3. Mutagenesis in Yeast,
  John F. T. Spencer and Dorothy M. Spencer 17
Ch. 4. Rare-Mating and Cytoduction in *Saccharomyces cerevisiae*,
  John F. T. Spencer and Dorothy M. Spencer 39
Ch. 5. Protoplast Fusion in *Saccharomyces cerevisiae*,
  Brendan P. G. Curran and Virginia C. Bugeja 45
Ch. 6. Meiotic Analysis,
  John F. T. Spencer and Dorothy M. Spencer 51
Ch. 7. Ascus Dissection,
  Carl Saunders-Singer 59
Ch. 8. Karyotyping of Yeast Strains by Pulsed-Field Gel Electrophoresis,
  John R. Johnston 69
Ch. 9. Chromosomal Localization of Genes Through Pulsed-Field Gel
  Electrophoresis Techniques,
  Graham R. Bignell and Ivor H. Evans 79
Ch. 10. Isolation, Purification, and Analysis of Nuclear DNA in Yeast
  Taxonomy,
  Ann Vaughan-Martini and Alessandro Martini 89
Ch. 11. Isolation of Yeast DNA,
  Peter Piper 103
Ch. 12. Isolation of Mitochondrial DNA,
  Graham R. Bignell, Angela R. M. Miller, and Ivor H. Evans 109
Ch. 13. Isolation of Yeast Plasma Membranes,
  Barry Panaretou and Peter Piper 117
Ch. 14. The Extraction and Analysis of Sterols from Yeast,
  Michael A. Quail and Steven L. Kelly 123
Ch. 15. Peroxisome Isolation,
  Ben Distel, Inge van der Leij, and Wilko Kos 133

CH. 16. Transformation of Lithium-Treated Yeast Cells and the Selection
of Auxotrophic and Dominant Markers,
*Robert C. Mount, Bernadette E. Jordan,
and Christopher Hadfield* 139
CH. 17. Biolistic Transformation of Yeasts,
Stephen A. Johnston and Michael J. DeVit 147
CH. 18. Genomic Yeast DNA Clone Banks: *Construction
and Gene Isolation,*
Graham R. Bignell and Ivor H. Evans 155
CH. 19. Yeast Colony Hybridization,
Matthew John Kleinman 189
CH. 20. Determination of Plasmid Copy Number in Yeast,
*Bernadette E. Jordan, Robert C. Mount,
and Christopher Hadfield* 193
CH. 21. Determination of Chromosome Ploidy in Yeast,
*Christopher Hadfield, Bernadette E. Jordan,
and Robert C. Mount* 205
CH. 22. Chromosome Engineering in Yeast with a Site-Specific
Recombination System from a Heterologous Yeast Plasmid,
Yasuji Oshima, Hiroyuki Araki, and Hiroaki Matsuzaki 217
CH. 23. Insertional Mutagenesis by Ty Elements
in *Saccharomyces cerevisiae,*
David J. Garfinkel 227
CH. 24. Reporter Gene Systems for Assaying Gene Expression in Yeast,
*Robert C. Mount, Bernadette E. Jordan,
and Christopher Hadfield* 239
CH. 25. Preparation and Use of Yeast Cell-Free Translation Lysate,
*Alan D. Hartley, Manuel A. S. Santos, David R. Colthurst,
and Michael F. Tuite* 249
CH. 26. Electrophoretic Analysis of Yeast Proteins,
Christopher M. Grant, Ian T. Fitch, and Michael F. Tuite 259
CH. 27. Preparation of Total RNA,
Alistair J. P. Brown 269
CH. 28. mRNA Abundance and Half-Life Measurements,
Alistair J. P. Brown and Francis A. Sagliocco 277
CH. 29. Polysome Analysis,
*Francis A. Sagliocco, Paul A. Moore,
and Alistair J. P. Brown* 297
CH. 30. Induction of Heat Shock Proteins and Thermotolerance,
Peter Piper 313
CH. 31. Rhodamine B Assay for Estimating Activity of Killer Toxins
Permeabilizing Cytoplasmic Membranes,
Vladimír Vondrejs and Zdena Palková 319

Contents

CH. 32. Nystatin-Rhodamine B Assay for Estimating Activity
of Killer Toxin from *Kluyveromyces lactis*,
Zdena Palková and Vladimír Vondrejs 325

CH. 33. Application of Killer Toxins in Stepwise Selection of Hybrids
and Cybrids Obtained by Induced Protoplast Fusion,
Vladimír Vondrejs, Zdena Palková, and Zuzana Zemanová ... 331

CH. 34. Killer Plaque Technique for Selecting Hybrids and Cybrids
Obtained by Induced Protoplast Fusion,
Zdena Palková and Vladimír Vondrejs 339

CH. 35. Genotoxicity Testing in *Schizosaccharomyces pombe*,
Pamela McAthey 343

CH. 36. Purification and Quantification of *Saccharomyces cerevisiae*
Cytochrome P450,
Ian Stansfield and Steven L. Kelly 355

CH. 37. Calorimetry of Whole Yeast Cells,
Linda J. Ashby and Anthony E. Beezer 367

CH. 38. Light Microscopy Methods,
Eva Streiblová and Jiří Hašek 383

CH. 39. Fluorescence Microscopy Methods,
Jiří Hašek and Eva Streiblová 391

CH. 40. Immunoelectron Microscopy,
Evert-Jan van Tuinen 407

Index 423

Contributors

HIROYUKI ARAKI • *Research Institute for Microbial Diseases, Osaka University, Osaka, Japan*

LINDA J. ASHBY • *Chemical Laboratory, The University, Canterbury, UK*

ANTHONY E. BEEZER • *Chemical Laboratory, The University, Canterbury, UK*

GRAHAM R. BIGNELL • *School of Biological and Chemical Sciences, University of Greenwich, London, UK; Current address: Section of Molecular Carcinogenesis, Institute of Cancer Research, Surrey, UK*

ALISTAIR J. P. BROWN • *Department of Molecular and Cell Biology, Marischal College, University of Aberdeen, Scotland*

VIRGINIA C. BUGEJA • *School of Natural Sciences, University of Hertfordshire, Hatfield, UK*

DAVID R. COLTHURST • *Biological Laboratory, University of Kent, Canterbury, UK*

BRENDAN P. G. CURRAN • *School of Biological Sciences, Queen Mary and Westfield College, London, UK*

MICHAEL J. DEVIT • *Department of Genetics, School of Medicine, Washington University, St. Louis, MO*

BEN DISTEL • *E. C. Slater Institute for Biochemical Research, University of Amsterdam, The Netherlands*

IVOR H. EVANS • *School of Chemical and Life Sciences, University of Greenwich, London, UK*

IAN T. FITCH • *Biological Laboratory, University of Kent, Canterbury, UK*

DAVID J. GARFINKEL • *NCI-Frederick Cancer Research and Development Center, ABL-Basic Research Program, Laboratory of Eukaryotic Gene Expression, Frederick, MD*

CHRISTOPHER M. GRANT • *Biological Laboratory, University of Kent, Canterbury, UK*

CHRISTOPHER HADFIELD • *Department of Genetics, University of Leicester, UK*

ALAN D. HARTLEY • *Biological Laboratory, University of Kent, Canterbury, UK*

JIŘÍ HAŠEK • *Laboratory of Cell Reproduction, Institute of Microbiology Academy of Sciences of the Czech Republic, Praha, Czech Republic*

JOHN R. JOHNSTON • *Department of Bioscience and Biotechnology, University of Stratclyde, Glasgow, Scotland*
STEPHEN A. JOHNSTON • *Southwestern Medical Center, University of Texas, Dallas, TX*
BERNADETTE E. JORDAN • *Department of Genetics, University of Leicester, UK*
STEVEN L. KELLY • *Krebs Institute for Biomolecular Research, University of Sheffield, UK*
MATTHEW JOHN KLEINMAN • *School of Biological Sciences, Birkbeck College, University of London, UK*
WILKO KOS • *E. C. Slater Institute for Biochemical Research, University of Amsterdam, The Netherlands*
ALESSANDRO MARTINI • *Dipartimento di Biologia Vegetale, Microbiologia Applicata, Universita' Degli Studi di Perugia, Italy*
HIROAKI MATSUZAKI • *Department of Biotechnology, Osaka University, Osaka, Japan*
PAMELA MCATHEY • *School of Life Sciences, The University of North London, London, UK*
ANGELA R. M. MILLER • *School of Biological and Chemical Sciences, University of Greenwich, London, UK*
PAUL A. MOORE • *Department of Molecular and Cell Biology, Marischal College, University of Aberdeen, UK*
ROBERT C. MOUNT • *Department of Genetics, University of Leicester, UK*
YASUJI OSHIMA • *Department of Biotechnology, Osaka University, Osaka, Japan*
ZDENA PALKOVÁ • *Department of Genetics and Microbiology, Charles University, Prague, Czech Republic*
BARRY PANARETOU • *MRC Laboratory of Molecular and Cellular Biology, University College London, UK*
PETER PIPER • *Department of Biochemistry and Molecular Biology, University College London, UK*
MICHAEL A. QUAIL • *Krebs Institute for Biomolecular Research, Department of Molecular Biology and Biotechnology, The University of Sheffield, UK*
FRANCIS A. SAGLIOCCO • *Laboratoire de Genetique, Unité Associée CNRS, Université de Bordeaux II, Talence, France*
MANUEL A. S. SANTOS • *Biological Laboratory, University of Kent, Canterbury, UK*
CARL SAUNDERS-SINGER • *Singer Instrument Co. Ltd., Watchet, UK*
DOROTHY M. SPENCER • *PROIMI, Genetics Division, San Miguel de Tucumán, Argentina*

Contributors

JOHN F. T. SPENCER • *PROIMI, Genetics Division, San Miguel de Tucumán, Argentina*
IAN STANSFIELD • *Biological Laboratory, University of Kent, Canterbury, UK*
EVA STREIBLOVÁ • *Laboratory of Cell Reproduction, Institute of Microbiology Academy of Sciences of the Czech Republic, Praha, Czech Republic*
MICHAEL F. TUITE • *Biological Laboratory, University of Kent, Canterbury, UK*
INGE VAN DER LEIJ • *E. C. Slater Institute for Biochemical Research, University of Amsterdam, The Netherlands*
EVERT-JAN VAN TUINEN • *Department of Molecular Cell Biology and Biotechnology, University of Amsterdam, The Netherlands*
ANN VAUGHAN-MARTINI • *Dipartimento di Biologia Vegetale, Microbiologia Applicata, Universita' Degli Studi di Perugia, Italy*
VLADIMÍR VONDREJS • *Department of Genetics and Microbiology, Charles University, Prague, Czech Republic*
ZUZANA ZEMANOVÁ • *Department of Genetics and Microbiology, Charles University, Prague, Czech Republic*

Chapter 1

Isolation and Identification of Yeasts from Natural Habitats

John F. T. Spencer and Dorothy M. Spencer

1. Introduction

Yeasts are relatively easy to isolate from natural habitats, the technique being to weigh out a sample of the substrate, suspend it in sterile water, and streak or spread a suitable dilution on either a nonselective medium (usually acidified malt extract agar to suppress bacterial growth) to obtain specimens of the total population, or on a selective medium to obtain cultures of particular species *(1,2)*. After the plates have been incubated at an appropriate temperature and colonies have developed, these are examined under a dissecting microscope to select at least one of each colony type. Unfortunately, many yeasts cannot be distinguished immediately by visual examination of either the vegetative cells or the colonial morphology, so that it is better to isolate as many colonies as can be handled, for identification.

In samples containing a low density of the yeast population, such as natural water, it may be necessary to concentrate the samples rather than dilute them, by passing the sample (usually 10–50 mL) through a sterile membrane filter, and then placing the filter on a plate of the isolation medium. Colonies are usually small and discrete and can be picked easily.

2. Materials

1. Acidified malt extract medium: Autoclave the malt extract agar, and let cool to about 60°C. Add approx 0.7 mL/100 mL of medium, of 1N HCl, to a final pH of 3.7–3.8, mix thoroughly, and pour plates *(1)*.

2. Yeast-nitrogen base, 0.67% (all percentages are w/v, unless otherwise stated), plus various selective carbon sources, 1% unless otherwise stated *(2)*.
   a. For *Cryptococcus* species, especially *Cryptococcus cereanus* from rotting cactus, *meso*-inositol.
   b. For *Candida sonorensis*, 0.5% methanol.
   c. For *Pichia cactophila, Pichia pseudocactophila,* and *Pichia norvegensis*, glucosamine.
   d. For *Pichia mexicana*, a species of *Clavispora, Debaryomyces hansenii*, and some species of *Cryptococcus, N*-acetylglucosamine.
3. Yeast-carbon base, plus different nitrogen sources *(2)*. Nitrate, to select for species of *Hansenula* (now, nitrate-positive species of *Pichia, Citeromyces,* and some species of *Candida*). This method can be used to detect the presence of *Candida utilis* in commercial baker's yeast. Other selective nitrogen sources include lysine, creatinine, ethylamine, and nitrite.
4. Millipore or other membrane filter apparatus, sterile filters, and flat tweezers for handling filters.
5. Dissecting microscope for examining colonies.
6. Other standard equipment for microbiological investigation.
7. Collecting vessels and tools. These can be sterile jars, flasks, or tubes for liquid samples, or small plastic bags for solids. (Whirlpack bags are suitable.) Collecting tools may vary according to the nature of the samples, and may be a spoon or knife, alcohol, and swabs for resterilizing the collecting tool, some kind of reaching pole and bottle handler for sampling small, shallow bodies of water, or some type of standard collecting bottle for taking samples at depth, for deep lakes or seas.

3. Method

1. Collect specimens in sterile containers. On returning to the laboratory (unless a mobile laboratory is available), use the following procedure.
2. Weigh out samples into tubes of sterile water.
3. Streak or spread on plates of acidified malt extract agar or YM agar (Difco, Detroit, MI), pH 3.7–3.8 *(1)*, and incubate for 3–6 d at approx 25°C, or at any temperature that is appropriate for the type of yeasts expected in the sample. Examine plates at frequent intervals, as the agar may be overgrown with filamentous fungi if left too long.
4. After incubation, examine colonies under the dissecting microscope, and pick at least one of each colony type, but preferably more. Count any distinctive types (form or color) to determine their frequency in the total population. If possible, determine the total colony count.
5. Restreak these colonies on malt extract agar or other appropriate medium, determine their morphological characteristics (colony morphology,

Identification of Yeasts

vegetative cells, and spores), and maintain the cultures on any suitable medium for more detailed study.
6. Attempt to identify isolates (*see* Note 2). For example, determine cellular morphology, presence or absence of spores, and in sporulating cultures, spore and ascus morphology. Also, determine fermentation and assimilation patterns for carbon sources *(1,3,4)* and assimilation of nitrogen sources. Attempt to find the species in a recognized key *(3,4)*.

4. Notes

1. A few yeast species from specialized habitats require particular procedures for isolation and handling.
2. *Brettanomyces* species form acid in quantity and die quickly in the common media. Addition of 0.5% calcium carbonate to yeast extract-glucose-peptone agar allows longer survival. Use of $CaCO_3$ in the isolation medium allows detection of acid-forming species. Clear zones form around the colonies.
3. *Cyniclomyces guttulatus* inhabits the rabbit digestive tract, and can be isolated from rabbit feces. It requires an array of amino acids, or else yeast autolysate in the isolation medium, an incubation temperature of 37°C, and a CO_2 concentration of 15% v/v in the atmosphere. 1% Proteose peptone, 1% dehydrated yeast autolysate, and 2% glucose, pH 3.5 (adjusted with $1N$ HCl), sterilized in a flowing steam, is satisfactory. This species sporulates at about 18°C, which is outside the temperature range permitting growth, so sporulation medium (YM medium or similar) must be inoculated with a heavy inoculum of vegetative cells, to obtain spores.
4. Osmotolerant species (*Zygosaccharomyces*, *Candida mogii*, *Schizosaccharomyces octosporus*) may require elevated concentrations of sugar or NaCl in the isolation medium. However, they can usually be adapted easily to the commonly used media, such as yeast extract-peptone-dextrose agar. Yeasts occurring in saline lakes, such as *Metschnikowia bicuspidata*, when isolated from infected brine shrimp in saline lakes, may require 2% NaCl in the isolation medium.
5. Identification may present some problems for the beginner. Detailed and extensive protocols for determining the morphological and physiological characteristics, used in identification of yeast species, are given in *The Yeasts, a Taxonomic Study*, by N. J. W. Kreger-van Rij *(4)*, and these should be followed exactly. The guidance of an experienced yeast taxonomist is highly desirable.

The volumes by Barnett et al. are either older (1983) *(3)* or very expensive, and are based only on physiological criteria, but contain a great deal of useful information. A computer program is available from Barnett.

Methods of identification based on molecular genetics (DNA reassociation, characterization, of ribosomal RNA (rRNA) *(7)* and DNA (rDNA) *(5,6)* and PCR studies of particular sequences of DNA *(8)*, are extremely useful for identification and also for determination of taxonomic position, but are in the province of the specialist.

6. Formation of pseudohyphae is no longer considered significant as a taxonomic criterion, and the genus *Torulopsis* has therefore been merged with *Candida*.

References

1. Phaff, H. J. and Starmer, W. (1987) Yeasts associated with plants, insects and soil, in *The Yeasts*, vol. 1, *Biology of Yeasts* (Rose, A. H. and Harrison, J. S., eds.), Academic, London, pp. 123–180.
2. Wickerham, L. J. (1951) *Taxonomy of Yeasts*. Techn. Bull. 1029, US Department of Agriculture, Washington, DC.
3. Barnett, J. A., Payne, R. W., and Yarrow, D. (1983) *Yeasts: Characteristics and Identification,* Cambridge University Press, Cambridge, UK.
4. Kreger-van Rij, N. J. W. (1984) *The Yeasts: A Taxonomic Study,* Elsevier, Amsterdam.
5. Lachance, M.-A. (1988) Restriction mapping of rDNA and taxonomy of *Kluyveromyces* van der Walt emend. van der Walt. *Yeast* **5**(Special Issue), S379–S383.
6. Shen, R. and Lachance, M.-A. (1993) Phylogenetic study of ribosomal DNA of cactophilic *Pichia* species by restriction mapping. *Yeast* **9**, 315–330.
7. Yamada, Y., Maeda, K., Nagahama, T., Banno, I., and Lachance, M.-A. (1992) The phylogenetic relationships of the genus *Sporopachydermia* Rodriguez de Miranda (Saccharomycetaceae) based on the partial sequences of 18S and 26S ribosomal RNAs. *J. Gen. Appl. Microbiol.* **38**, 179–183.
8. Fell, J. W. (1993) Rapid identification of yeast species using three primers in a polymerase chain reaction. *Mol. Marine Biol. Biotechnol.* **2(3)**, 174–180.

CHAPTER 2

Maintenance and Culture of Yeasts

John F. T. Spencer and Dorothy M. Spencer

1. Introduction

Yeasts generally are very easy to maintain, store, cultivate in almost any size of batch, and isolate from most habitats, but if a few elementary methods are understood, their isolation, storage, and culture, for day-to-day use in the laboratory is greatly simplified. Furthermore, the same methods can be used for maintenance of a stock collection for use in a production plant in a yeast-based industry. Being a group of unicellular organisms, yeasts can be handled in the same way as most bacteria, except that their nutritional and environmental requirements are, with few exceptions, much simpler. The techniques used will be considered under the following headings:

1. For short-term maintenance, for daily use. Rich, undefined media, such as yeast extract-peptone-glucose agar, YM agar, or malt agar, are commonly used (*see* Section 3.);
2. For medium-term storage;
3. For long-term preservation, as in a culture collection, where immediate access is less important, but maintenance of the characteristics of the species and strain is the primary objective; and
4. Methods for cultivation of yeasts, in small-, medium-, and large-scale volumes.

2. Materials
2.1. Growth Media (in g/L)

1. Malt agar: dehydrated malt extract (30 g), mycological peptone (0.5 g), agar (15 g).
2. MYGP (YM) agar: dehydrated malt extract (3 g), dehydrated yeast extract (3 g), peptone (5 g), glucose (10 g), agar (15 g).

From: *Methods in Molecular Biology, Vol. 53: Yeast Protocols*
Edited by: I. Evans Humana Press Inc., Totowa, NJ

3. Saboraud's glucose broth: glucose (40 g), mycological peptone (10 g), agar (15 g).
4. Yeast extract-peptone-dextrose agar: This medium is used extensively for maintaining and culture of yeast strains for genetic investigations. It contains: Glucose (dextrose) (20 g), yeast extract (10 g), peptone (10 g), and agar (15 g). The agar is omitted if the yeasts are to be grown in liquid culture. For some purposes, such as in fermentation tests, the peptone is omitted whether the yeasts are to be grown in liquid or on solid medium.
5. Wickerham's chemically defined medium *(1)*: This medium was introduced by L. J. Wickerham about 1950, and has become the standard chemically defined medium for investigations of yeasts of most species. It is available from Difco (Detroit, MI) in the dehydrated form, but can be made up in large batches in the laboratory if desired. The reagents must be of very high quality, as some grades and some brands contain toxic impurities that inhibit yeast growth.
    a. Carbon source (for yeast-carbon base): glucose, 10 g.
    b. Nitrogen source (for yeast-nitrogen base): ammonium sulfate, 3.5 g.
    c. Amino acids: L-Histidine (10 mg), DL-methionine (20 mg), DL-tryptophan (20 mg).
    d. Salts mixture: KH_2PO_4 (1 g), $MgSO_4 \cdot 7H_2O$ (0.5 g), NaCl (0.5 g), $CaCl_2 \cdot 6H_2O$ (0.5 g).
    e. Trace elements mixture: H_3BO_3 (500 µg), $MnSO_4 \cdot 4H_2O$ (400 µg) $ZnSO_4 \cdot 7H_2O$ (400 µg), $FeCl_3 \cdot 6H_2O$ (200 µg), $Na_2MoO_4 \cdot 2H_2O$ (200 µg), KI (100 µg), $CuSO_4 \cdot 5H_2O$ (40 µg).
    f. Growth factor mixture: Myo-Inositol (10 mg), calcium pantothenate (2 mg), niacin or nicotinic acid (400 µg), pyridoxine HCl (400 µg), thiamine-HCl (400 µg), *p*-aminobenzoic acid (200 µg), riboflavin (200 µg), biotin (20 µg), folic acid (2 µg).

The trace elements and growth factor mixtures can be made up as 10X or more concentrated mixtures and added in appropriate amounts if the medium is to be filter-sterilized, which is desirable for the most accurate determinations of growth and utilization of carbon and nitrogen sources. If it is to be autoclaved, then the vitamin mixture should be filter-sterilized and added aseptically after the rest of the medium has cooled.

For determination of assimilation of sugars and related compounds, the sugar solution should be sterilized separately, preferably by filtration. Separate sterilization of the yeast-carbon base is less important for determination of assimilation of nitrite and nitrate, though it may be desirable.

The commercially prepared media are available without amino acids, without vitamins, or without either, if desired, to allow greater flexibility in making special-purpose media.

Cultures may be stored in the above media in screw-capped tubes or bottles, such as McCartney bottles, though ordinary culture tubes with cotton plugs or plastic caps are satisfactory for periods of 3 or 4 wk or less. Cultures can be stored on Petri plates, on similar media, in the refrigerator, for a few weeks, if they are kept from drying out by placing in plastic bags or plastic boxes with tight-fitting covers.

2.2. Sporulation Media (2)

These are included, since the production of spores, by various yeast species, is an important type of culture of yeasts on solid or liquid media, in some investigations.

1. McClary's medium (g/L): potassium acetate (10 g), yeast extract (2.5 g), glucose (1.0 g).
2. Raffinose-acetate medium: raffinose (20 g), potassium acetate (10 g).
3. Sodium acetate medium: sodium acetate (5 g). Adjust pH to approx 6.8 with HCl.

    The described media are those most often used for sporulation of strains of *Saccharomyces cerevisiae*. They may be used as liquid media or solidified by addition of 1.5% (all percentages are w/v, unless otherwise indicated) agar.
4. Gorodkowa agar: Glucose (1 g), peptone (10 g), NaCl (5 g), agar (20 g). Glucose (2.5 g) and meat extract (10 g) are some times recommended for *Debaryomyces* spp.
5. Vegetable juice agar: Equal weights of potatoes, beets, carrots, and cucumbers are washed (not peeled), grated, and autoclaved for 10 min and pressed through cheesecloth. Dried yeast (2%) and agar (2%) are added and the medium is melted, dispensed, and autoclaved.

    Wickerham (1950) recommended a commercial vegetable juice preparation (V8 juice, 350 mL), diluted in water (350 mL) containing 5 g of moist yeast and 14 g of melted agar, and the pH adjusted to 6.8. The melted medium is dispensed and sterilized.
6. Dilute V8 agar. Commercial V8 juice is diluted 1:1 with demineralized water and adjusted to pH 5.5 with NaOH. Further dilutions are made to 1:2, 1:9, 1:19, and 1:29, agar is added to these dilutions, melted, and the medium is dispensed and sterilized.

    This medium is used to induce sporulation in *Metschnikowia* species and other yeasts that do not sporulate on other media.

7. Potato agar: Peel and grate 100 g of raw potatoes and soak in 300 mL of tapwater for several hours at 4°C. Filter the liquid through cloth, and autoclave for 1 h at 15 psi. To prepare the agar, add 230 mL of the extract to 770 mL tapwater containing 20 g glucose and 20 g agar. Dissolve the agar by heating, dispense the medium, and autoclave for 15 min at 15 psi.
8. Corn meal agar: Soak 12.5 g yellow corn meal in 300 mL of water, with stirring, in a waterbath at 60°C for 1 h and recover the liquid by filtering through paper. Make up the volume to 300 mL and add 3.8 g of agar. Autoclave the material for 15 min at 15 psi, filter the hot liquid through absorbent cotton wool, dispense, and sterilize for 15 min at 15 psi.
9. Vegetable wedges: With a cork borer, cut cylinders, about 1-cm diameter, from carrot, potato, beet, cucumber, or turnip, after thorough washing. Cut wedges obliquely through them. Rinse the wedges again, place them in culture tubes with a little water to prevent drying out, and autoclave for 15 min at 15 psi.
10. Gypsum blocks: Make by mixing eight parts of gypsum with three parts of water. Cast the paste into small cylinders or cones, 3–4 cm high and allow to set. Place the blocks in covered jars or dishes and hold at 110–120°C for 2 h. Before use, add sterile water or a solution of a mannitol-phosphate (18 mL of 0.5–2.0% mannitol + 2 mL of 5% K_2HPO_4), 1 cm deep, to the dishes. Spread a thin layer of freshly harvested yeast cells on the gypsum. Or, prepare gypsum slants by pouring the paste into sterile, plugged tubes, slant them, and harden them by holding at 50°C for 24–48 h.
11. Rice agar: Put 20 g of white rice into 1 L of tapwater and bring to the boil. Let simmer for 45 min in a covered pot, and then filter the liquid through a cloth. Let cool (discard the solids), make up the volume to 1 L, and add 20 g agar. Melt the agar by steaming, dispense in tubes or bottles, and autoclave for 15 min at 15 psi.
12. Oatmeal agar: Add 4 g of oatmeal to 100 mL of tapwater, boil for 1 h, and filter through cheesecloth. Discard the solids, make up the filtrate to 100 mL, and add 1.5 g of agar. Melt the agar, mix well, and dispense in tubes. Autoclave 15 min at 15 psi.

2.3. Media for Germination of Teliospores of Basidiomycetous Yeasts

1. Hay infusion agar: Add 50 g of decomposing hay to 1 L of demineralized water, autoclave for 30 min at 15 psi, and filter off the solids. Dissolve 2 g of K_2HPO_4 and 15 g of agar in 1 L of the filtrate, adjust the pH to 6.2, dispense the material in tubes or bottles if desired, and autoclave for 20 min at 15 psi.
2. Sucrose-yeast extract agar: Add 1 g of KH_2PO_4, 0.5 g of $MgSO_4 \cdot 7H_2O$, 0.1 g of $CaCl_2$, 0.1 g of NaCl, 5.0 µg of biotin, 20 g of sucrose, 0.5 g of

yeast extract, and 40 g of agar to 1 L of demineralized water, and dissolve these constituents by heating. Autoclave the medium for 15 min at 15 psi.

2.4. Materials for Medium-Term Storage

1. Storage on agar slants, liquid medium, or under mineral oil: Slants of yeast extract-peptone-dextrose agar or bottles or tubes of yeast extract-peptone-dextrose broth, as described previously; sterile mineral oil; pipeter with sterile tips, set to deliver 5–10 mL of oil.
2. Drying on filter paper *(3)*: Whatman (Maidstone, UK) No. 4 filter paper or equivalent, cut into 1-cm squares, wrapped in aluminum foil and autoclaved; yeast cultures, grown in a heavy patch on a plate of YEPD agar, for 2–3 d at 30°C; suspending medium (evaporated milk, commercial, sterile); sterile toothpicks (blunt-ended); desiccator.

2.5. Materials for Long-Term Storage

2.5.1. Lyophilization

1. Cultures grown on YEPD slants for about 48 h, at 25–30°C.
2. Suspending media: Skim milk (20%), DMSO (5%), sodium glutamate (10%), or inactivated horse serum no. 5, containing 7.5% glucose, sucrose, or inositol (Wellcome Reagents Ltd., Beckenham, UK). (Skim milk is probably better and is cheaper. Reconstituted skim milk powder may be used.)
3. Freeze-drying apparatus, if available.
4. Acetone + dry ice freezing bath, if a commercial freeze-drying apparatus is not at hand.
5. Ampules, standard.
6. Filter paper labels, sterile, to fit inside the ampules. The labels are numbered, placed inside the ampules, the ampules are plugged loosely with nonabsorbant cotton, and sterilized.
7. Sterile Pasteur pipets for filling ampules.
8. Small cotton plugs for ampules.
9. Torch, for sealing ampules under vacuum.
10. Where a commercial drier is not available:
    a. Vacuum pump in good condition.
    b. Acetone-dry ice bath, for freezing.
    c. Manifold for attaching ampules for vacuum drying.

2.5.2. Storage Under Liquid Nitrogen

1. In ampules:
    a. Plastic ampules, 2 mL, screwcap, sterile.
    b. Inoculum: yeast cells, grown in liquid YM medium (Difco), grown at 25°C for 72 h with shaking, to a cell density of 10^6–10^7 cells/mL.

c. 10 or 20% v/v Cryoprotectant, sterile glycerol solution, 10% v/v DMSO, YM broth (Difco), 5–10% hydroxyethyl starch.
d. Liquid nitrogen refrigerator, and also a domestic freezer (–30°C) if available, or a cooling bath.
e. Dewar flask for transporting frozen ampules.
f. Racks for holding the ampules in the freezer, and other ancillary equipment.

2. In straws:
   a. Polypropylene "straws," prepared from 2.5-cm lengths of drinking straws of this material. (The sections are held tightly, 1 mm from the end, in a pair of unridged forceps, and this end is held about 1 cm from the flat flame of a fishtail burner. The polypropylene melts very quickly and forms a seal. The melted part sets firmly in a few seconds when removed from the heat and the end is thus sealed. The straws are autoclaved at 121°C in a glass Petri dish.)
   b. Inoculum and cryoprotectant are prepared as in method 1.

2.5.3. Freezing in Glycerol Solution (3)

1. Inoculum: yeast cells, grown for 48 h in YE, YM, or other normal medium, to approximately 10^7–10^8 cells/mL.
2. Cryoprotectant: glycerol, sterile.
3. Small culture tubes or vials, 3–4 mm diameter, sterile.
4. Freezer, capable of operation at –70°C.

3. Methods

3.1. Medium-Term Storage

3.1.1. Storage on Agar Slants, in Liquid Medium, or Under Mineral Oil

Two tubes or slants of each culture are inoculated and incubated at 25°C for 48 h. One tube is reserved for starting a new stock culture when required; the other is used for daily transfers as required. The tubes or slants are closed tightly and refrigerated. If it is desired to keep the cultures under oil, after incubation, the tubes are filled with sterile mineral oil to a level just above the level of the agar and then refrigerated. The cultures normally remain viable for 6–12 mo, although different species have different storage lives.

3.1.2. Drying on Filter Paper (3)

1. Place three or four drops of sterile evaporated milk in separate locations in a sterile Petri dish.

Maintenance and Culture of Yeasts

2. Scoop up a toothpick load of cells from the patch and resuspend in one of the drops of sterile milk, in as heavy a suspension as possible.
3. Using sterile forceps, pick up one of the pieces of sterile filter paper out of its foil wrapping, soak in the cell suspension and return it to the foil wrapping. Handling the filter paper and foil in this way, aseptically, requires some practice.
4. Leave the foil wrapping unsealed around three of the edges, to facilitate drying.
5. Dry in a desiccator for 2–3 wk. Handle carefully so that the papers do not fall out of the wrappings.
6. Remove the packets from the desiccator and fold over the edges of the foil to seal.
7. Store in a dry container at 4°C (shelf life 2–3 yr). The method is satisfactory for *Saccharomyces cerevisiae*; less so for other species.
8. The cultures can be revived by removing the paper aseptically from the foil and placing on a plate of the appropriate medium. Incubate at 23–30°C for several days and pick off colonies.

3.2. Long-Term Storage

3.2.1. Lyophilization (3)

1. Suspension: Mix equal amounts of inoculum (not washed) and suspending medium, aseptically in a sterile tube or bottle.
2. Inoculation of ampules: Six drops of the mixture (0.2 mL) are transferred to the bottom of each ampule with a sterile Pasteur pipet. Do not touch the side of the ampule. Save the original inoculum for determination of viability of the culture.
3. If a commercial freeze-drier is used, place the ampules in the centrifuge head, mount the latter in the drying chamber, switch on the centrifuge and vacuum pump, and dry for 3 h. Follow the manufacturers' instructions at all times.
4. Take ampules out of the centrifuge, trim off any projecting cotton, and push the plug halfway down the ampule. Pull a neck on the ampule, using the torch.
5. Mount the constricted ampules on the manifold, and install the manifold on the freeze-dryer. Pull a vacuum on the ampules and dry again overnight if possible.
6. Seal the ampules with the torch, test with a spark tester and store at 4°C in the dark. Shelf life may be up to 30 yr. Most yeast species can be stored by this method.

3.2.2. Storage Under Liquid Nitrogen (3)

1. Ampules:
   a. Mix equal quantities and the glycerol solution (or other cryoprotectant) in a sterile tube, so that the final concentration of glycerol is 5% v/v. Transfer 1 mL of the mixture to each of the ampules.
   b. Freeze the preparations in a domestic freezer or cooling bath, to −30°C, at a rate of about 5°C/min, and allow to dehydrate for 2 h.
   c. Transfer the frozen ampules, without thawing, to the liquid nitrogen refrigerator. Use the chilled Dewar flask if necessary.
   d. Maintain the level of liquid nitrogen to where the ampules are completely submerged. Shelf life is up to 4 yr.
   e. Cultures are revived by rapid thawing in a waterbath at 35°C, with shaking.
2. Storage in straws:
   a. Inoculum and cryoprotectant are prepared as in step 1 above.
   b. Filling of straws. Inoculum and cryoprotectant are mixed as in Section 3. One straw is taken out of the Petri dish with sterile forceps and filled with inoculum mixture using a Pasteur pipet. Place the tip of the pipet close to the bottom of the straw, and fill the straw to within 3 mm of the upper end (about 0.1 mL of inoculum mixture).
   c. Heat seal the straw as described in Section 2.
   d. Freeze the straws in ampules as described in Section 3.
   e. Store in the liquid nitrogen refrigerator as before.
   f. Revival: Thaw in a water bath at 35°C as before, resuspend the cells by squeezing the straws, wipe the straws with 95% v/v alcohol and cut off the end with sterile scissors.
   g. Make sure the cells are all resuspended, by repeated pipeting, and transfer two drops to 0.5 mL of sterile water, giving an approx 1:10 dilution. Plate out and make isolations as required.

3.2.3. Storage by Freezing in Glycerol Solution (3)

1. Take an aliquot of the inoculum and make up to a glycerol concentration of 15–50% v/v. Different workers prefer different concentrations of glycerol in the final mixture.
2. This mixture is transferred to a small sterile culture tube (2–5 mL), and is then slanted and frozen at −70°C, and stored at this temperature.
3. Routine transfers are made by scraping a little of the culture from the surface of the frozen medium, and transferring to fresh medium.
4. If the freezer should fail, the culture can be transferred and a fresh subculture frozen later.
5. Survival (shelf life) is for several years, cultures stored at −70°C surviving longer than those kept at −20°C. Several tubes of each strain should be stored.

3.3. Culture of Yeasts in the Laboratory

3.3.1. Culture on Solid Medium

Yeasts can be cultivated in almost any size vessel, from a test tube on up. Although for some purposes, they may be grown on the surface of solid medium in a Petri dish and scraped off, this method yields only a relatively small quantity of cells and is not generally used. Yeasts may be sporulated on McClary's medium, sodium acetate agar, raffinose-acetate agar, malt agar, Gorodkowa agar (for *Debaryomyces* spp), gypsum blocks, carrot, or potato wedges, and similar media, but these are special instances, when evidence of sporulation, or the spores themselves, are required for taxonomic or genetic investigations.

3.3.2. Culture in Liquid Medium

1. Small cultures: These may be grown in culture tubes of the desired size. The aeration can be improved if the tubes are shaken in a rack, sloped at an angle of about 30°.
2. Culture in Erlenmeyer flasks: Any desired liquid medium may be used. Usually a v/v ratio of about 5:1 is used, and the flasks are normally incubated on a shaker operated at 200–250 rpm, and having an eccentricity of 1" (2.5 cm), which gives more or less adequate aeration. Incubation temperatures range from 25–30°C.
3. Culture in stirred vessels *(4)*: A number of makes of commercial equipment for pilot plant scale culture of yeasts are available. Most are reliable. Usually an inoculum is grown in shaken flasks, and added to the prepared fermentor at rates up to 10% v/v, though rates of less than 1% have been used successfully. Fermenter volumes in the laboratory are usually in the range 2–30 L (working volume), though vessels of 100–150 L are fairly common. At the larger volumes, mass and heat transfer problems become significant, in particular, oxygen transfer, and adequate monitoring equipment is essential.

   At one time, it was necessary to calibrate the vessels in advance, usually by the sulfite oxidation method, but reasonably accurate oxygen electrodes, for determination of the actual dissolved oxygen level, are now available. For operation of the equipment, follow the manufacturer's instructions. The microbiological techniques are the same as for laboratory culture.

   Inoculation techniques for stirred vessels range from opening a port in the top of the vessel and pouring in the culture to be used, to blowing the culture into the fermentor using sterile compressed air, from a specially constructed bottle that can be filled under aseptic conditions in a sterile room. The same techniques are employed whether the cells are to be grown in batch, fed-batch, or continuous culture.

For yeasts grown in small vessels, the pouring technique is usually satisfactory.

3.4. Fed-Batch, Continuous, and High-Density Culture (5)

Fed-batch cultures, as used in the manufacture of baker's yeasts, can be done in the same type of stirred vessels as for stirred batch cultures, and using similar media. The cell yield is enhanced by further addition of media as the fermentation progresses, until the desired cell density is reached. For best results, automatic controls linked to continuous sensors are desirable.

High cell-density cultures can also be carried out on a small-scale in ordinary small fermentors, though larger equipment should be designed for the purpose. In particular, adequate baffling and stirring, to ensure proper heat transfer and oxygen transfer is essential. The cell density is controlled by use of an appropriate concentration of nutrients, particularly the carbon source, in the feed, and by choice of the optimum feed rate. Cell densities, using yeast, may reach 200 g/L or higher. In production of biomass, this has the advantage of eliminating at least one stage in the recovery system, as cultures at that density can be sent directly to the drier or filter press, without centrifuging or washing, depending on the final use for the product. High cell-density continuous processes are also well suited to the production of heterologous proteins using methylotrophic yeasts. These are grown in a medium containing methanol and glycerol, and have been used for production of tumor necrosis factor and other proteinaceous pharmaceutical products.

4. Notes

1. Drying on silica gel has been used for storage of yeasts, including genetic stock cultures. The method was proven unreliable and is no longer used, but investigators may find it mentioned in the literature.
2. Cultures will survive in closed tubes or bottles, in liquid medium or on agar slants, as described earlier, for surprising lengths of time. However, they can be kept under sterile mineral oil, in the refrigerator, as an added precaution. They are generally viable for 6–12 mo under these conditions. For routine work, when the cultures are transferred, two new tubes or bottles should be inoculated, and these are incubated for 24–48 h and then closed tightly and refrigerated. Only one tube is used for daily transfers, and the other is reserved for starting new cultures for storage.

3. Other similar media can be used if desired. Corn meal agar frequently is used also for detection of ballistospore formation in *Sporobolomyces* and *Bullera* species, and formerly, for inducing pseudomycelium formation in *Candida* species and others. Rice agar and potato-dextrose agar may also be used for inducing formation of pseudomycelium.
4. Hay infusion agar and sucrose-yeast extract agar are used for inducing germination of teliospores of basidiomycetous yeasts, since spore germination and subsequent development of basidial structures are decisive criteria for the classification of these yeasts.
5. A recent taxonomic study of the genus *Saccharomyces* has been published by Barnett *(6)*.

References

1. Wickerham, L. J. (1950) *Taxonomy of Yeasts,* Tech. Bull 1029, US Dept. Agriculture, Washington, DC.
2. Kreger-van Rij, N. J. W. (1984) *The Yeasts, a Taxonomic Study,* Elsevier, Amsterdam.
3. Kirsop, B., Painting, K., and Henry, J. (1984) *Yeasts: Their Identification, Preservation and Use in Biotechnology,* Bangkok MIRCEN, Thailand Institute of Scientific and Technological Research, Bangkok, Thailand.
4. McNeil, B. and Harvey, L. M., (eds.) (1990) *Fermentations, a Practical Approach,* IRL Press, Oxford University Press, Oxford.
5. Spencer, J. F. T., Spencer, D. M., and Hayen, G. (1990) High Cell-Density Fermentations, in *Yeast Technology* (Spencer, J. F. T. and Spencer D. M., eds). Springer-Verlag, Heidelberg.
6. Barnett, J. A. (1992) The taxonomy of the yeast *Saccharomyces* Meyen ex Reess: a short review for non-taxonomists. *Yeast* **8,** 1–23.

CHAPTER 3

Mutagenesis in Yeast

John F. T. Spencer and Dorothy M. Spencer

1. Introduction

Mutations of numerous types can be induced in yeast. The basic principle is to bring the yeast in contact with the mutagen (UV light, X-rays, EMS, MMS, nitrous acid, nitrosoguanidine [NNG], ICR-170, nitrogen mustard, and so on), for long enough to bring about 50–95% killing, after which the mutagen is removed. The yeast strain may be grown in nutrient medium for a round of cell division to fix the mutation in the cells (a "recovery" step), and the cells are plated out on complete medium. Finally, the cells are replica plated to minimal medium, if auxotrophic mutants are desired, or on various other media designed to detect mutations such as resistance to a number of compounds (canavanine, cycloheximide, heavy metals, and similar toxic compounds). Recessive mutations are generally induced in haploid strains, but dominant mutations, including many of the mutations to resistance, can be induced in cells of higher ploidy. Mutations in the mitochondrial (mt) DNA form a special class of mutations in *Saccharomyces cerevisiae*, the classic *petite colonie* mutation, which manifests itself as a small colony on agar, and which lacks the ability to grow on nonfermentable substrates. The mtDNA in this mutation is nonfunctional or absent, giving rise to the respiratory deficiency, and the mutation occurs spontaneously, or is induced by manganese, and by many carcinogenic drugs, such as acriflavin, ethidium bromide, adriamycin, daunomycin, and benzidine, to name a few. Mutations conferring resistance to such antibiotics as chloramphenicol, erythromycin, oligomycin, and paromomycin, affecting the mitochondrial system, also occur in mtDNA.

The primary effect of mutagenesis by UV irradiation is the formation of pyrimidine dimers, which in wild-type yeasts are removed by various repair systems, including photoreactivation by certain wavelengths of visible light. Since this interferes with recovery of mutants induced by UV, induction of mutations by this method should be done under yellow light to avoid this effect. Nevertheless, the method for mutation induction remains approximately the same, and UV irradiation is widely used to induce mutations in yeasts.

2. Materials

2.1. Induction by Radiation

1. UV or X-ray source.
2. Yeast culture, in broth, approx 48 h old (early stationary phase)
3. Water blanks, 9.0- and 9.9-mL, sterile.
4. Petri dishes, empty, glass or plastic, sterile.
5. YEPD (1% yeast extract, 1% peptone, 2% glucose, all percentages are w/v, unless otherwise stated) agar plates (or other complete medium).
6. Plates of minimal medium (YNBD [yeast nitrogen base without 0.67% amino acids, 2% glucose] or equivalent).
7. Dropout media or other selective media (YNBD, plus all amino acids or purine and pyrimidine bases required, except the one of interest).

2.2. Induction by Chemical Mutagens (EMS, MMS, NNG, NA, ICR-170, and so on)

1. Mutagenic agents (ethyl methane sulfonate [EMS], methyl methane sulfonate (MMS), N-Methyl-N'-nitro-N-nitrosoguanidine [MNNG] and nitrous acid [NA]) are some of the more commonly used mutagens for yeast.
2. Yeast cultures, normally haploid, grown to stationary phase (48 h at 30°C).
3. Sterile phosphate buffer, $0.1M$, pH 7.1.
4. $Na_2S_2O_3$, 5% sterile, for terminating the reactions where MNNG, MMS, or EMS are used.
5. Water blanks, 9.0- and 9.9-mL, sterile (in tubes or bottles).
6. Complete and minimal media (YEPD and YNBD), solid.
7. Sterile pipets, Petri dishes, and other laboratory glassware routinely used in microbiological procedures, and tubes or flasks of YEPD broth, if the "recovery" step in Section 3.1.7. is to be used.

2.3. Isolation of Particular Classes of Mutants
2.3.1. Induction of Mitochondrial Mutations
2.3.1.1. PETITE COLONIE MUTANTS
1. Tubes of YEPD broth, 5 mL/tube.
2. Plates of YEPD agar.
3. Acriflavin solution, 0.4%, or ethidium bromide solution, 1% (stock solution).
4. Plates of YEPG (1% yeast extract, 1% peptone, 4% glycerol v/v) agar.

2.3.1.2. MIT⁻ MUTANTS
1. Yeast culture, haploid, carrying the *pet9* mutation (a nuclear mutation conferring respiratory deficiency) and one other auxotrophic marker, but otherwise being *rho⁺mit⁺* (i.e., wild-type with respect to mtDNA), and also a strain of a *pet9 rho°* tester strain (*rho°* meaning lacking detectable mtDNA).
2. Medium containing 10% glucose, 1% yeast extract, 1% peptone. Both broth and agar.
3. 350 mM Sterile MnCl$_2$ solution.
4. YEPG agar (ethanol may be substituted for glycerol).
5. YNBD agar.

2.3.1.3. ANTIBIOTIC-RESISTANT MUTANTS
1. Yeast cultures, respiratory competent, 48 h, in YEPD broth. These may be haploid or diploid, but must be respiratory competent.
2. YEPG plates, containing antibiotics, such as chloramphenicol or erythromycin (1.5–2.5 mg/mL) or oligomycin (2.5 µg/mL).
3. 350 mM Sterile MnCl$_2$ solution.
4. Tubes of 5 mL sterile YEPD.

2.3.2. Mutator Mutants
1. Yeast cultures (auxotrophic for lysine), 48 h, in YEPD medium. The presence of other auxotrophic markers is desirable.
2. Plates of synthetic complete medium (YNBD, plus amino acids as required), containing 1 and 20 µg/mL of lysine.
3. Sterile water for washing cells.

2.3.3. Determination of Revertant Frequency
1. Flask of limiting medium, at least 1 L (the low-lysine medium described above; 1 µg/mL lysine) containing all other requirements for the yeast strain used. Place a magnetic stirring bar in the medium before sterilizing.
2. Yeast strain, grown in normal medium.
3. Multiwell plates, presterilized, 2 mL/well capacity, 1000 wells/yeast strain.
4. Automatic pipetor, to deliver 1 mL medium/well.
5. Magnetic stirrer.

2.3.4. Temperature-Sensitive Mutants

1. Yeast strains, mutagenized according to previous protocols.
2. Complete medium (YEPD, as described previously).

2.3.5. Mutants Defective in Nuclear Fusion: The karl-1 Mutation

1. Yeast strains carrying selectable mutations, including resistance to canavanine (*can1*) and/or cycloheximide (*cyh^r*) and possibly *ade2* (red). Strain JC1 *(MATα his4 ade2 canl cyh^r rho$^-$)* and GF4836-8C *(MATa leul thrl RHO$^+$)* were originally used. (RHO$^+$ means that the mtDNA lacks any of the deletion mutations that cause the *petite* phenotype.)
2. Equipment and reagents for mutagenesis.
3. Media:
   a. Minimal medium (YNBD);
   b. Supplemented glycerol minimal medium (yeast nitrogen base without 0.67% amino acids, 4% glycerol v/v, supplemented with histidine (0.3 mM) and adenine (0.15 mM) buffered at pH 6.5, plus canavanine (60 mg/L) and cycloheximide (2 mg/L) added after autoclaving.

2.3.6. Cell Wall Mutants: Mutants Having Altered Mannans

1. Yeast strain, haploid. Both *Saccharomyces* and *Kluyveromvces* spp. have been used.
2. Antisera raised against the mannans or against particular chemical structures in the cell wall (e.g., pentasaccharide side chains in *Kluvveromyces lactis* walls). These can be obtained commercially from specialist laboratories.
3. Equipment and reagents for mutagenizing the yeast with the desired mutagen (including plates).

2.3.7. Cell Wall Mutants: "Fragile" Mutants, Lysing at Normal Osmotic Pressures

1. Haploid yeast strains.
2. Equipment and reagents for mutagenesis.
3. Plates of osmotically stabilized medium (containing 1.2M sorbitol or 0.6M KCl.
4. Plates of normal YEPD medium.

2.3.8. Cell Wall Mutants: Mutants Having Easily Digested Cell Walls

1. Yeast strains.
2. Equipment and reagents for mutagenesis, including YEPD plates.
3. YEPD plates containing snail enzyme (β-glucuronidase), pepsin, lipase, or trypsin (approx 1 mg/mL).

Mutagenesis in Yeast 21

2.3.9. Antibiotic-Sensitive Mutants: "Kamikaze" Mutants (7)

1. YEPD plates and YEPD broth.
2. Yeast strain BL15 (kamikaze), which grows normally at 32°C but dies after 40 min exposure to 42°C.
3. Antibiotics, as mentioned earlier, and trimethoprim if desired.

2.3.10. PEP4 Mutants

1. Haploid yeast strain.
2. Solutions of N-acetylphenylalanine-β-naphthyl ester (APE) and fast garnet (GBC), which gives a red color with the naphthol released.
3. Hide powder azure, for detection of proteinase B-deficient mutants.

2.3.11. Inositol-Requiring Mutants

1. Mutagenized (EMS) haploid yeast strain, having any required auxotrophic or other mutations to be used as markers.
2. Media: YEPD agar plates, and minimal medium (YNBD), plus any amino acids and other factors required for growth of the yeast strain, but without inositol.
3. Minimal medium as described, + inositol, 2 µg/mL.

2.3.12. Inositol-Excreting Mutants

1. Mutagenized yeast strain, as described.
2. Media: YEPD, minimal medium (YNBD plus adenine and any amino acids required by the indicator yeast strains).
3. Indicator yeast strains, auxotrophic for both adenine and inositol.

2.3.13. Sterol-Requiring Mutants

1. Wild-type, haploid cells, grown for 48 h at 30°C in YEPD.
2. YEPD with 40 µg/mL ergosterol and 15–20 µg/mL nystatin.
3. Equipment for UV-irradiation.

2.3.14. Mutants Auxotrophic for 2'-Deoxythymidine 5'-Monophosphate (dTMP)

1. Haploid yeast strain, wild-type.
2. Minimal medium (yeast nitrogen base without 0.67% amino acids) containing either glucose (2%) or glycerol (4% v/v) as sole carbon source, plus sulfanilamide (5 mg/mL), aminopterin (100 µg/mL), vitamin-free casamino acids (0.15%), adenine (30 µg/mL), and dTMP (100 and 10 µg/mL).
3. YEPD plates, with and without dTMP (100 µg/mL if used).
4. Phosphate buffer, 0.1M (pH 7.1), for washing cells.

2.3.15. Glycolytic Pathway Mutants

2.3.15.1. ALCOHOL DEHYDROGENASE (ADH) MUTANTS

1. Graded series of YEPD plates containing allyl alcohol, 20–100 mM in steps of 20 mM or continuous culture apparatus (turbidostat). Use of a turbidostat is specified because it indicates cell growth directly and hence is a measure of adaptation of the yeast to the compound. A chemostat depends on the use of a metabolizable growth-limiting constituent of the medium (not an inhibitor) and cell growth is not measured in this way.
2. Yeast strains *(petites)*.

2.3.15.2. PYRUVATE CARBOXYLASE (PYC) MUTANTS

1. Equipment and reagents for mutagenizing with EMS.
2. YNBD agar plates.
3. YEP-ethanol (2%) and YNB-ethanol (2%) agars.

2.3.15.3. PHOSPHOENOLPYRUVATE CARBOXYKINASE (PEPCK) MUTANTS

1. YEPD, YEPG, and YEP-ethanol (2%) agar. Concentration of the carbon source, 2%.
2. Medium containing nystatin (3 µg/mL) for enrichment of the culture in mutants.
3. Assay medium for PEPCK (per 0.5 mL, 25 µmol imidazole, 25 µmol NaHCO$_3$, 1 µmol MnCl$_2$, 1 µmol ADP, 0.25 µmol NADH, 1 µmol phosphoenolpyruvate, 80 µkat malate dehydrogenase, 40 µkat hexokinase.

3. Methods

3.1. General Methods for Mutation Induction by UV and X-Irradiation (see Notes 1 and 2) and Isolation of Auxotrophs

1. Switch on the UV source, if this is to be used, and allow time enough for it to stabilize.
2. Dilute the culture 1:10 (1 mL of culture in a 9.0-mL water blank).
3. Transfer the diluted culture to a sterile Petri dish.
4. Place the Petri dish in the irradiation apparatus and remove cover (*see* Notes 1 and 2).
5. Remove the dish cover and irradiate, agitating the dish as continuously as possible (*see* Notes 1 and 2).
6. Take samples at intervals, beginning after 5–10 s, depending on the dose rate of the source. (This permits isolation of mutants at the best percentage killing for obtaining the greatest number of mutants among the survivors, and at the same time, provides data for construction of a killing curve.)

Mutagenesis in Yeast

7. Place the samples in a series of water blanks, and when the irradiation procedure is complete, dilute the samples appropriately, to give 100–150 cells in 0.25 mL, for spreading on the YEPD plates. (Record all dilutions.)
   Alternatively, the samples may be placed in tubes or flasks of YEPD broth and incubated for 2–3 h, to allow a round of cell division to take place, to fix the mutations, and the culture may then be diluted and plated.
8. Incubate the plates at 25–30°C, for 4–5 d, until the colonies have grown well.
9. Screen the colonies for the desired mutant phenotype. For example, to isolate auxotrophic mutants, replica plate the colonies to minimal medium and again incubate. Retain the master plates. Examine the plates of minimal medium and look for colonies that have not grown. Locate these colonies on the master plates, pick and restreak, and repick to complete media and to dropout and other diagnostic media as desired.

3.2. General Method for Mutation Induction Using Chemical Mutagens (see Note 3)

1. Prepare culture as for UV irradiation (*see* Section 3.1.2.).
2. Add desired mutagen (ethylmethane sulfonate, EMS; methyl methane sulfonate, MMS; MNNG, nitrous acid, NA) typically, to 3% v/v.
3. Incubate for desired time (10–40 min).
4. If EMS, MMS, or MNNG are used, dilute the culture into 5% thiosulfate solution to stop the reaction. For NA, dilute into $0.1M$ phosphate buffer (pH 7.1).
5. Make dilutions, plate out, and isolate mutants as for mutagenesis by irradiation, according to phenotype (*see* Notes 4–6).

3.3. Mitochondrial Mutations: Petite Mutants (see Notes 7–9)

3.3.1. Acriflavin-Induced Petites

1. Add one loopful of the acriflavin solution to a tube of YEPD broth and mix.
2. Inoculate with a light inoculum of the desired yeast strain. This may be haploid, diploid, or of higher ploidy, or an industrial yeast of unknown ploidy.
3. Incubate in the dark at 30°C, without agitation.
4. After 2–4 d, streak culture out on a plate of YEPD agar and incubate at the same temperature for 3–5 d.
5. Pick small colonies (restreak if desired) to YEPD and YEPG agar, and again incubate. Retain those colonies that do not grow on glycerol as the sole carbon source.
6. Ethidium bromide-induced *petites* may be induced using the same procedure, using a final concentration of 20 µg/mL of ethidium bromide in the medium.

3.3.2. Mit⁻ Mutants

These are mutants having point mutations in the mtDNA rather than large deletions. The frequency of their occurrence is very low, compared with that of rho^- and $rho^°$ mutants, and if they are to be isolated, it is necessary to use mutagenesis with $MnCl_2$ (3) to increase their frequency, and to eliminate the rho^- and $rho^°$ mutants. The latter is done by using a strain carrying the *op1 (pet9)* nuclear mutation, when inducing the *mit⁻* mutations (mtDNA deletion mutation in an *op1/pet9* genetic background are inviable). The procedure is as follows:

1. Grow a haploid strain carrying the *pet9* mutation and one other-auxotrophic-marker, but otherwise being rho^+mit^+, in a medium containing 10% glucose, 1% yeast extract, and 1% peptone.
2. Recover and wash the cells and resuspend in sterile water. Dilute to a density of 10^7 cells/mL.
3. Inoculate 5 mL of the same medium with 0.1 mL of the cell suspension, and add 0.1 mL of sterile $MnCl_2$ solution, 350 mM, to give a final concentration of 7 mM of $MnCl_2$. Do not exceed a concentration of 10 mM.
4. Incubate at 28°C for 24 h with shaking.
5. Dilute the cell suspension to 2×10^3 cells/mL.
6. Plate 0.1 mL of the suspension on agar having the same composition as above. Incubate at 28°C for 3 d.
7. Isolate small colonies, which will be *mit⁻*. The other classes of *petite* mutants are also induced by Mn^{2+}, but are eliminated. The visible colonies will be either non-mutant or the desired *mit⁻* mutants.
8. Pick the *mit⁻* colonies to an 8 × 8 grid, again on the 10% glucose, 1% yeast extract, 1% peptone agar, and incubate for 3 d.
9. Replica plate to media containing glycerol or ethanol, and to plates of 0.67% YNB (without amino acids) + 2% glucose, covered with a lawn of a *pet9 rho°* tester strain of the opposite mating type and complementary auxotrophic requirements, and incubate for 3 d at 28°C.
10. Look for:
    a. Absence of growth of haploids on the glycerol-containing and YNB-glucose media.
    b. Confluent growth of diploids on the YNB-glucose medium.
11. Replica plate the diploid colonies to YEPG medium, note those that fail to grow, and isolate the corresponding haploid clones from the original grid.
12. Suspend these cells in sterile water, at 10^3 cells/mL, and plate aliquots of 0.1 mL on 10% glucose, 1% yeast extract, 1% peptone agar. Incubate at 28°C and repeat steps 8–11 on subclones.
13. Isolate a single subclone from each mutant.

Mutagenesis in Yeast

14. Conditional *mit⁻* mutants (temperature-sensitive) can be isolated by the same procedure, to step 8, after which the grids are replica plated to duplicate YEPG plates and one plate is incubated at the desired restrictive temperature and one at the permissive temperature.
15. Diploid colonies that do not grow at the restrictive temperature but do grow at the permissive one are noted, and the corresponding haploid clones are selected on the original grid, after which steps 12 and 13 are followed, to make the final selection.

3.3.3. Antibiotic-Resistant Mutants (see Notes 10–12)

1. Recover the cells and dilute to 10^3 cells/mL with sterile water. Spread 0.1 mL of this suspension, or streak on plates of YEPD medium. Incubate at 28–30°C for 2–3 d.
2. Pick single colonies, to give as many subclones as the number of mutants desired.
3. If it is desired to mutagenize the cultures with Mn^{2+}, inoculate the subclones into individual tubes of YEPD medium, and add 0.1 mL of $MnCl_2$ solution to each tube. If mutagenesis with Mn^{2+} is not required, use YEPG and omit the $MnCl_2$ solution. Incubate the cultures on a roller drum operated at about 25–30 rpm, or on a slanted rack on a rotary shaker, for up to 1 wk, at 28–30°C.
4. Recover the cells from each tube and resuspend in 1 mL of sterile water. Plate the cells on YEPG agar containing the antibiotic desired (chloramphenicol, erythromycin, oligomycin, or other).
5. Incubate at 28–30°C until resistant colonies appear. This may require at least 1 wk, depending on the strain and on the antibiotic used.
6. Pick and restreak colonies on the same medium. Ideally, only one colony should be taken from each yeast culture, to be certain that the mutations are independent.
7. Grow the cells in 5 mL of YEPG at 28°C for 2 d. Then dilute in sterile water to a density of 10^3 cells/mL, and plate out on YEPG agar. This ensures that the clone is a genuine resistant mutant.
8. Pick a single colony from each subclone and store on YEP-glycerol agar containing the antibiotic.

3.3.4. Mutator Mutants (5)

These mutants show an increased rate of spontaneous mutation.

3.3.4.1. Isolation of Mutator Mutants

1. Wash mutagenized cells thoroughly by centrifugation, about 4000g for 10 min, and resuspend in sterile water.
2. Spread on both media (high- and low-lysine).

3. Incubate at 30°C for several days, until colonies appear on the low-lysine plates. Pick colonies from plates showing more revertants per plate (6–20 times as many).

3.3.4.2. Determination of the Mutation Rate

1. Grow the culture in liquid medium, recover, and wash the yeast cells.
2. Inoculate the flask of medium with approx 5×10^3 cells/mL and allow to equilibrate on the magnetic stirrer.
3. Fill the wells, using the automatic pipetor and aseptic techniques.
4. Plate aliquots on solid medium to determine the number of revertants in the inoculum.
5. Seal the boxes of multiwell plates and incubate at room temperature. Do not agitate the wells.
6. Count the number of wells having no visible revertant colonies. These can be seen readily with the naked eye. Count also the total number of revertant colonies per strain.
7. Count the total number of cells (two wells/plate) by hemocytometer.
8. Determine whether the reversions were at the desired (lysl) locus or were owing to the presence of a super-suppressor, by plating one revertant colony on complete, synthetic complete, minimal, and omission medium.
9. Calculate the mutation rate from the formula

$$e^{-m} = N_0/N$$

where N is the total number of wells in the experiment, N_0 is the number of wells without revertants, and m is the average number of mutational events/compartment (= well). In addition, m_b is the average number of mutants/well, determined by direct plating of the original inoculum, and m_g is the corrected number of mutational events, so that $m_g = m - m_b$, and the mutational events/cell/generation, $M = m_g/2C$, where C is the number of cells/well after growth has ceased in the limiting medium. The number of cell generations in the history of a culture is approximately twice the final number of cells.

10. If the mutants are grouped into categories, the mutation rate is $M_i = f_i M$, where M is the mutation rate per cell per generation, and f_i is the fraction of mutants tested, which were found in the ith category,

   This procedure is for determination of the spontaneous mutation rate. Mutator strains may also be induced by mutagenic agents.

3.3.5. Temperature-Sensitive Mutants (see Notes 13 and 14)

1. Treat yeast strain with mutagen according to previous procedures.
2. Dilute suspension of mutagenized cells appropriately and plate on complete medium, two plates/dilution.

3. Incubate one plate at the restrictive temperature and one at the permissive temperature (usually 36°C and 23–25°C for heat-sensitive mutants).
4. Colonies that do not grow at the restrictive temperature are isolated and tested.

3.3.6. Mutants Defective in Nuclear Fusion: The karl-1 Mutation (see Note 15 and ref. 6)

These mutants are very useful for transferring cytoplasmic elements (mitochondria, killer virus-like particles; VLPs) between cells, and for single-chromosome transfer. The method described here is that used by Condé and Fink (1976) (6).

1. Mutagenize strain JC1 by standard methods.
2. Mate with strain GF4836-8C by mass mating and spread the mating mixture on YEP-glucose plates.
3. Incubate 20–50 h at 30°C to allow mating.
4. Resuspend the cells in 1 mL of water, dilute 1:10, and spread 0.2 mL/plate on selective media (supplements).
5. Isolate colonies growing on the selective plates. These will mostly be Mata his– ade– can^r cyh^r (RHO$^+$), having the nucleus from JC1 and the mitochondria from GF4836-8C. The parental strains will not grow on the selective medium.
6. Backcross the putative *kar* mutant strains to GF-4836-8C and determine the frequency of heteroplasmon formation, according to the numbers of clones having the JC1 nucleus (red color, auxotrophic for histidine, canavanine, and cycloheximideresistant).

Cells having diploid nuclei or two nuclei will be white, as the *ade*– mutation in strain JC1 will be complemented. Cross the unmutagenized GF4836-4C strain to unmutagenized JCl and observe the number of red colonies as a control; the colonies resulting from this cross should be white.

3.3.7. Cell Wall Mutants: Mutants Having Altered Mannans (7)

1. Mutagenize the cells according to standard procedures (*see* Section 3.2.).
2. Grow cells for 2 d, wash, and agglutinate with the antiserum.
3. Discard agglutinated cells and recover and plate out the cells remaining in the supernatant. Repeat two or three times.
4. Isolate cells and test for (lack of) agglutination by the antiserum.

3.3.7.1. Cell Wall Mutants: "Fragile" Mutants, Lysing at Normal Osmotic Pressures (8)

1. Mutagenize yeast cells according to standard methods, using UV irradiation or chemical mutagens.

2. Dilute and plate on osmotically stabilized medium.
3. After growth, replica plate to normal YEP-glucose medium. Isolate colonies growing only on osmotically stabilized medium.

3.3.7.2. CELL WALL MUTANTS:
MUTANTS HAVING EASILY DIGESTED CELL WALLS (SEE NOTES 16–18 AND REF. 9)

1. Mutagenize yeast strains, dilute and spread aliquots on YEP-glucose plates (see Sections 3.1. and 3.2.).
2. After growth, replica plate to the media containing lytic enzymes. Isolate colonies from the master plates, which do not grow on the plates containing enzymes.

3.3.8. Antibiotic-Sensitive Mutants: "Kamikaze" Mutants (see Notes 19 and 20 and ref. 10)

Strain BL15 of *S. cerevisiae* is temperature-sensitive (restrictive temperature 42°C) and is used for isolation of mutants sensitive to emetine, cycloheximide, amecitin, fusidic acid, and other drugs that inhibit protein synthesis. Normal yeast strains are generally impermeable to these compounds, so the first step in isolation of strains that are resistant to these drugs because of changes in the protein synthesizing system, is to isolate strains that are permeable to these antibiotics and hence sensitive to them.

1. Mutagenize the culture and dilute into YEP-glucose broth (1:5). Incubate overnight.
2. Dilute to 4×10^7 cells/mL in YEP-glucose broth, containing the desired antibiotic (100 μg/mL emetine; 300 μg/mL trimethoprim) and incubate 1 h at 32°C.
3. Raise the temperature to 42°C and hold for 7 h, which reduces the viable count to 7×10^5 cells/mL (see Note 2).
4. Wash the cells with water to remove the antibiotic, dilute and plate on YEP-glucose agar.
5. After growth, replica plate to plates containing the antibiotic to detect colonies sensitive to the drug.
6. Use sensitive strains to isolate mutants, resistant to these antibiotics because of alterations in the mechanisms for protein synthesis (Littlewood, 1972 [9]).

3.3.9. PEP4 Mutants (11)

These mutants are recessive and pleiotropic, and cause deficiencies in carboxypeptidase Y, proteinase A, RNases, and a repressible alkaline phosphatase. The PEP4 gene product may be required for maturation of several vacuolar hydrolases.

1. Mutagenize the yeast strain according to standard methods, dilute, and plate on YEP-glucose agar. Incubate 2–3 d.

2. To 3 mL molten 0.6% agar at 50°C, add 2 mL APE solution, mix, and pour over an individual plate.
3. When the agar has solidified (10 min), gently pour over 3.5–4 mL of GBC solution. Within a few minutes, wild-type colonies turn red, whereas mutant colonies turn red more slowly.
4. Pour off the GBC solution as soon as the colonies start to turn red.
5. Stab mutant columns with a sterile toothpick, streak on YEPD, and retest.

3.3.10. Membrane Mutants (12)

These include inositol-requiring mutants, inositol-secreting mutants, and sterol-requiring mutants, and are obtained by standard methods for mutagenizing the desired yeast strain. Inositol-secreting mutants may be detected by replica plating the colonies, after mutagenesis, to an inositol-deficient medium covered with a lawn of an inositol-requiring, ade1 or ade2 strain. Colonies secreting inositol are surrounded by a halo of red satellite colonies. Inositol is a precursor of a number of membrane phospholipids, essential for metabolism and growth of yeasts.

3.3.10.1. INOSITOL-REQUIRING MUTANTS

1. Dilute suspension appropriately and spread an amount sufficient to give 100–150 colonies per plate, on several plates of YEP-glucose agar.
2. Incubate at 30°C for 3–5 d.
3. Replica plate to minimal medium with and without inositol, and incubate for 3–7 d.
4. Select colonies growing on minimal medium with inositol but not on minimal medium without inositol.

3.3.10.2. INOSITOL-EXCRETING MUTANTS (SEE NOTE 21)

1. Spread mutagenized, diluted yeast strain on YEP-glucose agar as before, and incubate for 3–5 d at 30°C.
2. Spread a lawn of the indicator strain on plates of the minimal medium and let dry.
3. Replica plate the colonies from the master plate (YEP-glucose) to the plates having the lawn. Incubate for 3–7 d at 30°C.
4. Look for white colonies on the selective medium, surrounded by a red halo, formed by the cells of the inositol- and adenine-requiring mutant.

3.3.10.3. STEROL-REQUIRING MUTANTS (SEE NOTES 22–24)

Sterol-requiring mutants *(13)* are obtained by mutagenizing with UV-light, and selecting for nystatin-resistant colonies that are temperature-sensitive at 36°C, by selecting for strains resistant to nystatin (15–28 mg/L) in the presence of cholesterol (40 mg/L).

Sterol-requiring mutants can also be obtained by selecting for strains requiring heme for growth.

1. Plate approx 10^7 cells/plate, on YEP-glucose agar containing cholesterol, and irradiate with a UV germicidal lamp, at 45-cm distance, for 30 s.
2. Incubate plates for 2 d at 22°C.
3. Harvest the cells. Spread on plates of YEP-glucose, containing cholesterol and nystatin, approx 10^6 cells/plate.
4. Incubate for 2 wk at 22°C.
5. Replica plate to YEP-glucose plates and incubate 24 h at 36°C, to identify thermosensitive mutants.
6. Test strains resistant to nystatin at 22°C and unable to grow at 36°C, from steps 4 and 5, for ergosterol and/or cholesterol requirements at 22, 25, 28, 30, and 36°C.

3.3.11. Mutants Auxotrophic for 2'-Deoxythymidine 5'-Monophosphate (dTMP) (see Notes 25 and 26 and ref. 14)

These mutants are useful in the study of DNA replication. Since normal wild-type yeast cells are not permeable to thymidine monophosphate, the first step in obtaining the desired mutants is to obtain mutants which are permeable to dTMP. These mutants can then be used for investigation of the incorporation of compounds such as 5-bromo-deoxyuridine 5-monophosphate into the DNA.

1. Grow the wild-type yeast to exponential phase and wash cells in phosphate buffer.
2. Plate 2×10^7 cells/plate on minimal medium containing sulfanimamide and aminopterin as described earlier, and incubate at 34°C for 4 d.
3. Pick colonies appearing on the plates, test, and discard petites (90%). If glycerol is used as the sole carbon source, *petite colonie* mutants should not appear. If the carbon source is glucose, it may be possible to identify the RC colonies visually by their size.
4. Plate grande colonies on SG medium containing sulfanilamide, aminopterin, and dTMP (10 g/mL). These strains should be respiratory-competent and highly permeable to dTMP.

To obtain mutants requiring dTMP:

1. Mutagenize these strains by standard methods using EMS, dilute, and plate on medium (YEP-glucose or YEP-glycerol agar), containing 100 μg/mL of dTMP. Incubate 5 d at 30°C.
2. Replica plate colonies on these plates to YEP-glucose and SD medium with and without dTMP for 3 d at 30°C, and isolate clones auxotrophic for dTMP.

3.3.12. Glycolytic Cycle Mutants (15): Alcohol Dehydrogenase Mutants

These include alcohol dehydrogenase tADH), pyruvate carboxylase (pyc), and phosphoenolpyruvate carboxykinase (PEPCK) mutants, and also hexokinase and phosphofructokinase mutants and mutants affecting galactose utilization and phosphogluconate dehydrogenase mutants. Most of the mutants are isolated by methods based on failure to grow on glucose, and growth on glycerol or other nonfermentable substrates.

3.3.12.1. ALCOHOL DEHYDROGENASE MUTANTS

These are selected for their resistance to allyl alcohol. Yeasts producing normal amounts of alcohol dehydrogenase convert allyl alcohol to acrolein, which is toxic, and are killed. Mutants lacking the gene encoding this enzyme do not convert allyl alcohol and survive.

1. Either:
   a. Mutagenize the yeast strain (petite) and spread 10^7 cells/plate on medium containing the lowest concentration of allyl alcohol, followed by plating the colonies obtained on the medium having the next higher concentration, until the highest is reached; or
   b. Grow the culture in YEP-glucose medium in the turbidostat, and increasing the concentration of allyl alcohol from zero, first to 10–12 mM, and increasing the concentration stepwise until the desired concentration is reached.
2. Isolate strains resistant to allyl alcohol, grow, break the cells, and test for altered forms of ADH using SDS-PAGE electrophoresis (*see* Note 27).

3.3.12.2. PYRUVATE CARBOXYLASE MUTANTS (PYC) *(16)*

These mutants have nutritionally complex requirements, which makes it desirable to isolate them in otherwise wild-type strains. They affect the anapleurotic reactions which supply intermediates to the Kres cycle reactions. This may make it possible to investigate the interface between reactions taking place within the mitochondria and those going on in the cytoplasm. The mutation involved in a deficiency in pyruvate carboxylase is a recessive single-gene nuclear one. These mutants resemble biotin-deficient mutants in having an aspartic acid requirement; pyruvate carboxylase requires biotin as a cofactor.

1. Mutagenize the yeast with EMS (1 h in 3% EMS), recover the cells, wash first with 6% $Na_2S_2O_3$, and then with phosphate buffer.
2. Dilute the suspension and plate on YNB-glucose agar, 100–200 viable cells/plate, to eliminate amino acid auxotrophs.

3. Replica plate the colonies to YEP-ethanol and YNB-ethanol. Isolate colonies growing on the former medium but not on the latter.
4. Mate these colonies to a wild-type haploid and isolate segregants, unable to grow on YEP-ethanol.
5. Cross these segregants to wild-type strains again, and continue isolation of segregants until strains giving high spore germination and 2:2 segregation of the character are obtained.
6. Cross this strain to an ADH-negative strain, to show that it has intact Krebs cycle and oxidative phosphorylation systems. Only ADH-negative strains having these systems, survive. If these pathways are inhibited in ADH-negative strains, the cells die. If reactions involving Krebs cycle enzymes are to be investigated using these mutants, they must, of course, be present.

3.3.12.3. PHOSPHOENOLPYRUVATE
CARBOXYKINASE (PEPCK) MUTANTS (*SEE* NOTE 28 AND REF. *17*)

These mutants are selected from colonies growing on media containing glycerol, but not growing on ethanol as C source.

1. Mutagenize and wash the culture, and transfer the cells to medium containing ethanol as sole carbon source, and nystatin (3 µg/mL) to kill growing cells.
2. The cells are diluted as required and plated on YEP-glucose agar, and incubated 3–4 d at 25–30°C.
3. Colonies are replica plated to YEP-glycerol and YEP-ethanol media.
4. Colonies growing on glycerol but not on ethanol are selected, the cells grown to yield 100 mg wet wt of cells, broken by vortexing with glass beads, and tested using the assay system described earlier.

3.3.13. Transport Mutants (18)

Transport processes can involve the entire range of compounds and elements that must cross the boundaries of the cell itself, and the membranes that separate the organelles within the cell; the nucleus, the mitochondria, the vacuole, Golgi bodies, endoplasmic reticulum, and such microbodies as peroxisomes, glyoxysomes, and others, where reactions producing compounds toxic to the cell (hydrogen peroxide, for instance) are sequestered until they can be detoxified. These substances can be grouped into small molecules and large molecules, the latter including mostly proteins such as enzymes and also heterologous proteins excreted by engineered yeast strains and having therapeutic (hormones, antigens such as Heptatitis B surface antigen) and technological value (industrial enzymes). Small molecules include H^+-ions, other cations and anions

Mutagenesis in Yeast

required by the cell, neutral molecules such as sugars and amino acids, and extracellular metabolites such as ethanol and polyhydroxy alcohols and glycolipids. All of these must be transported in and out of the cell or across membranes within it. The study of transport of ions and small molecules, through specific channels in the membranes, is a whole field of research in itself, so in this chapter, it is possible only to mention a few general principles used in the investigation of these phenomena. The transport of proteins from the site of their biosynthesis, and their excretion into the vacuole or secretion into the external medium, is also too complex a system to discuss in detail here, and will likewise be discussed only in very general terms.

Two types of transport mutants exist: In the first, the phenotype is of a recessive mutation to resistance to inhibitors such as L-ethionine, 5-fluorouracil, L-canavanine, 4-aminopyrazolepyrimidine, and pyrithiamine. These compounds are often analogs of the compound whose transport is inhibited. In the other group, a transport system in inhibited but there is no resistance to an inhibitor. Mutations affecting sugar transport usually fall into the second group.

The *petite* mutation may also affect sugar transport, galactose transport in particular being affected. In general, the first evidence that a mutation in the transport process has occurred is that the substance that is normally transported, remains outside the cell or compartment, instead of being found within it.

3.3.14. Secretory Mutants *(see Notes 29 and 30 and ref. 19)*

These are defined as mutants that normally produce an extracellular or intraperiplasmic enzyme (invertase or acid phosphatase, for instance), and in which the mutant continues to produce the enzyme, but it is not released into the periplasmic space or the environment. These mutants can be isolated by searching for strains that no longer carry out the normal wild-type function, and testing them to determine the presence of the enzyme in the cytoplasmic extract, and its absence from the normal.

4. Notes
4.1. Mutation Induction by Irradiation

1. Irradiation chambers: For X-rays and other high-energy systems delivering penetrating radiation (X-rays [gamma-irradiation], α-particles, β-particles, various atomic nuclei), follow the operator's instructions, put your

hands in your pockets, and stay out of the way. Do not attempt to use the equipment yourself unless you are a trained operator.
2. For UV irradiation, there are a variety of sources available. Some are almost fully automatic, so that when the dish is in place, one can start the timer and shaker, to agitate the culture, and take samples at the desired intervals. In its simpler form, one may hold the dish under the source after removing the lid, and agitate by hand, removing the dish for sampling. You may find that one hand is more deeply tanned than the other, but you will not, with good fortune, acquire skin cancer. If there is any danger of eye exposure, use UV-protective goggles.

4.2. Chemical Mutagens

3. Chemical mutagens are highly toxic and dangerous. Procedures consistent with local and national safety guidelines should be used, and followed exactly.
4. The culture may be enriched in mutants impairing growth by transferring the treated cells to medium containing nystatin, which kills growing cells but not nongrowing ones. For example, cultures can be enriched in auxotrophs by growth in minimal medium containing nystatin. Prototrophs will grow in this medium and are killed, whereas auxotrophs do not grow, and survive. After suitable incubation, the culture is plated out as before.
5. Auxotrophic mutants may be identified and isolated by plating a culture, preferably enriched, on a complete medium containing Magdala Red. The auxotrophs take up more of the dye and so are more deeply colored. The dye is nontoxic to yeasts and the colonies can be isolated and the phenotypes determined (*see* Horn and Wilkie, 1966 *[1]*).
6. When a mutation confers a selectable phenotype, such as resistance to killing by heavy metals, the selective agent can be used in an initial enrichment step, to enhance the mutant yield.

4.3. Petite *Mutants*

7. *Petites* induced by acriflavin may retain nonfunctional mtDNA in the mitochondrial genome that will stain with DAPI (4',6-diamidino-2-phenylindole; a fluorescent dye [19]), and that may carry genes for antibiotic resistance in cryptic form. *Petites* induced with ethidium bromide, which destroys mtDNA in growing and nongrowing DNA, have no DNA at all (rho^0 *petites*). Interestingly, mutants of *Schizosaccharomyces pombe* entirely lacking mtDNA have been isolated, and these have been found to exhibit alkaloid resistance *(21)*.
8. Cultures incubated in the presence of either acrivflavin or ethidium bromide, for induction of *petite* mutants, must be kept in the dark.

9. Obtaining spontaneous *petites* by starvation: Cultures are inoculated into 5 mL of Schöpfer's medium (containing, per L, 30 g glucose, 1 g asparagine, 1.5 g KH_2PO_4, and 0.5 g $MgSO_4 \cdot 7H_2O$), in culture tubes or small bottles, and incubated 6–10 d at 30°C. The cells are then recovered by centrifugation and plated on YEP-glucose agar. Small colonies are picked to YEP-glucose and YEP-glycerol agars, and any *petite* mutants found are isolated and retained.

4.4. Antibiotic-Resistant Mutants

10. These mutants usually occur spontaneously at a relatively high frequency and mutagenesis is not necessary. However, mutagenesis with Mn^{2+} or other agents can be used if a higher mutation frequency is desired.
11. These mutants are resistant to relatively high levels of antibiotic, and can be used to demonstrate recombination in mtDNA.
12. Where possible, the strains should be tested genetically to be sure that the mutation is mitochondrial and not nuclear. This cannot be done with nonsporulating industrial yeast strains. However, since these strains are not haploid, the mutations are unlikely to be nuclear.

4.5. Temperature-Sensitive Mutants

13. The method has been used in the isolation of cell division cycle mutants, which may be detected by Giemsa staining of cells grown at the restrictive temperature to show the point at which nuclear division and other events in the cell division cycle have been halted.
14. Osmotic-remedial mutants may be isolated by replica plating to a third plate containing elevated concentrations of KCl ($1–3M$) and incubating at the restrictive temperature along with the second plate, isolating colonies that grow on high-KCl medium and not on normal medium.

4.6. Mutants Defective in Nuclear Fusion (karl-1 Mutants)

15. Other strains carrying the *karl-1* mutation, of both mating types and with other auxotrophic requirements, are in the collection of the Carlsberg Laboratories (Copenhagen, Denmark).

4.7. Mutants Having Easily Lysed or Easily Digested Cell Walls

16. These mutants are often unstable and revert rapidly to growth on normal media.
17. They may also show increased sensitivity to increased or decreased temperatures, antibiotics, or other drugs.

18. Mutants with easily digested cell walls were originally intended for use in livestock feeds and human food.

4.8. "Kamikaze" Mutants

19. Emetine sensitivity results from two mutations, sensitivity to trimethoprim, from one.
20. In step 3, wild-type cells (carrying only the particulare mutation to temperature sensitivity), are killed, whereas the mutants are not.

4.9. Inositol-Excreting Mutants

21. Most of the inositol-excreting mutants obtained by this method were constitutive for inositol-1-phosphate synthetase, but the original authors found one nonconstitutive mutant in which the synthesis of phosphatidylcholine was affected.

4.10. Sterol-Requiring Mutants

22. The original authors *(12)* obtained 120 isolates of this sort, of which, 12 grew better in the presence of ergosterol.
23. Selection of strains requiring supplementation of the medium with heme for growth, also yielded mutants in the sterol biosynthetic pathway. At first, these strains grew poorly in medium containing ergosterol or cholesterol instead of α-aminolevulinic acid. After several generations, spontaneous mutants in the sterol biosynthetic pathway appeared (Taylor and Parks, *14*).
24. Mutants requiring unsaturated fatty acids can be isolated by normal methods, by replica plating mutagenized cells on media with and without different fatty acids. These are usually incorporated into the medium as fatty acid esters.

4.11. dTMP-Requiring Mutants

25. The frequency of dTMP-requiring mutants is 4–11 per 11,000 colonies tested, which is rather low.
26. At least one mutation in the PHO80 gene, controlling phosphatase expression, is allelic to one or more mutations conferring permeability to dTMP on yeast.

4.12. Mutants Deficient in Alcohol Dehydrogenase

27. Cell extracts made by vortexing a small amount of cells mixed with ballotini beads in phosphate buffer (pH 6.0) are sufficient for electrophoretic determination of changes in alcohol dehydrogenase.

4.13. PEPCK Mutants

28. Mutants deficient in fructose-1,6-bisphosphatase, though mentioned as belonging in this group, are not isolated by this procedure *(17)*.

4.14. Secretory Mutants

29. The general pathway for secretion of proteins, including enzymes, is from the site of synthesis (endoplasmic reticulum) to the Golgi body, where it is packaged into vesicles, which are then directed to the vacuole, the mitochondrion, or the cytoplasmic membrane. The vesicle then fuses with the membrane (at least in the latter case) and the protein is released into the other compartment, for instance, the periplasmic space. The mechanism by which the protein is transported across the cell wall and capsule, if any, is not known; this may be by simple diffusion, although this implies the presence of pores of a size that enzymes such as invertase or acid phosphatase are retained in the periplasmic space because of being too large to pass through these channels.
30. There is an unusual mutant of this class, in which the *grande* form of the yeast utilizes sucrose, but the *petite* mutant does not. Tests of the cell extract shows that invertase is still present in the cell extract, but is not transported into the periplasmic space *(20)*.

4.15. Sulfite-Resistant Mutants

31. Interesting new examples of this class of mutant have been isolated in *S. cerevisiae*, following EMS treatment, MET genes being possibly implicated *(22)*.

References

1. Horn, P. and Wilkie, D. (1966) Use of magdala red for the detection of auxotrophic mutants of *Saccharomyces cerevisiae. J. Bacteriol.* **91**, 1388,1389.
2. Rickwood, D., Dujon, B., and Darley-Usmar, V. M. (1988) Yeast mitochondria, in *Yeast: A Practical Approach* (Campbell, I. and Duffus, J. H., eds.), IRL, Oxford, pp. 185–254.
3. Putrament, A., Baranowska, H., and Prazmo, W. (1973) Induction by manganese of mitochondrial antibiotic resistance mutations in yeast. *Mol. Gen. Genet.* **126**, 357–366.
4. Williamson, D. H. and Fennell, D. J. (1975) The use of fluorescent DNA-binding agent for detecting and separating yeast mitochondrial DNA, in *Methods in Cell Biology*, vol. 12 (Prescott, David M., ed.), Academic, New York, pp. 335–351.
5. Von Borstel, R. C., Cain, K. T., and Steinberg, C. M. (1971) Inheritance of spontaneous mutability in yeast. *Genetics* **69**, 17–27.
6. Condé, J. and Fink, G. R. (1976) A mutant of *Saccharomyces cerevisiae* deficient in nuclear fusion. *Proc. Natl. Acad. Sci. USA* **73**, 3651–3655.

7. Cohen R. E., Ballou, L., and Ballou, C. E. (1980) *Saccharomyces cerevisiae* mannoprotein mutants. Isolation of the *mmn5* mutant and comparison with the *mmn3* strain. *J. Biol. Chem.* **255,** 7700–7707.
8. Venkov, P., Hadjiolov, A. A., Battaner, E., and Schlessinger, D. (1974) *Saccharomyces cerevisiae*: sorbitol dependent fragile mutants. *Biochem. Biophys. Res. Commun.* **56,** 599–604.
9. Mehta, R. D. and Gregory, K. F. (1981) Mutants of *Saccharomyces cerevisiae* and *Candida utilis* with increased susceptibility to digestive enzymes. *Appl. Env. Microbiol.* **41,** 992–999.
10. Littlewood, Barbara S. (1972) A method for obtaining antibiotic-sensitive mutants in *Saccharomyces cerevisiae. Genetics* **71,** 305–308.
11. Jones, E. W. (1977) Proteinase mutants of *Saccharomyces cerevisiae. Genetics* **85,** 23–33.
12. Culbertson, M. R. and Henry, Susan A. (1975) Inositol-requiring mutants of *Saccharomyces cerevisiae. Genetics* **80,** 23–40.
13. Karst, P. and Lacroute, F. (1977) Ergosterol biosynthesis in *Saccharomyces cerevisiae. Mol. Gen. Genet.* **154,** 269–277.
14. Taylor, F. R. and Parks, L. W. (1980) Adaptation of *Saccharomyces cerevisiae* to growth on cholesterol: selection of mutants defective in the formation of lanosterol. *Biochem. Biophys. Res. Commun.* **95,** 1437.
15. Little, J. G. and Haynes, R. H. (1979) Isolation and characterization of yeast mutants auxotrophic for 2'-deoxythymidine 5'-monophosphate. *Mol. Gen. Genet.* **168,** 141–151.
16. Ciriacy, M. (1975) Genetics of alcohol dehydrogenase in *Saccharomyces cerevisiae.* 1. Isolation and genetic analysis of adh mutants. *Mutat. Res.* **29,** 315–326.
17. van de Poll, K. W. and Schamhart, D. H. J. (1977) Characterisation of a regulatory mutant of Fructose 1, 6-Biophosphate in *Saccharomyces carlbergensis. Mol. Gen. Genet.* **154,** 61–66.
18. Cooper, T. G. (1981) In *The Molecular Biology of the Yeast* Saccharomyces. *Metabolism and Gene Expression.* Cold Spring Harbor Laboratory, Cold Spring Harbor, NY, pp. 399–461.
19. Novick, P., Field, C., and Shekman, R. (1980) Identification of 23 complementation groups required for post-translational events in the yeast secretory pathway. *Cell 21,* 205–215.
20. Spencer, J. F. T., Spencer, D. M. and Miller, R. (1983) Inability of *petite* mutants of industrial yeasts to utilize various sugars, and a comparison with the ability of the parent strains to ferment the same sugars microaerophilically. *Z. Naturforsch.* **38c,** 405–407.
21. Massardo, D., Mauna, F., Sachöfer, B., Wolf, K., and DelGiudice, L. (1994) Complete absence of mtDNA in the *petite*-negative yeast *Schizosaccharomyces pombe* leads to resistance towards the alkaloid lycorine. *Curr. Genet.* **25,** 80–83.
22. Xu, X. Wightman, J. D., Geller, B. L., Arram, D., and Bakalinsky, A. T. (1994) Isolation and characterisation of sulfite mutants of *S. cerevisiae. Curr. Genet.* **25,** 488–496.

CHAPTER 4

Rare-Mating and Cytoduction in *Saccharomyces cerevisiae*

John F. T. Spencer and Dorothy M. Spencer

1. Introduction

Rare-mating is an adaptation of techniques used for normal classical mating in yeasts. It was first used to study the switching of mating types in *Saccharomyces cerevisiae* by Gunge and Nakatomi *(1)*, but was later adapted to the investigation of industrial yeasts that have low mating frequencies and low spore viabilities, or that do not mate or sporulate at all *(2)*. The method depends on the fact that occasional mating-type switching occurs in industrial yeasts, which are normally diploids or of higher ploidy, and may display aneuploidy as well. This results in the occurrence at low frequency of mating cells, that can then conjugate with a known laboratory mating strain of either **a** or α mating type (or **aa** or αα mating type), and whose progeny then carry the auxotrophic or other markers present in the laboratory strain. In addition, the ability to sporulate, in the progeny, is often enhanced, so that genetic analysis can be carried out on the strain, and some idea of the nature of the genome of the industrial strain can be obtained.

Cytoduction is the transfer of subcellular organelles, such as mitochondria and/or killer virus-like particles (VLPs) from one strain of yeast to another, without transferring any nuclear genes. This can be done by fusing or mating a strain carrying a *kar1-1* mutation, with the strain that is to be used as the donor of the organelles, isolating strains carrying the *kar1-1* nucleus and the organelles, and fusing cells of this strain with the desired recipient strain. The *kar1-1* strains are defective in nuclear fusion and the nuclei segregate during mitosis, so that strains originating from cells carrying only the *kar1-1* nucleus can be isolated, and these cells also carry the various subcellular organelles of the donor strain. If killer VLPs are to be transferred, a respiratory-competent *kar1-1* mutant strain can be used, but if mitochondria are to be transferred, both the *kar1-1* mutant and the recipient industrial yeast strain must be converted to the *petite* form by treatment with ethidium bromide, to destroy the original mitochondrial DNA before replacing it with that obtained from the donor strain. This will be described later in the Methods section.

2. Materials

2.1. Obtaining Petite Mutants by Acriflavin Treatment

1. Acriflavine solution: 0.4% w/v in sterile water. Keep the stock solution foil-wrapped to exclude light, at 4°C.
2. YEP-glucose broth: 1% w/v yeast extract, 0.5% w/v peptone, 2% w/v glucose, dispensed as 5-mL aliquots in tubes or bottles. Solidify, if required, with 2% w/v agar.
3. YEP-glycerol plates (for testing putative *petites*): 1% w/v yeast extract, 1% w/v peptone, 2% v/v glycerol, 2% agar.

2.2. Obtaining Petite Mutants by Ethidium Bromide Treatment

1. YEP-glucose-ethidium bromide broth: Add ethidium bromide solution (10 mg/mL stock in sterile water, kept foil-wrapped at 4°C) to YEP-glucose broth to give a final concentration of 20 µg/mL (e.g., 0.01 mL stock added to a 5-mL aliquot of YEP-glucose broth).
2. YEP-glycerol plates (for testing putative *petites*): As in Section 2.1.

2.3. Mating and Selecting Hybrids

1. YEP-glucose broth: As in Section 2.1.2.
2. YEP-glucose broth: Aliquots in vessels allowing reasonable aeration, e.g., 5-mL vol in 50-mL conical flasks.
3. YNB-glycerol-ethanol plates: 0.67% w/v yeast nitrogen base (without amino acids), 2% v/v glycerol (or potassium or sodium lactate), 3% v/v ethanol (to inhibit sporulation).
4. Sterile capped plastic centrifuge tubes (5-mL minimum capacity) suitable for low-speed centrifugation.

2.4. Cytoduction

1. Media as in Section 2.3.
2. A yeast strain carrying the *kar1-1* mutation and, preferably, other mutations as markers, especially on *ade1* or *ade2* mutation to give a red color to the colonies. This can be constructed *(3)*, but it is usually quicker and easier to obtain a strain from a private or public commercial collection.
3. YEP-glucose plates freshly inoculated with a killer-sensitive strain of *S. cerevisiae*, so as to produce continuous lawns on incubation.

3. Methods

3.1. Obtaining Petites by Acrifavine Treatment

1. Add one loopful of acriflavin solution to 5 mL of YEP-glucose broth, then inoculate with the industrial yeast strain.
2. Incubate the culture without agitation at 30°C for 2–3 d *in the dark*, and streak out a loopful on YEP-glucose agar.
3. Pick small colonies and test for respiratory deficiency: Do this by subculturing similar sized inocula (sterile toothpick or loop) into known positions on YEP-glucose and YEP-glycerol plates. Incubate c. 2 d at 30°C and compare the growth of the inocula on the two plates: True cytoplasmic *petites* are unable to grow on a nonfermentable (e.g., YEP-glycerol) medium. *Petite* cultures on YEP-glucose plates are very white in color, whereas respiratory-competent *wt* cultures tend to be creamy-colored.

3.2. Obtaining Petites by Ethidium Bromide Treatment (for Use in Transfer of Subcellular Organelles by Cytoduction) (see Note 2)

1. Inoculate liquid YEP-glucose medium containing 20 µg/mL of ethidium bromide, and grow at 30°C, in the dark (vessels can be foil-wrapped, to exclude light), as in the method for induction of *petites* with acriflavin, but with agitation *(4)*.
2. Reinoculate the culture into fresh medium containing ethidium bromide, and reincubate.
3. Repeat step 2 to ensure that 100% of the cells are converted to *petites*.
4. Streak out the culture after 2–3 d and pick and test small colonies (*see* Section 3.1.3.).

3.3. Rare-Mating (see Note 3 and Ref. 2)

1. Grow the laboratory mating strains, carrying any desired auxotrophic markers, and also the *petites* of the industrial strains, separately in YEP-glucose broth for 48 h.
2. Recover the cells centrifugally (500g for 5 min room temp) and mix the cells—resuspended aseptically in the residual supernatant—in one of the centrifuge tubes. Inoculate aliquots of the mixed pairs of strains into fresh YEP-glucose broth, either in shallow layers in flasks or bottles, or in slanted tubes for better access of air. Incubate at 30°C without agitation, for approx 3–7 d (the time is not critical, but may depend on the strains used).
3. Again, recover the cells centrifugally, wash in sterile dH$_2$O, and spread a heavy suspension on YNB-glycerol-ethanol plates. Incubate at 30°C.
4. Prototrophic, respiratory-competent colonies usually appear after 3–5 d. Isolate these and purify by restreaking on the same medium. They may then be subjected to genetic analysis (*see* Chapter 6 and Notes 4 and 5).

3.4. Cytoduction (see Note 6)

3.4.1. Transfer of Killer VLPs

1. Mate the donor strain (a known killer) with the *kar1-1* strain, carrying either the *ade1* or *ade2* mutation (giving a red color) and isolate red colonies by restreaking.
2. Test these isolates for the killer character by patch-inoculation onto the spreads of killer-sensitive cells; incubate 1–2 d at 30°C and look for the clearing zones that indicate that toxin has been released, killing and lysing sensitive cells (*see* ref. 5).
3. Mate one of these *kil*$^+$ strains with the chosen recipient strain (fusion can also be used; *see* Note 6).
4. Isolate white strains having the killer character. If the strains sporulate, dissect asci from these strains and look for red segregants (*see* Chapter 6 on meiotic mapping). If there are none, the nuclei carrying the *kar1-1* mutation have segregated out and presumably the new killer strain has no genes originating from it.

3.4.2. Transfer of Mitochondria Between Industrial Yeast Strains

The technique is the same as in Section 3.4.1., except that both the *kar1-1* vector strain and the proposed recipient strain must be converted to the *petite* form, using ethidium bromide to destroy all the mtDA (*see* Note 2).

1. Mate the donor strain and the *kar1-1 petite* strain.
2. Isolate red colonies: The *kar1-1 petite* is white, and strains which have acquired mitochondria from the donor, will be red, which makes the task easier.
3. Mate these red, respiratory-competent strains with the *petite* mutant obtained from the desired recipient strain.
4. Isolate white respiratory-competent colonies. Test in any way possible for the presence of genes from the *kar1-1* mutant strain (*see* Note 7).

4. Notes

1. All growth media are, of course, sterilized by conventional methods—usually autoclaving at 15 lb (sq in.) (121°C) for 15 min. The slight caramelization that may occur to sugars (especially glucose) in these media can be ignored for most purposes.
2. The method given herein is for liquid cultures, but *petites* can also be conveniently induced on solid medium, by streaking cultures on YEP-glucose plates containing 10–50 µg/mL ethidium bromide. After incubation at 30°C (2–4 d), the small colonies arising are tested as in Section 3.2. Ethidium bromide may cause complete destruction of the mtDNA in yeast, and converts both growing and nongrowing cells to the *petite* form, whereas *petites* made using acriflavin typically contain nonfunctional mtDNA.
3. A variant of the protocol given here involves mixing the two freshly grown cultures, and then filtering the mixture through a sterile 0.45-µm filter—this places the cells in close proximity, thereby encouraging mating. Enclose the filter in a moist chamber, to prevent drying, for 2–3 h at 20–30°C, then transfer the filter to a tube or bottle of YEP-glucose broth and proceed as in Section 3.3.2.
4. Yeast hybrids obtained by rare-mating may be sporulating or nonsporulating. Sporulating strains may be analyzed by dissection of asci and tetrad analysis in the normal way, provided that spore viability is high enough. Spore viability may be higher in strains obtained by mating with **aa** or $\alpha\alpha$ mating diploids, than with **a** or α haploid maters, or vice versa, depending possibly on the ploidy of the original strain. Nonsporulating strains could be analyzed using mitotic recombination, induced by low doses of UV irradiation.
5. Hybrids between laboratory mating strains and industrial yeasts either marked with a mutation to mitochondrial antibiotic resistance or by the ability to utilize starch, not present in the laboratory strain, have been obtained in this way. The mutation to antibiotic resistance (chloramphenicol, erythromycin, or oligomycin) is cryptic in the *petite* mutant, but is again expressed in the hybrid, so the hybrid/sporulating cultures are not the result of a mutation in or contamination of the auxotrophic laboratory strain.

6. Cytoduction can be effected via protoplast fusion protocol, *see* Chapter 5.
7. If the recipient strain sporulates, the presence of genes from the vector strain *(kar1-1)* can be detected by dissecting asci as before and testing the segregants. If it does not, sometimes the presence of genes from the *kar1-1* nucleus can be detected by inducing mitotic recombination, with a light dose of UV irradition, and looking for recombinant sectors, red in the first instance. Appearance of red sectors shows that the strain carries either the entire nucleus of the *kar1-1* strain, or that the nuclei of the two strains have fused and the nucleus of the hybrid carries genes from the *kar1-1* mutant. The latter condition can only be verified if the hybrid sporulates and yields viable spores.

References

1. Gunge, N. and Nakatomi, Y. (1972) Genetic mechanisms of rare matings of the yeast *Saccharomyces cerevisiae* heterozygous for mating type. *Genetics* **70,** 41–58.
2. Spencer, J. F. T. and Spencer, Dorothy M. (1977) Hybridization of nonsporulation and weakly sporulating strains of brewer's and distiller's yeast. *J. Inst. Brewing* **83,** 287–289.
3. Condé, J. and Fink, G. R. (1976) A mutant of *Saccharomyces cerevisiae* defective for nuclear fusion. *Proc. Natl. Acad. Sci. USA* **73,** 3651–3655.
4. Rickwood, D., Dujon, B., and Darley-Usmar, V. M. (1988) Yeast mitochondria, in *Yeast: A Practical Approach* (Campbell, I. and Duffus, J., eds.), IRL, Oxford, pp. 185–254.
5. Woods, D. R. and Bevan, E. A. (1968) Studies on the nature of the killer factors produced by *Saccharomyces cerevisiae*. *J. Gen. Microbiol.* **51,** 114–126.

CHAPTER 5

Protoplast Fusion in *Saccharomyces cerevisiae*

Brendan P. G. Curran and Virginia C. Bugeja

1. Introduction

Saccharomyces cerevisiae can exist as either a haploid cell of either *Mat a* or *Mat α* mating type, a diploid cell or a polyploid cell, but the transfer of genetic material is normally restricted to conjugation between haploids of opposite mating type. It is possible, however, to isolate genetically stable yeast hybrids from strains that are unable to mate by using protoplast fusion. In this process, the yeast cell walls are enzymatically removed, the resulting protoplasts are fused together using polyethylene glycol (PEG) and calcium ions (Ca^{2+}), before being embedded in agar under appropriate selective conditions. Colonies of genetically stable hybrids are normally visible within 5–10 d at 25°C.

The mechanism by which PEG and Ca^{2+} bring about protoplast fusion is unknown, but it is thought that the PEG has the dual role of diminishing the electrostatic field between the lipid membranes *(1)* and removing water from the protoplasts. This allows extensive areas of lipid from different protoplasts to come into contact with one another. The calcium ions appear to perturb the membrane structure *(2)*, resulting in the intermingling of lipids from both membranes, and small cytoplasmic connections are made. Finally, as the protoplasts are rehydrated, they swell and the cytoplasmic connections coalesce, generating a heterokaryon.

The fused protoplasts can regenerate their cell wall if they are embedded in osmotically stabilized agar. Approximately 5% of heterokaryons generated in this way will undergo karyogamy to give a stable hybrid, whereas the other 95% will divide without karyogamy resulting in the stable inheritance of cytoplasmic markers, such as mitochondria *(3)*.

2. Materials

1. Strains: The strains to be fused must carry appropriate markers that complement, so that fusants may be selected for and parental cells against (*see* Note 1).
2. Liquid growth medium (YEPD) (2 × 25 mL): 1% w/v yeast extract, 2% w/v bactopeptone, 2% w/v glucose.
3. Regeneration medium (400 mL): 0.67% w/v yeast nitrogen base without amino acids, 2.0% w/v glucose, 18.2% w/v sorbitol, 3.0% agar (*see* Note 2).
4. Appropriate amino acid solutions, as stock solutions, 200X concentration needed in medium (the amino acids are those required by the auxotrophic parent strains).
5. MP buffer (500 mL): $1.0M$ sorbitol, $0.1M$ sodium chloride, $0.01M$ acetic acid, adjust to pH 5.5 with sodium hydroxide (IM) (*see* Note 2).
6. 60% w/v Polyethylene glycol (PEG) 4000 (50 mL).
7. $1M$ Calcium chloride (10 mL).
8. Sterile water (200 mL).
9. $1M$ β-Mercaptoethanol (10 mL).
10. Suc d'Helix Pomatia (Industrie Biologique Française, Paris, France), supplied sterile 1-mL glass vials.
11. Novozyme SP234 (Novo Enzyme Products Ltd., Windsor, UK).
12. Protoplasting solution: 6 mL MP buffer, 18 mg Novozyme SP234, 100 μL Suc d'*Helix pomatia*, 100 μL $1M$ β-mercaptoethanol; prepared immediately before use and filter-sterilized (0.45 μM) (*see* Note 4).

3. Method

Use sterile glassware and sterile techniques throughout. All procedures are carried out at room temperature unless stated otherwise.

3.1. Day 1

1. Autoclave regeneration medium and aliquot 19 × 10 mL samples into sterile universal bottles.
2. Pour the remainder into 19 Petri dishes.
3. Inoculate each of the yeast strains into 25 mL YEPD and grow overnight at 25°C.

Protoplasts in S. cerevisiae 47

3.2. Day 2

4. Dissolve the 10-mL aliquots of medium by steaming and then keep molten in a 45°C water bath. Add 0.1-mL of the appropriate 200X concentration amino acid solution(s) to two sets of six tubes (for use as regeneration medium for the parental strains).
5. Harvest the cells in mid exponential growth phase ($5 \times 10^6 - 2 \times 10^7$ mL) by spinning at 3000g for 5 min (*see* Note 5).
6. Wash once by resuspending the pellet in 10 mL of sterile water and harvesting again.
7. Resuspend in protoplasting solution (*see* Note 2). Incubate for 30 min at room temperature. Protoplast formation can be monitored by diluting 0.1 mL of the suspension in 0.9 mL of water and observing a decrease in cloudiness as the protoplasts burst.
8. Harvest the protoplasts at 3000g (5 min) and wash twice in 10-mL MP buffer by resuspending and harvesting.
9. Finally, resuspend in 3 mL of MP buffer and count a suitable dilution using a hemocytometer (1/80 dilution in MP buffer is normally sufficient).
10. Mix 2×10^7 protoplasts from each strain together and divide into two tubes.
11. Pellet as before, but resuspend one pellet in 2 mL of 60% PEG and 0.2 mL of 1M CaCl$_2$. Add 2 mL of MP buffer and 0.2 mL of 1M CaCl$_2$ to the other tube as a control. Allow to stand for 3 min at room temperature.
12. Add 6 mL of MP buffer, mix, and allow to stand at room temperature for 6 min.
13. Harvest as before, wash once in 10-mL MP buffer by resuspending the pellet and harvesting again. Resuspend in 1 mL of MP buffer.
14. Plating out (*see* Note 6).
    a. Add 3×0.1 mL aliquots of the fusion solution to 3×10 mL of molten minimal medium without amino-acids (*see* Note 1) and plate immediately.
    b. Dilute 0.1 mL into 9.9 mL of MP buffer and plate as in a. This dilution ensures discrete colony formation in cases of very efficient fusion/regeneration.
    c. Plate 0.3 mL of the control into one molten agar to check for any cross feeding.
    d. To calculate the percentage regeneration of the parental strains make an additional 1/100 dilution in MP buffer and plate 3×0.1 mL aliquots from the 1/100 dilution (b) and from this 1/10,000 dilution into minimal medium containing the requirements for the individual parental strains.
15. Incubate the plates for 4–6 d at 25°C (*see* Notes 7–9).

4. Notes

1. This protocol is written assuming that the parent strains carry complementary auxotrophic markers such that hybrid fusants are prototrophic. For parent strains that share an auxotrophic deficiency, the selective minimal medium must, of course, be appropriately supplemented.
2. Protoplasts from different strains can vary in their osmotic sensitivity, but 0.8 to 1.2M sorbitol is adequate for most *S. cerevisiae* strains.
3. The fusion efficiency of different PEG batches varies considerably. If the fusion efficiency is low but regeneration is adequate, try different sources of PEG.
4. This particular combination of enzymes and β-mercaptoethanol works extremely well, but any enzyme or combination of enzymes which produce yeast protoplasts can be used, e.g., Zymolase, Glusulase, and so on.
5. The speed at which different strains protoplast varies to some extent, but the most important aspect of protoplasting is the physiological state of the cells. Early to midexponentially growing cells are the most suitable. If there is difficulty in protoplasting, use younger cells.
6. It is of vital importance to embed the protoplasts because yeast protoplasts (unlike other fungi) cannot regenerate on the surface of the agar.
7. Fusion and regeneration frequencies vary considerably from strain to strain.
8. The transfer of cytoplasmic markers is extremely efficient during protoplast fusion, but 3% glycerol replaces the 2% glucose in the regeneration medium when wild-type and *petite* mutants are fused *(3)*.
9. The percent of fused protoplasts that can undergo karyogamy appears to be restricted to protoplasts containing nuclei at a specific point in G1 of the cell cycle *(4)*. Thus, when asynchronous populations of cells are used in a fusion, the percentage of fused protoplasts that undergo karyogamy is very low. Procedures that increase the proportion of cells at this point in G1 greatly improve hybrid formation frequencies *(4)*.

The most efficient means of preparing a population of cells at this point in G1 is to treat *Mat a* cells with 5 U of α factor (Sigma, St. Louis, MO) for 3–4 h before protoplasting them *(5)*. This can increase the frequency of hybrid formation 20-fold (depending on the strains used). A similar procedure using factor *a* on *Mat α* cells is not as efficient *(6)*.

Acknowledgments

The authors wish to thank Gail Cohen, Ruth Graham, and Jean Smith for preparing this manuscript.

References

1. Maggio, B., Ahkong, Q. F., and Lucy, J. A. (1976) Poly(ethylene glycol) surface potential and cell fusion. *Biochem. J.* **158,** 647–650.
2. Papahadjopoulos, D., Vail, W. J., Pangborn, W. A., and Poste, G. (1976) Studies on membrane fusion II. Induction of fusion in pure phospholipid membranes by calcium ions and other divalent metals. *Biochem. Biophys. Acta* **448,** 265–283.
3. Curran, B. P. G. and Carter, B. L. A. (1986) α-factor enhancement of hybrid formation by protoplast fusion in the yeast *S. cerevisiae* II. *Curr. Genet.* **10,** 943–945.
4. Curran, B. P. G. and Carter, B. L. A. (1989) The α-factor enhancement of hybrid formation by protoplast fusion in *S. cerevisiae* can be mimicked by other procedures. *Curr. Genet.* **15,** 303–305.
5. Curran, B. P. G. and Carter, B. L. A. (1983) α-factor enhancement of hybrid formation by protoplast fusion in the yeast *S. cerevisiae. J. Gen. Microbiol.* **129,** 1539–1591.
6. Bugeja, U. C., Whittaker, P. A., and Curran, B. P. G. (1988) The effect of α-factor on hybrid formation by protoplast fusion in *S. cerevisiae. Fems. Microb. Letts.* **51,** 101–104.

Chapter 6

Meiotic Analysis

John F. T. Spencer and Dorothy M. Spencer

1. Introduction
1.1. Tetrad Analysis

Diploid strains of asci, placed on suitable medium, sporulate and form four haploid spores/ascus, which may be arranged as tetrahedra, diamonds, or linearly. Strains forming linear arrays of spores in the ascus are rare, dissection of the asci with the micromanipulator is difficult, so the method is now little used. Hawthorne *(1)* used it extensively in his early studies of yeast genetics. Generally, the yeast spores are considered unordered, and the ascus walls are dissolved enzymatically and the spores separated with a micromanipulator and grown to form colonies, which may consist of haploid or diploid cells, depending on whether or not the strain is heterothallic. The segregation of auxotrophic or other markers can then be determined, and map distances, in centimorgans, estimated. Allelic relationships can also be investigated.

Tetrad analysis is used to determine the distance between two markers on the same chromosome, according to the number of crossovers between them, as indicated by the segregation of the aforementioned markers. If no crossing-over has occurred, the asci will all be parental ditype, PD (AB AB ab ab), where A and B are the genes being investigated. If one crossover has occurred, only tetratype asci, T, will be formed (AB Ab aB ab). Multiple crossovers will result in the formation of three ascus types: PD, T, and NPD, or nonparental ditype (Ab Ab aB aB). The map distance can be calculated from the formula $X = 100/2 \, [(T + 6 \, NPD)/(PD + NPD + T)]$, and is the average number of crossovers per chromatid. If the map distance is greater than 35–40 cM, the equation overestimates the true map distance, however *(2)*.

1.2. Random Spore Isolation (3)

Where a micromanipulator is not available, random spore isolation will yield a certain amount of genetic information. The method is based on disruption of the asci, usually enzymatically, and partition of the spores in a biphasic system, usually with mineral oil as one component. The spores are then spread on a plate for germination and growth, and the clones can then be replica plated to drop-out other selective media to determine the average segregation ratios. Recombination, linkage, and gene conversion can be determined in this way, though tetrad analysis *per se* is not possible. Postmeiotic segregation can be observed as sectored colonies.

2. Materials
2.1. Tetrad Analysis
2.1.1. Sporulation

1. Presporulation medium: 1% Yeast extract (all percentages are w/v, unless otherwise indicated), 1% peptone, 2% glucose, plus 1.5% agar.
2. Sporulation medium: McClary's medium, containing 1% K acetate, 0.25% yeast extract, and 0.1% glucose *(3)*, or raffinose-acetate medium (2% raffinose plus 1% K acetate).

2.1.2. Dissection

1. Dissection agar (YEPD): 1% Yeast extract, 1% peptone, 2% glucose, plus 2–2.5% agar, dispensed in 10-mL aliquots in tubes or bottles.
2. Dissecting chamber (*see* Note 1).
3. Microscope slides, 1.5 × 3", and a jar of alcohol to keep them in.
4. Dissecting needles, usually flat-ended (*see* Note 2).
5. Enzymes for dissolution of the ascus wall; Helicase (snail gut enzyme), Novozyme 234, Zymolyase, mushroom enzyme (preparation from common *Agaricus* from the market: The mushrooms are blended in a Waring blender in buffer, approx pH 6, filtered through cheesecloth, and then through an asbestos filter pad (R. A. Woods, personal communication).

2.1.3. For Determination of the Auxotrophic Requirements of the Single-Spore Clones

1. YEPD plates (*see* Section 2.1.2.).
2. Drop-out media containing 0.67% yeast-nitrogen base, plus all amino acids (usually, 20 mg/L) required by the original parental strains in the diploid, minus one, to identify the requirement for a particular amino acid.
3. Sterile velvet pads, filter paper, or other material for replica plating. Filter paper gives more clearly defined replicas.

2.2. Random Spore Isolation
2.2.1. Grinding Method
1. Sporulated yeast culture (*see* Section 3.1.1.2. and Note 3).
2. Snail enzyme or other enzyme for dissolution of ascus walls (*see* Section 2.1.2.).
3. Dounce or Potter-Elvehjem homogenizer, glass or Teflon pestle, motor-driven at approx 100 rpm *(3)*.
4. Gelatin solution, 15%.
5. Mineral oil, sterile.
6. Glass powder, sterile.
7. YEPD and drop-out plates *(see above)*.

2.2.2. Enzymatic Method (see *Notes 7 and 8*)
1. Sporulated yeast culture as before.
2. Enzyme preparation; Helicase, Zymolyase, Novozyme 234, mushroom enzyme, or other.
3. Other reagents and equipment as for spore dispersal by grinding (*see* Section 2.2.1.).
4. YEPD and drop-out plates *(see above)*.

3. Methods
3.1. Tetrad Analysis
3.1.1. Sporulation
1. Inoculate the desired culture (diploid or of higher ploidy) on to presporulation medium (tube, 5 mL, liquid, or agar, plate, or slant) and incubate at 25–30°C for 24–48 h.
2. From the presporulation culture, inoculate the sporulation medium (tube, 2 mL, liquid, or plate). Incubate at 20–30°C for 48–72 h, or until spores appear.

3.1.2. Dissolution of the Ascus Walls
1. Dilute the selected enzyme into sterile water, so that the final dilution is from 1:4–1:40.
2. Suspend a small loopful of the spore preparation in 0.2 mL of the enzyme solution in a sterile culture tube or Bijou bottle.
3. Incubate at 30–37°C for 25–30 min, or until the ascus walls have just dissolved. One's shirt pocket is often a satisfactory incubator.

3.1.3. Microdissection
1. Swab out the dissecting chamber with alcohol and let dry. If the atmosphere is very dry, place a piece of moist filter paper in the chamber, where it will not interfere with the use of the microscope.
2. With tweezers, take a microscope slide (approx 75 × 37 mm) from the alcohol jar and flame it. Place it on top of the dissecting chamber.

3. Place a plate of dissection agar on the template, and with a flamed microspatula, cut out a slab of agar about 2 × 4 cm and transfer it to the sterile slide on the dissecting chamber.
4. Cut a narrow strip (about 2 mm wide) from one side of the slab, and leave in place.
5. With a small (1 mm) loop, streak a loopful of the spore suspension along the edge of the narrow strip. Leave a very narrow strip (about 1 mm) clear, for a working area.
6. *Invert* the slide over the dissection chamber, and place on the microscope stage, using the 10X objective and a 20X eyepiece if available. Focus on the cells and dissolved asci.
7. Mount a dissecting needle in the holder of the micromanipulator and adjust the height to approximately the correct position.
8. Locate the tip of the needle in the center of the field of view, with the coarse controls, but do not touch the agar.
9. Find a well-isolated ascus, and with the fine controls, touch the agar beside the ascus. Lift the needle away from the agar. The ascus should adhere to the needle. If not, then repeat the procedure until it does.
10. Transfer the ascus to the larger slab, in the position where the first spore is to be left, set it down, and mark the agar with the needle.
11. Separate the spores: Touch the needle to the agar, beside the spores, and tap the bench or the base of the micromanipulator, enough to make the needle vibrate gently. The spores should separate. They can then be picked up, one at a time, and set out in a line across the slab.
12. Repeat until the slab is full, 8 or 10 asci.
13. With the sterile microspatula, transfer the slab from the chamber to a fresh plate of YEPD agar.
14. Clean up, put the dissection chambers away, and hide all the good needles.

3.1.4. Testing for Segregation of Auxotrophic Markers

1. Transfer the colonies of single-spore clones to two YEPD plates, using sterile toothpicks and a template. Up to 32 clones can be accommodated on one plate, in groups of four.
2. Replica plate the colonies to plates of dropout media for determination of the auxotrophic requirements.
3. Calculate the segregation patterns and map distances.

3.2. Random Spore Isolation
3.2.1. Grinding Method

1. Suspend the sporulated culture in water and add glass powder.
2. Grind the preparation in the homogenizer, using the minimum force necessary to break open the asci and destroy vegetative cells (up to 10 min).

3. Shake the suspension with an equal volume of mineral oil (e.g., paraffin). The spores aggregate in the oil layer. Separate the phases by centrifuging (low g in a bench centrifuge is sufficient).
4. Repeat if necessary, until the suspension is free of vegetative cells.
5. Add gelatin solution (to 15% w/v final concentration) and shake thoroughly.
6. Spread aliquots on YEPD agar and incubate until colonies are well developed.
7. Replica plate to drop-out media and other selective media, and score when colonies are again developed.

3.2.2. Enzymatic Method

With this technique, the ascus walls are dissolved using snail enzyme or other similar one, as listed previously, after which, similar procedures can be used for separating the spores (*see* Note 4).

1. Dilute enzyme appropriately (from 1:4–1:40; determine by experiment if necessary) and treat the sporulated culture with it until the ascus walls are dissolved (30–60 min should be sufficient).
2. Partition the spores by shaking in a 1:1 mixture of sterile mineral oil and water; adjustment of the solution to 15% w/v gelatin can then aid in dispersing clumps of spores (*see* Section 3.2.1.5.).
3. Spread on YEPD agar and incubate at the desired temperature until colonies are well grown.
4. Replica plate to drop-out media as before.

3.3. Mapping
(see *ref.* 6 and Note 6)

1. Cross the strain carrying the unknown gene to one carrying a centromere-linked marker such as *trpl*, *pet18*, *met14*, and so on.
2. If the unknown gene shows less than 2/3 tetratype asci relative to the centromere-linked marker, cross the strain to a set of centromere tester strains. (These can be obtained from the Yeast Genetic Stock Center at the University of California, Berkeley).
3. Determine whether the gene shows linkage to one of the centromere markers.
4. If the gene is not centromere-linked, determine the chromosome where it is located. Use either trisomic analysis or the chromosome-loss method (*see* Note 7).
5. Localize the unmapped gene to the right or left arm of the chromosome by mitotic crossover analysis (*see* Note 7).
6. Map the gene to its site by tetrad analysis using known markers on this arm of the chromosome.

4. Notes

1. Dissection Chamber: Make a 75- × 37-mm microscope slide, to which is cemented a strip of alloy or copper, bent from a strip approx 1 cm in width and 18 or 19 cm long. Araldite or other epoxy cement is satisfactory.

   Dissection may also be done on a plate of YEPD agar directly. In the Queen Mary College system (Developed by E. A. Bevan), dissection is with a microloop, which must be made in a microforge, rather than a flat-ended needle, and is done under a reasonably high-powered stereomicroscope. Dissection must be done in a very clean environment, to avoid contamination by airborne fungal spores. In the Seattle-Berkeley system, the plate is help upside-down in a sophisticated holder, and special long-focus lenses may be required. In the Goldsmith system, the same effect is obtained using a relatively simple holder, but this is held in the normal slide holder on the microscope stage, and the range of movement is limited. The single-spore clones can be easily replica plated to selective media, for detection of postmeiotic segregation (pms).

2. Flat-ended needles: These are drawn from a 5-mm borosilicate (Pyrex) glass rod (soda glass has different melting characteristics and is not satisfactory). The needles are drawn in two stages; the first, over a normal burner flame, to a diameter of about 0.2 mm, when the thin portion is bent sharply at right angles. This section is then drawn out to the desired thickness (just visible to the naked eye is about right) over a very small flame, such as the pilot light of an ordinary Bunsen burner. The needle is then held against a wood block or a similar material, and cut off with a razor blade or sharp scalpel. The end should be inspected to see if there is a spur of glass that will interfere with microdissection.

   G. R. Fink draws out the thin portion separately and cuts it into short sections, which are then glued to the 0.2-mm section of the needle, thus avoiding the more difficult part of the process *(4)*. Sections of optical fibers also have been suggested.

3. Yeasts may be sporulated in liquid medium (McClary's medium or raffinose acetate medium *[3]*), if desired. The yeast strains are grown in YEPD broth as presporulation medium, in volumes of 5 mL or more, with vigorous aeration, for 48 h, and used to inoculate the sporulation broth at a ratio of 10% of the volume of the latter. For normal genetic investigations, 2 mL of sporulation medium are inoculated at this ratio, and the culture is again incubated with vigorous aeration, until spores appear. The cells are washed and stored under a little water or buffer in the refrigerator, where they normally retain their viability for some weeks. The advantage of this system is that any volume of uniform culture can be produced and stored until required.

Temperature for sporulation: Yeasts do not necessarily sporulate best, nor yield spores of high viability, at 30°C. Better sporulation is often obtained at 25 or even 20°C.

4. A disadvantage of the enzymatic method is that the vegetative cells are not destroyed. However, if cultures are used that are prototrophic in the diploid state, and that sporulate well, the method can be used successfully. The diploids will grow on minimal medium, whereas those haploid segregants bearing auxotrophic markers will not. So, a reasonable sample of meiotic products can be identified and characterized by replica plating. If *ade1* or *ade2* markers are present in the diploid, then haploids can conveniently be identified simply as pink colonies (the pigment arising from the mutational blocks in the biosynthetic pathway for adenine).

A more elegant way of isolating the spores is to treat the spores with cell wall-dissolving enzymes under protoplasting conditions. In suitable strains, this treatment removes the walls from the vegetative cells as well as dissolving the ascus walls. The culture is then diluted with water, if necessary, and the protoplasts of the vegetative cells burst, leaving only the tetrads of ascospores, which may either be separated by micromanipulation or treated as random spores.

5. Random spore isolates may be further enriched in spores by density gradient centrifugation on a urografin gradient to concentrate the asci *(5)*. This can be done in a laboratory centrifuge. Some workers have reported that the method is not satisfactory for all strains of *S. cerevisiae*, however.

6. The mapping procedure given in Section 3.3. is broadly applicable, for example, to yeast much less genetically well known than *S. cerevisiae*. However, in *S. cerevisiae*, an easier approach is now possible, using specially constructed mapping strains.

Acquire the set of nine strains, each of both **a**- and α-mating types, carrying a total of 66 markers spaced approx 50 cm apart over the entire yeast genome. These strains are crossed with the tester strain carrying the unmapped gene, the resulting hybrids are sporulated, and tetrad analysis is done. In general, examination of four or five crosses should allow the mapping of most genes. The set of strains carrying the appropriate markers can be obtained from the Yeast Genetics Stock Center (Donner Laboratory, Department of Genetics, University of California, Berkeley, CA). An elegant alternative method—for *S. cerevisiae*—involves using specially constructed *cir*° tester strains (available from the aforementioned lab) in which the 2-μm plasmid is integrated near a centromere. In a cross with a (*cir*°) strain carrying the unmapped marker, specific mitotic chromosome loss is induced, so identifying the relevant chromosome *(6)*.

Genes may also be mapped to a particular chromosome by cloning the unmapped gene, separating the chromosomes of the yeast strain, and hybridizing the cloned gene to a Southern blot. The chromosomes of *S. cerevisiae* can be identified quite accurately and the gene located in that way (*see* Chapter 9 by Bignell and Evans, and ref. *6*). A refinement of this approach, again for *S. cerevisiae*, is to use the physically mapped clone bank (large *Not*1, *Sfi*1 chromosomal DNA segments) of M. V. Olson, available from the YGSC.

7. In trisomic analysis for assignment of a gene to a particular chromosome, the haploid strain carrying the unmapped gene is crossed with a strain carrying a marked chromosome known to be in the disomic condition. If the unmapped gene is on the disomic chromosome, the segregation for it will be 4:0, 3:1, and 2:2; otherwise, it will be 2:2 only.

In mapping with super-triploids, specially constructed triploids carrying a marker on each chromosome in the duplex condition are mated with a haploid carrying the unmapped gene. Several strains are required to locate the unmapped gene.

In mitotic mapping, induction of mitotic recombination (e.g., by UV) is used to demonstrate linkage to a particular marker. (Further information on these methods and the results they yield are given in refs. *6–9*.)

References

1. Hawthorne, D. C. (1955) *Genetics* **40**, 511.
2. Mortimer, R. K. and Hawthorne, D. C. (1969) Yeast Genetics, in *The Yeasts* (Rose, A. H. and Harrison, J. S., eds.), Academic, New York, pp. 385–460.
3. Fowell, R. R. (1969) Sporulation and Hybridization of Yeasts, in *The Yeasts* (Rose, A. H. and Harrison, J. S., eds.), Academic, New York, pp. 303–383.
4. Fink, G. R. (1986) *Manual of Methods in Yeast Genetics,* Cold Spring Harbor Laboratory, Cold Spring Harbor, New York.
5. Dawes, I. W., Wright, J. F., Vezinhet, F., and Ajam, N. (1980) Separation on urografin gradients of subpopulations from sporulating *Saccharomyces cerevisiae* cultures. *J. Gen. Microbiol.* **119**, 165–171.
6. Sherman, F. and Wakem, P. (1991) "Mapping yeast glues," in guide to yeast genetics and molecular biology. *Meth. Enzymol.* **194**, 38–57.
7. Mortimer, R. K. and Schild, D. S. (1983) Genetic mapping in *Saccharomyces cerevisiae,* in *The Molecular Biology of the Yeast* Saccharomyces*: Life Cycle and Inheritance* (Strathern, J. N., Jones, E. W., and Broach, J. R., eds.), Cold Spring Harbor Laboratory, Cold Spring Harbor, NY, pp. 11–26.
8. Sherman, F., Fink, G. R., and Hicks, J. L. (1986) *Laboratory Course Manual for Methods in Yeast Genetics,* Cold Spring Harbor Laboratory, Cold Spring Harbor, NY.
9. Mortimer, R. K., Contopolou, C. R., and King, J. S. (1992) Genetic and physical maps of *Saccharomyces cerevisiae. Yeast* **8**, 817–902.

CHAPTER 7

Ascus Dissection

Carl Saunders-Singer

1. Introduction

Many yeast researchers are reluctant to embrace micromanipulation because of its reputation as a somewhat testing black art. However, as strain construction, meiotic mapping, cell and zygote isolation, pedigree studies, and other experiments are powerful weapons in the armory of the geneticist, microtechniques are attractive.

The technique that remains the most infamous is tetrad analysis, and it is not difficult to understand the reason for this if a little of the background is appreciated.

Before the use of digesting enzymes *(1)*, tetrads of *Saccharomyces cerevisiae* were dissected painstakingly by expressing the spores from the ascus by pinching them between needles or a needle and a slide and then transferring the liberated spores onto agar droplets. This lengthy and highly skilled procedure took many hours to master and, with equipment that was often homemade, dissection rates were likely to be very slow.

Even today methods differ greatly from country to country, from coast to coast, and even from adjacent laboratory to laboratory. Variations include dishes open at the top or inverted, with dissecting tools approaching from above or below the plate and even by the unaided hand *(2)*. The glass tools may be loops, pointed, ball-ended, or flat-ended needles, all made in a variety of ways. Dissection may take place in Petri dishes or on thin sheets of agar on glass slides. There are many permutations of dish layout, and, a wide variety of microscopes, mechanical stages, and micromanipulators, both commercial and homemade, and of varying compatibility and performance. Most of the equipment, even that supplied by major manufacturers, was not designed for work on yeast in the first place and has seen little refinement since conception. Indeed, some combinations are so difficult to use that it is a testament to the persistence of the originators that they are in use at all, and many would-be ascus analysts have been put off for life by trying them.

The author is indebted to all those who have shared their methods. The techniques described in the following are a distillation of the most successful. The instrumentation is the latest and best available and is specially designed for ascus dissection and the like. Dissection has the reputation of being difficult to master. This is not the case; anyone can learn it in about 10 min.

2. Materials

1. Singer MSM System series 200 (Watchet, Somerset, UK), with dissecting needles (*see* Note 1 and ref. *3*) (Fig. 1).
2. Sporulated culture.
3. YPD dissection, Petri plates: 10 g yeast extract, 20 g peptone, 20 g glucose, 20 g agar, 1000 mL distilled water (Difco, Detroit, MI) (*see* Note 2).
4. 3-mm loop.
5. Enzyme: β glucuronidase type H-3AF, zymolyase or lyticase (Sigma, St. Louis, MO) (*see* Note 3).

3. Methods

3.1. Digesting the Sporulated Culture

1. Prepare dissecting plates (*see* Note 2).
2. Inspect the sporulated culture microscopically to ensure the presence of four spore asci and to assess the sporulation rate (*see* Note 4).
3. Prepare enzyme and pipet about 0.2-mL into a small centrifuge tube.
4. Pick up a pinhead size sample of the culture using a sterile toothpick or flamed loop and suspend it in the enzyme (*see* Note 5).
5. Incubate the suspension for 10–20 min at room temperature (*see* Note 6).
6. Carefully add sterile water to the top of the tube. Allow the solids to fall to the bottom of the tube and then carefully aspirate the liquor from the top down to the original volume (*see* Note 7).

3.2. Inoculating the Dish

1. Mark the bottom of the Petri dish with a lab pen with a small mark at each end of where the inoculum should be (*see* Note 8).
2. Inoculate the dish with the digested sample (*see* Note 9).
3. Raise the overarm of the microscope (*see* Note 10).

Ascus Dissection

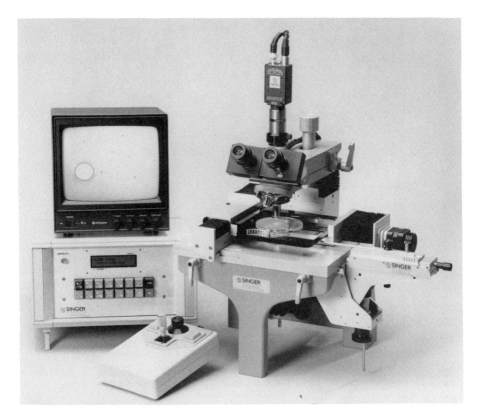

Fig 1. The Singer MS System series 200 is a complete workstation for micromanipulation in yeast genetics. Comprising microscope with focusing limb, stage-mounted micromanipulator, and microprocessor-controlled motor-driven stage. The instrument comes with factory made glass dissecting needles and a special holder. The equipment may be fitted with CCTV and a video printer (invaluable for teaching).

3.3. Loading the Petri Dish
1. Place the Petri dish, inverted, in the plate holder with inoculum marks on the left and lying north–south (*see* Note 11).
2. Carefully lower the microscope overarm and move the stage to the streak (*see* Note 12).

3.4. Searching the Inoculum
1. Search the inoculum north–south for a tetrad (*see* Note 13).
2. Center the tetrad in the field.
3. Pick up the tetrad on the needle (*see* Note 14).

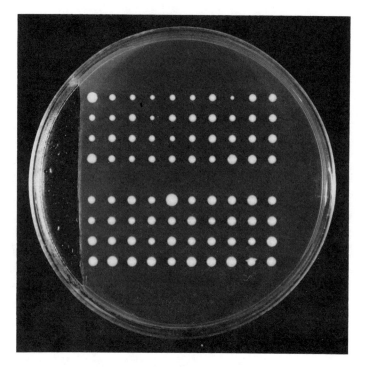

Fig 2. Photograph of an incubated dissecting plate with colonies at 6-mm centers. The inoculum has been removed.

3.5. Move the Tetrad to the Matrix

1. Move the stage to the matrix. Drop the tetrad at A1 (*see* Note 15).
2. Separate the spores (*see* Note 16).
3. Leave one spore at A1 and deposit the others at B1, C1, and D1 (*see* Note 17).
4. Return to the inoculum and search for another tetrad (*see* Note 18).
5. Dissect this at A2, B2, C2, and D2 (*see* Note 19).
6. Repeat the process for 20 tetrads (*see* Note 20).
7. Remove the Petri dish and incubate (*see* Note 21 and Fig. 2).

3.6. Alternative Techniques

3.6.1. Micromanipulating Other Structures

The skills learned during ascus dissection are directly applicable to other micromanipulation techniques. As well as spores, vegetative cells, zygotes, and other structures can be isolated and moved to recordable coordinates on a Petri dish. If two or more structures are picked up on the needle, they will automatically come together and can be deposited, touching, on the agar: This can be used for spore-spore or spore-cell pairing.

The matrix can be changed to a 1-mm × 1-mm grid giving 3600 addressable and, moreover, accurately recoverable locations on a Petri dish. Each generation can be assigned to an address for pedigree analysis or the plate can be scanned for mutants: The Petri dish holder is very precise and so the dish can be removed, incubated, replaced, and addresses can be revisited to record cell progression.

3.6.2. Other Yeast(s)

The techniques described herein can just as easily be used for other yeasts such as *Schizosaccharomyces pombe*. *S. pombe* tetrads appear completely different from those of *S. cerevisiae* in that they can look like bananas or peanuts in the shell.

Sporulation techniques are different, and conveniently with this yeast, the ascospores liberate spontaneously from the ascus on incubation. *S. pombe* tetrads are slightly more "sticky" than those of *S. cerevisiae*, but may be picked up with facility as can the spores.

4. Notes

1. The MSM System comes factory configured and ready to use. It is a good idea to read the handbook thoroughly, but this can be left until later if you wish. Set up the MSM as described in the handbook. Switch on the MSM so that the microscope light works (the LCD screen will display a scrolling message and a tone will sound). Familiarize yourself with the micromanipulator controls. Ensure that the needle is vertical and centered in the field. Lower it until the holder is just above the glass window in the stage. High-specification needles made by the company are supplied with the instrument, and are available for resupply (part no. MSM/N/SO/10). The needles, comprising a pipet-mounted glass fiber cleaved flat at one end, also function as a light conductor for specimen illumination.
2. It is important to prepare high-quality dissecting plates. Autoclave all components together for 15 min at 120°C and 1 bar. For the clearest medium, autoclave the components separately. Allow the mixture to cool slightly before pouring 20-mL into each Petri dish with the dishes on a *horizontal* surface. Dissection plates should be dried at room temperature for 2–3 d or it may be difficult to pick up the tetrads, if necessary dry the plates in a laminar flow cabinet.

3. Enzyme may be diluted 1:5 to 1:40, depending on the enzyme, with sterile water and stored for weeks in the refrigerator.
4. You should be able to see the tetrads in their tetrahedral form.
5. It should appear very slightly milky.
6. A proper digestion is vital for easy dissection. Inspect the digesting asci on a slide under 400X magnification. Properly digested tetrads have the four spores in a cruciform shape in a single plane. The spores should be easy to see and each have a strong, black outline. The best way to assess a sample is by a trial dissection. The asci should stay together for streaking, but break easily on dissection. There should be minimal signs of random spores.
7. The concentration of the inoculum is important: Too dense and you will have to "dig" out the tetrads, too dilute and you may have trouble finding a tetrad in a poorly sporulated strain. Remember, you can always adjust the concentration by adding water or aspirating it from the suspension.
8. The MSM is factory programmed to reproduce the Petri dish layout shown in Fig. 3. This design produces colonies at a convenient spacing for replica plating and makes use of the whole of the Petri dish. Note that the area on the dish where the inoculum should be spread is shown hatched on the left (this segment may be precut with a flamed spatula so that the whole of the inoculum can be removed after dissection).
9. A single, light stroke is best. Try not to disrupt the agar surface.
10. This is convenient as it allows a lot of room when loading the Petri dish and minimizes needle breakage.
11. The Singer MSM System automatically goes to the streak if it is placed in the correct location.
12. Press one of the joystick keys: The stage will go automatically to the inoculum. Focus on the streak using the coarse and fine focus.
13. Move the stage by displacing the joystick: It will move in all directions in a series of quick, small steps. Smaller movements may be made by first holding down the small button on the top of the joystick and at the same time moving the joystick in the desired direction. "Walk" the tetrad to the center of the field.
14. Rotate the micromanipulator vertical drive control clockwise to raise the needle toward the agar. Remember, you cannot lose the needle from the field if it is properly centered, so just turn until it appears. You will first see the needle as an out of focus shadow, becoming gradually sharper as it approaches the agar surface. Carefully raise the needle, and as it touches the agar surface, the liquid on the surface will suddenly "jump" up the sides of the needle and a black, circular meniscus clearly will be seen round the needle tip. This sudden appearance of the meniscus is the vital clue to the plane of the agar surface and can be perceived even when the

Ascus Dissection

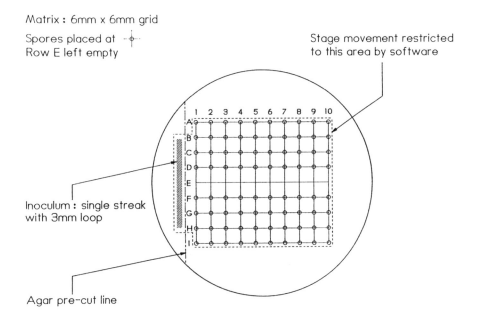

Fig. 3. Factory set Petri dish layout for tetrad dissection.

needle tip is not in focus. The diameter of the needle may be a surprise at first in looking very large, but its size and shape enable it to pick up anything from an ascospore to a zygote. To pick up the tetrad, raise the needle until you see the meniscus, then with the micromanipulator joystick move the needle to the tetrad. Move the tetrad around a little, and then briskly turn the micromanipulator vertical drive anticlockwise, lowering the needle, and with any luck taking the tetrad with it. You may need to practice this a little until it becomes easy. Try using the outer edge of the needle and different places on the circumference. The secret is to keep on trying until it happens! When you have your tetrad on the needle, lower it away from the agar surface by four or five full turns of the micromanipulator knob (in case the agar is not flat).

15. Press one of the keys (the "arrow" keys) on the joystick unit. The stage will go automatically to the top left position—A1. (The stage does not go to A1 automatically each time, but rather to the place it was in last before the MSM was switched off.)
16. Rotate the micromanipulator control clockwise to raise the needle to the agar. Be careful: Look for the meniscus! Focus the microscope with the fine focus knob so that the meniscus is as sharp as possible. You may be able to

see the tetrad on the tip of the needle. Dropping the ascus is the reverse of picking it up: You need to "dab" the needle tip on the agar until the tetrad comes off. After each "dab" it is a good idea to move the needle to one side, as this may reveal that it has been dropped. Break the ascus by touching it with the needle and oscillating the micromanipulator joystick, or by tapping the far end of the micromanipulator on the side with the index finger. Make certain that for breaking up the tetrad the needle is just at the "meniscus" stage: If it is pressing too strongly on the agar, you may tear it and lose a spore. Be confident in this tapping and/or oscillation of the micromanipulator joystick. A sharp tap will cause the needle to oscillate and with luck the ascospores will fly apart. Sometimes this takes patience and persistence. Tapping the bench does not work with the MSM. When you have broken the ascus apart, pick up any combination of spores using the same technique as for the tetrads. Of course, you want to leave one on A1.

17. You can move the stage up and down the matrix (grid) column, row A through row I, by displacing the stage joystick in a north-south direction. Each movement of the joystick will move on matrix square of 6-mm. The stage position will be shown on the console LCD. Dissect the ascospores, placing them at A1, B1, C1, and D1. *Do not forget to drop the needle away from the agar every time you move from point to point!*

18. After you have dropped the last ascospore of the four, move the stage back to the top row, A, and lower the needle. The "arrow" keys on the joystick unit are like the "Enter" key on a computer. Pressing either now causes the stage to "shuttle" back and forth between the inoculum and the matrix. Do not forget to lower the needle away from the agar each time before pressing "arrow" or you will lose the tetrad. You will have noticed that the stage joystick operates in different ways, depending on what "mode" (search or matrix) it is in. The stage will return automatically to the exact place in the inoculum from where you picked up the ascus. Search for another tetrad by moving the stage with the stage joystick as before. Do not forget the "nudge" button on the top of the stick. Pick up another ascus. Lower the needle. You can clean the needle tip by repeatedly dabbing it on a clean part of the agar, and, more severely, by digging it into the surface and stirring it around.

19. Press one of the joystick keys; the stage will go back to A1. Focus on the spore at A1 and then, using the joystick, move to A2. Dissect this tetrad at A2, B2, C2, and D2.

20. When you have completed the top half of the dish by placing and dissecting ten tetrads, and before you return to the inoculum, key F1 on the console keyboard followed by =. The stage will go automatically to point F1. Now key an "arrow" key to return to the inoculum. Pick up a tetrad, key

the "arrow" key, and the stage will return to F1. Dissect the tetrad F1, G1, H1, and I1. Carry on until you finish the bottom of the Petri dish.
21. When you come to the end of the last column, it is good "housekeeping" to key A1 and then =.

Acknowledgments

The author thanks the late Seymour (Sy) Fogel for an introduction to tetrad dissection, the author's late father, A. E. Saunders-Singer, for a thorough grounding in instrument design, and Trevor Clarke.

References

1. Johnston, J. R. and Mortimer, R. K. (1959) *J. Bacteriol.* **78,** 292.
2. Munz, P. A., (1973) Dissecting yeast asci without a micromanipulator, *Experientia* **29,** 251.
3. Sherman, F. and Hicks, J. (1991) Micromanipulation and dissection of asci. *Methods Enzymol.* **194,** 21–37.

CHAPTER 8

Karyotyping of Yeast Strains by Pulsed-Field Gel Electrophoresis

John R. Johnston

1. Introduction

Pulsed-field (gradient) gel electrophoresis (now abbreviated to PFGE or PF) was first described by Schwartz and Cantor *(1)* and Carle and Olson *(2)* in 1984. It allows separation of DNA molecules in the size range 50 kb to approx 10 Mb in agarose gels to which electric fields are applied in different directions. The chromosomes of *Saccharomyces cerevisiae* lie within the approximate size range of 200 kb to 3 Mb and can therefore be separated by this method to provide "electrophoretic karyotyping" of both laboratory-bred *(3)* and industrial *(4)* strains of this and related species. Several other yeasts of various genera, such as *Candida albicans* and *Schizosaccharomyces pombe*, have also been karyotyped by this method *(4–8)*.

The original systems of apparatus have been extensively modified, largely in the design of electrodes and the consequent electric fields. Several of these recently have been reviewed *(9,10)*, so only a few of the more commonly used types are briefly described:

1. OFAGE (orthogonal field alternation gel electrophoresis) *(2)*: This is a simple construction with two pairs of electrodes, diagonally placed, resulting in curved tracks of outside lanes owing to inhomogeneous fields; this system has now largely been replaced by improved types.
2. FIGE (field inversion gel electrophoresis) *(11)*: The polarity of a single field is switched (i.e., 180°C) each cycle, which comprises a longer pulse

time (or higher voltage) for the forward direction, i.e., toward the anode. An advantage is that a conventional horizontal gel box may be used. However, the relationship between size and migration rate is complex *(11)*, usually necessitating use of programmed ramping, i.e., increasing of the switching cycle during the run. Variations in ramping have recently been more extensively investigated *(12)*. Without ramping, karyotyping of several yeasts other than *Saccharomyces*, which possess only larger chromosomes (>1.5 kb), has been achieved *(8)*. Straight lanes are obtained, but bands generally are more diffuse than with other systems.

3. CHEF (contour clamped homogeneous electric fields): Here the angle between the alternating electric fields is more than 90°C. Most commonly, a hexagonal array of small electrodes connected by a series of resistors results in electric fields angled at 120°C *(13)*. The clamping of electrode potentials results in highly uniform fields and straight tracks. Full instructions for construction of CHEF systems have recently been given *(14,15)*. Another form of "homogeneous crossed fields" is rotation of the gel through more than 90°C at each switching cycle in a single uniform electric field *(16,17)*.

4. TAFE (Transverse alternating field electrophoresis) *(18)*: Here the electric fields are transverse to the gel and DNA molecules therefore change directions across the thickness of the gel. Although the fields are homogeneous across the depth of the gel, therefore producing straight lanes, field strength, and orientation angles vary along the length of the gel from top to bottom. Because the gel is vertical, its size is limited and handling the gel is more difficult.

5. PACE (Programmable autonomously controlled electrode): This system provides variable and precise control of all parameters and therefore allows study of the relationships between DNA mobility and switch time, voltage, field angles, agarose concentration, and temperature *(19)*. Presently, it represents the most flexible and sophisticated of the PF systems.

In recent years, modifications to apparatus and other novel aspects of design of systems have been regularly reported. These include pneumatic rotation of gels *(20)*, conditions for higher field strengths and mobility inversion *(21)*, feedback clamping of voltages in a modified CHEF apparatus *(22)*, simultaneous electrophoresis in a double-decker gel arrangement *(23)*, and faster separation by "secondary" PFGE *(24)*. Commercial versions of various systems are increasingly available, e.g., "CHEF-DRII" (Bio-Rad, Hercules, CA); "Hex-a-field" (BRL, Gaithersburg, MD), also a CHEF system; "Geneline" (Beckman, Fullerton, CA), a TAFE system and "RAGE" (Stratagene, La Jolla, CA), a rotating system.

The other important development, besides PFGE, which has permitted the karyotyping of yeasts electrophoretically, is the development of protocols for releasing the chromosomes from a yeast cell as intact DNA molecules: This essentially involves enzyme and detergent treatment of yeast cells incorporated into agarose blocks so that mechanical stresses on the large fragile DNA molecules, both during and after cell disruption, are minimized: Effective suppression of DNase activity (through EDTA and proteinase K) is also important. However, it has been reported recently that cell wall lytic enzymes are not needed for preparing chromosomal DNA for PFGE *(25)*, and a radically different freeze-grinding method has been published *(26)*. In the Methods section, a protocol for the preparation of intact chromosomal DNA is agarose blocks is given (Section 3.1.), followed by the method for pulsed field gel electrophoresis (Section 3.2.).

2. Materials

2.1. DNA Preparation

1. YPD culture medium: 1% w/v yeast extract, 2% w/v peptone, 2% w/v dextrose (glucose).
2. EDTA: $0.5M$ as stock:
   a. pH 7.5,
   b. pH 7.8,
   c. pH 9.0,
   d. pH 9.5.
3. $0.125M$ EDTA, pH 7.8.
4. SCE: $1M$ sorbitol, $0.1M$ sodium citrate, pH 5.8, 10 mM EDTA.
5. Lyticase (Sigma): 10 U/mL in 50% w/v glycerol-TE 50, storage at $-20°C$.
6. Lysis buffer: 3% *N*-lauryl sarcosine, 0.5 Tris-HCl, $0.2M$ EDTA (pH 9.0).
7. SE: 75 mM NaCl, 25 mM EDTA, pH 7.5.
8. ES: 0.5 EDTA (pH 9.5), 1% *N*-lauryl sarcosine.
9. Proteinase K (Sigma): 10 mg/mL in 50% w/v glycerol-TE 50, storage at $-20°C$ (or made fresh as required).
10. TE 50: 10 mM Tris-HCl, pH 7.5, 50 mM EDTA, pH 7.5.
11. PMSF (phenylmethylsulfonyl fluoride): 10 mM PMSF in isopropanol, storage at $-20°C$. Resuspend precipitate before use. *Caution:* PMSF is toxic and should be handled using disposable gloves in a fume cupboard.
12. LMP agarose: Low melting point agarose (*see* Section 3.1.).
13. Plastic molds (inserts): Pharmacia (Uppsala, Sweden) LKB, cat. no. 80-1102-55.

2.2. Gel Electrophoresis

1. Selected electrophoresis system (CHEF, PACE, and so on).
2. Glass plates, size determined by apparatus (e.g., 5 × 5 in., 3-mm thick for CHEF gel shown in Fig. 1). Small areas of velcro glued to two corners of the plates prevent the gel sliding off the plate during the run. Open, or slotted, gel comb or wells. Suitably sized gel box.
3. Agarose (high purity, molecular biology grade): Typically 1% w/v aqueous, but may be varied. (Agarose solution can be prepared quickly in a microwave oven.) Volume dependent on apparatus, e.g., 100 mL for plates above (step 2).
4. TBE buffer: Stock solution as 10X, i.e., $0.9M$ Tris-base, $0.9M$ boric acid, 20 mM EDTA. Dilute to 0.5X, using cooled distilled water, if you wish to save buffer cooling time.
5. Ethidium bromide: Stock solution 10 mg/mL. Stain gel with approx 0.5 μg/nmL. *Caution:* Ethidium bromide is toxic and possibly carcinogenic, handle and dispose of with due caution (*see* e.g., Chapter 18).
6. Ultraviolet transilluminator; Polaroid camera.

3. Methods
3.1. DNA Preparation

1. Grow cells of the chosen strains (*see* Note 1) in 5–10 mL YPD, generally overnight, e.g., cultures can be in 50-mL disposable, screwcap centrifuge tubes, vigorously shaken.
2. Estimate final cell densities by either cell chamber (hemocytometer) counts or spectrophotometrically (optical density of the chosen wavelength having been previously calibrated for cell concentration). Flocculence may cause a problem in cell counting. This can usually be overcome by EDTA treatment (*see* Note 2).
3. To make a maximum of ten 100-μL plugs, remove volumes containing approx 5×10^8 cells, transferring to Eppendorf (1.5 mL microfuge) tubes.
4. Centrifuge the samples in a microfuge (few seconds), then wash with $0.125M$ EDTA (pH 7.8).
5. Resuspend each cell pellet in 1 mL SCE, add 200 U Lyticase, and incubate at 30°C for 30 min without shaking (*see* Note 3).
6. The efficiency of spheroplasting may be checked either during or after the previous step by adding 20 μL of suspension to each of two glass tubes, one containing 1 mL lysis buffer and the other 1 mL SCE. When the former tube is clear, compared to the latter tube as control, most cells have spheroplasted (*see* Notes 3 and 4).
7. Centrifuge the spheroplasts gently, e.g., 600*g* for 5 min, and remove the supernatant with a Pasteur pipet.

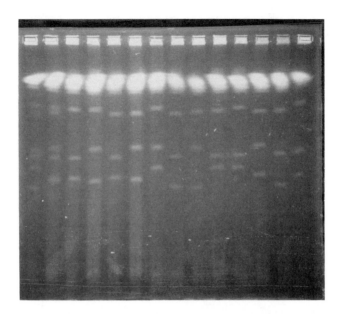

Fig. 1. A CHEF gel (1% agarose) run at 160V for 24 h at 12°C with 18-s pulses throughout run. These conditions were selected to optimize resolution of chromosome I bands (the lowest). Only bands for chromosomes I, VI, and III (the shortest chromosomes) are discreet (bands for all other, larger chromosomes congregate together between the plugs and chromosome III). Lane 1: diploid strain EM93 (different length homologs of chromosomes I, VI, and III are resolved); lane 2: diploid strain EM93-X1 (cross of two EM93) meiotic segregants, each carrying a chromosome I of size intermediate between the long and short homologs of EM93; lanes 3–6, 7–10, 11–14: segregants of 3 meiotic tetrads of strain EM93-X1. These tetrads are, respectively, parental ditype (lanes 3–6), nonparental ditype (lanes 7–10) and tetratype (lanes 11–14) for the proposed two distinct chromosome-length polymorphisms (CLP) between the long and short homologs of chromosome I of EM93 *(30)*.

8. Process the pelleted spheroplasts into agarose plugs, one strain at a time by:
   a. Resuspending by adding 430 µL $1M$ sorbitol (bringing the total volume to approx 500 µL);
   b. Adding and quickly mixing 500 µL of 1.5% LMP agarose, dissolved in SE, from a tube kept at 48°C in a temperature-block or waterbath;
   c. Pipeting 100 µL aliquots into plastic moulds (Pharmacia), sealed by taping backs, and refrigerating for rapid setting;
   d. Removing plugs from molds into five plug volumes of ES, containing 1 mg/mL proteinase K (placed in six-well culture dishes).

The plugs will slide out of the molds by applying pressure to the untaped backs with a rubber teat. Preparation of a few plugs may be sufficient, but 9–10 plugs should be prepared from strains to be analyzed by PFGE many times (*see* Note 7).

9. Shake the plugs overnight very gently on a platform shaker at either room temperature or 30°C (proteinase K treatment should clear the plugs) (*see* Notes 6 and 7).
10. On the next day, rinse the plugs repeatedly with TE50. Change the TE three times to dilute the *N*-lauryl sarcosine, then gently shake plugs in the TE, changing every 1–2 h.
11. The plugs are now ready for use and may be stored in TE50 (or, for longer keeping, $0.5 M$ EDTA [pH 7.8] at 4°C). If, however, the DNA is to be cut with restriction enzymes, then wash the plugs twice with 0.1 mM PMSF in TE50 for 2 h, following the initial three rinses with TE50 (step 10).

As indicated in the introduction, a nonenzymic procedure for preparing intact chromosomal DNA has been reported (*see* Note 8).

3.2. Gel Electrophoresis

1. Pour the hot agarose gel onto a *level* glass plate, using either an open or slotted comb, and allow to set (*see* Notes 9 and 12).
2. Load the plugs, usually half pieces, either into slotted wells or against the bottom and front surfaces of an open well. If using the latter, fill the remainder of the well, therefore overlaying the plugs, with melted agarose; a prewarmed wider-bore pipet is useful here to prevent blocking by setting agarose (*see* Notes 5, 10, and 11).
3. Fit the glass plate with the gel into the correct orientation in the PFGE apparatus (movement of DNA is to the positive electrodes). Check that the gel is level in the apparatus and covered to depth of a few mm by pumped, cooled 0.5X TBE buffer. Check that all tubing and connectors are secure, to avoid any buffer leaks (*see* Note 13).
4. Ensuring that safety features are functioning, switch on the power pack, and adjust the voltage (current) and the timer and/or ramping programs. Check that the buffer is at the desired running temperature (frequently 10–12°C).
5. During and/or at the end of a timed run, check the parameters of voltage, current, pulse-time, and buffer temperature.
6. Switch off the power pack and observe safety features before removing the gel from the apparatus. Cut between the gel and velcro areas with a razorblade and slide the gel from its glass plate into a box/tray containing buffer/distilled water with ethidium bromide (0.5 µg/mL).

Karyotyping of Yeast Strains by PFGE

7. Stain for the required period, usually 20–30 min. Rinse in distilled water, view, and photograph if wished at this stage. Generally, a period of destaining (overnight or 1–2 d, in distilled water) produces improved contrast for photographing bands.

For a brief discussion of interpretation, *see* Note 14.

4. Notes

1. Yeast strains may differ in their cell surface characteristics, which may affect their behavior during culturing or spheroplasting.
2. If strains are flocculent, i.e., dispersed cells aggregate into flocs during growth, then centrifuge and resuspend in 50 mM EDTA (pH 7.5) to redisperse cells (Section 3.1., step 2). Occasionally, a strongly flocculent strain may require additional one to two washes with 50 mM EDTA (pH 7.5).
3. Different strains may require different lengths of incubation to produce high levels of spheroplasting. Occasionally, a strain requires a high concentration of lyticase (Section 3.1., step 5).
4. The method described omits 2-mercaptoethanol (ME). However, its use, or preferably that of the substitute dithiothreitol (DTT), may improve spheroplasting of particular strains. If either of these are used, then wash spheroplasts afterward with 1M sorbitol (Section 3.1., step 5). *Note:* 2-Mercaptoethanol is toxic and should be handled carefully and only in a fume cupboard.
5. Half-sized pieces of plugs are normally used when many lanes per run are required. (Section 3.2., step 2).
6. Some workers prefer proteinase K treatment at 50°C (Section 3.1., step 9).
7. Cost-saving methods have been used in which proteinase K is omitted *(16)* (Section 3.1., step 8). One of these (Maren Bell, personal communication) is also very quick, without any overnight step. For comparison it is given in the following:
   a. Pick up yeast cells from overnight growth patches with a bent 200 µL pipet tip.
   b. Resuspend in 125 µL SEBZ (*see* h) in an Eppendorf tube by vortexing.
   c. Shake at approx 300 rpm at 37°C for 30–60 min.
   d. Add 125 µL 1% w/v low-melting temperature agarose in SEZ (SEBZ without ME) at 45°C to the spheroplast solution and mix gently by pipeting.
   e. Rapidly pipet this mixture into plug molds, sitting on ice.
   f. After 5–10 min on ice, remove plugs to 3 mL of EST (*see* h), and shake at 500 rpm at 37°C for 30–60 min.
   g. Plugs are now ready for use. One plug can be used in four CHEF gels. Remaining plug portions are stored in EST or in 0.5M EDTA at 4°C.
   h. SEBZ is 1M sorbitol, 50 mM EDTA (pH 7.8), 28 mM 2-mercaptoethanol 1 mg/mL zymolyase 100-T; EST is 100 mM EDTA, 10 mM Tris-HCl, pH 8.0 1% w/v sarcosyl.

8. A freeze-grinding method, in which liquid-nitrogen-frozen cells are homogenized with a mortar and pestle, recently has been reported *(26)*.
9. Special agarose preparations for PFGE, such as Beckman LE and Bio-Rad Chromosomal grades, are now available (Section 3.2., step 1).
10. For sizing smaller yeast chromosomes, lambda ladders such as wild-type (48.5-kb increments) and deletion mutant Δ39 (39-kb increments) should also be loaded (Section 3.2., step 2).
11. An overloaded concentration of DNA may lead to slower migration of bands; if so, the sizes of their DNA would be overestimated *(9)*.
12. Spurious results (retardation of bands) have been reportedly owing to a specific batch of agarose, used for the gel *(9)*. A new batch of agarose should therefore be tested in a trial run with controls.
13. The electrophoresis buffer 0.5X TBE may be replaced by 0.5X TAE. A stock solution of 5X TAE contains (per L) 24.2 g Tris-base, 5.71 mL glacial acetic acid, 20 mL of $0.5M$ EDTA (pH 8.0). TAE is preferable to TBE if gels are to be blotted onto nitrocellulose filters.
14. Karyotyping involves determining the number and sizes of chromosomes per cell. Electrophoretic karyotyping of *S. cerevisiae*, *S. pombe*, and a few other yeasts requires the identification of DNA bands with chromosomes already known in terms of genetic maps. This is accomplished by Southern gel hybridizations of cloned genes, mapped on particular chromosomes, to specific DNA bands (*see* Chapter 9). In the case of a few haploid strains of *S. cerevisiae*, the pattern of bands, from lowest to highest, has been clearly identified with the range of 16 chromosomes, from shortest (CHR I) to longest (CHR XII). Individual chromosomes are, however, often of different lengths and this has been termed chromosome length polymorphism (CLP). Karyotypes including translocations have also been obtained *(27)*.

    In a natural diploid isolate of *S. cerevisiae*, strain EM93 *(28)*, extensive CLP has been shown between homologs of the four shortest chromosomes (CHR I, VI, III, and IX) *(29)*.

Figure 1 shows an example of a CHEF gel used to analyze EM93 and meiotic tetrads. Most industrial strains of *S. cerevisiae* have more complex karyotypes, featuring degrees of polyploidy and aneuploidy. A convenient representation of their karyotype is a scanning densitometer plot of their bands *(30;* De Zoysa and Johnston, unpublished results).

Acknowledgment

The assistance of Maren Bell is gratefully acknowledged.

References

1. Schwartz, D. C. and Cantor, C. R. (1984) Separation of yeast chromosome-sized DNAs by pulsed field gradient gel electrophoresis. *Cell* **37,** 67–75.
2. Carle, G. F. and Olson, M. V. (1984) Separation of chromosomal DNA molecules from yeast by orthogonal field alternation gel electrophoresis. *Nucleic Acids Res.* **12,** 5647–5664.
3. Carle, G. F. and Olson, M. V. (1985) An electrophoretic karyotype for yeast. *Proc. Natl. Acad. Sci. USA* **82,** 3756–3760.
4. Johnston, J. R. and Mortimer, R. K. (1986) Electrophoretic karyotyping of laboratory and commercial strains of *Saccharomyces* and other yeasts. *Int. J. Syst. Bacteriol.* **36,** 569–572.
5. DeJonge, P., Dejonge, F. C. M., Meijers, R., Steensma, H. J., and Scheffers, W. A. (1986) Orthogonal field alternation gel electrophoresis banding patterns of DNA from yeasts. *Yeast* **2,** 193–204.
6. Vollrath, D. and Davis, R. W. (1987) Resolution of DNA molecules greater than 5 megabases by contour-clamped homogeneous electric fields. *Nucleic Acids Res.* **15,** 7865–7876.
7. Smith, C., Matsumoto, T., Niwa, O., Kleo, S., Fan, J. B., Yanagida, M., and Cantor, C. R. (1987) An electrophoretic karyotype for *Schizosaccharomyces pombe* by pulsed field gel electrophoresis. *Nucleic Acids Res.* **15,** 4481–4489.
8. Johnston, J. R., Contopoulou, C. R., and Mortimer, R. K. (1988) Karyotyping of yeast strains of several genera by field inversion gel electrophoresis. *Yeast* **4,** 191–198.
9. Anand, R. and Southern, E. M. (1990) Pulsed field gel electrophoresis, in *Gel Electrophoresis of Nucleic Acids, A Practical Approach,* 2nd ed., (Rickwood, D. and Hames, B. D., eds.), IRL, Oxford, pp. 101–123.
10. Eby, M. J. (1990) Pulsed-field separations: continued evolution *Biotechnology* **8,** 243–245.
11. Carle, G. F., Frank, M., and Olson, M. V. (1986) Electrophoretic separation of large DNA molecules by inversion of the electric field. *Science* **232,** 65–68.
12. Heller, C. and Pohl, F. M. (1990) Field inversion gel electrophoresis with different pulse time ramps. *Nucleic Acids Res.* **18,** 6299–6304.
13. Chu, G., Vollrath, D., and Davis, R. W. (1986) Separation of large DNA molecules by contour-clamped homogeneous electric fields. *Science* **234,** 1582–1585.
14. Chu, G. (1989) Pulsed field electrophoresis in contour-clamped homogeneous electric fields for the resolution of DNA by size or topology. *Electrophoresis* **10,** 290–295.
15. Meese, E. and Meltzer, P. S. (1990) A modified CHEF system for PFGE analysis. *Technique* **2,** 26–42.
16. Southern, E. M., Anand, R., Brown, W. R., and Fletcher, D. S. (1987) A model for the separation of large DNA molecules by crossed field gel electrophoresis. *Nucleic Acids Res.* **15,** 5925–5943.
17. Serwer, P. (1987) Gel electrophoresis with discontinuous rotation of the gel: an alternative to gel electrophoresis with changing direction of the electric field. *Electrophoresis* **8,** 301–304.

18. Gardiner, K. and Patterson, D. (1989) Transverse alternating field electrophoresis and applications to mammalian genome mapping. *Electrophoresis* **10**, 296–302.
19. Birren, B. W., Hood, L., and Lai, E. (1989) Pulsed field gel electrophoresis: studies of DNA migration made with the programmable, autonomously-controlled electrode electrophoresis system. *Electrophoresis* **10**, 302–309.
20. Sutherland, J. C., Emrick, A. B., and Trunk, J. (1989) Separation of chromosomal length DNA molecules: pneumatic apparatus for rotating gels during electrophoresis. *Electrophoresis* **10**, 315–317.
21. Gunderson, K. and Chu, G. (1991) Pulsed-field electrophoresis of megabase-sized DNA. *Mol. Cell. Biol.* **11**, 3348–3354.
22. Chu, G. and Gunderson, K. (1991) Separation of large DNA by a variable-angle contour-clamped homogeneous electric field apparatus. *Anal. Biochem.* **194**, 439–446.
23. Nagy, A. and Choo, K. H. (1991) Pulsed field gel electrophoresis using a double-decker gel system. *Nucleic Acids Res.* **18**, 5317.
24. Zhang, T. Y., Smith, C. L., and Cantor, C. R. (1991) Secondary pulsed field gel electrophoresis: a new method for faster separation of larger DNA molecules. *Nucleic Acids Res.* **19**, 1291–1296.
25. Gardner, D. C. J., Heale, S. M., Stateva, L. I., and Oliver, S. G. (1993) Treatment of yeast cell with wall lytic enzymes is not required to prepare chromosomes for pulsed-field gel analysis. *Yeast* **9**, 1053–1055.
26. Kwan, H., Li, C., Chiu, S., and Cheng, S. (1991) A simple method to prepare intact yeast chromosomal DNA for pulsed field gel electrophoresis. *Nucleic Acids Res.* **19**, 1347.
27. Mortimer, R. K., Game, J. C., Bell, M., and Contopoulou, C. R. (1990) Use of pulsed-field gel electrophoresis to study the chromosomes of *Saccharomyces* and other yeasts. Methods: A companion to *Methods Enzymol.* **1(2)**, 169–179.
28. Mortimer, R. K. and Johnston, J. R. (1986) Genealogy of principal strains of the yeast genetic stock center. *Genetics* **113**, 35–43.
29. Johnston, J. R., Curran, L., and Mortimer, R. K. (1991) Chromosome length polymorphisms in ancestral diploid strain EM93 of *S. cerevisiae*. *Abstr. Gen. Soc. Amer. Yeast Gen. Mol. Biol.*, San Francisco, p. 115.
30. Johnston, J. R. (1987) Electrophoretic karyotyping of strains of *Saccharomyces* and other yeasts. *Found. Biotechnol. Ind. Ferment. Res.* **5**, 43–55.

CHAPTER 9

Chromosomal Localization of Genes Through Pulsed-Field Gel Electrophoresis Techniques

Graham R. Bignell and Ivor H. Evans

1. Introduction

The development of electrophoretic techniques for separating intact chromosomal DNA from lower eukaryotes has provided a powerful system for analyzing the karyotypes of these organisms. Before the advent of pulsed field gel electrophoresis *(1)*, the study of lower eukaryotic chromosomal organization, by conventional genetic and cytogenetic methods, had proven to be unproductive, but now the genome of these organisms can be analyzed with relative facility. Electrophoretic karyotypes have been determined for filamentous fungi such as *Aspergillus nidulans (2)*, *Mucor cirinelloides (3)*, and *Schizophyllum commune (4)*, and for yeasts such as *Schizosaccharomyces pombe (5,6)* and *Saccharomyces cerevisiae*. These chromosomal separations by pulsed-field gel electrophoresis (PFGE) have, of course, allowed sizing of individual chromosomes and hence the nuclear genomes of these species, usually using the chromosomes of *S. cerevisiae* and *S. pombe* as size markers (*see* Chapter 8, by J. R. Johnston), but they have also enabled cloned genes to be assigned to their chromosomes by hybridization with suitable labeled probes, so bypassing the much more laborious procedures of conventional genetic mapping.

This hybridization method of assigning genes to specific yeast chromosomes is used increasingly. For example, the genes for the extracellular glucoamylases of diastatic strains of *S. cerevisiae* (*STA1, STA2,* and

From: *Methods in Molecular Biology, Vol. 53: Yeast Protocols*
Edited by: I. Evans Humana Press Inc., Totowa, NJ

STA3), as well as the intracellular sporulation-specific glucoamylase gene, have been shown to be present on chromosomes IV, II, XIV, and IX, respectively *(7,8)*. These approaches can also be valuable in evolutionary studies as exemplified by the work of Adams and colleagues *(9)*. They studied karyotypic changes in thirteen populations of *S. cerevisiae* when grown for 1000 generations under conditions of limiting organic phosphate, using chromosome-specific probes and found karyotypic changes in eight of the populations. In one of the latter, they found that *PHO10*, an acid phosphatase structural gene important for growth in media limiting in organic phosphate, had been deleted from its original locus on chromosome VIII, and that amplified copies had been translocated to other chromosomes.

2. Materials

2.1. Preparation of Karyotypic Southern Blots

1. Depurination solution: $0.25M$ HCl.
2. Denaturation solution: $1.5M$ NaCl, $0.5M$ NaOH.
3. Neutralizing solution: $1.5M$ NaCl, $1M$ Tris-HCl, pH 8.0.
4. Whatman (Maidstone, UK) 3MM filter paper.
5. 20X SSC: $3M$ NaCl, $0.3M$ trisodium citrate, pH 7.0.
6. 2X SSC: Made by diluting 20X SSC stock.
7. Two glass or perspex plates, slightly larger than the gel.
8. Nitrocellulose paper.
9. Four rubber bungs, ≈4 cm tall, and suitable flat-bottomed glass or plastic dish.
10. Acetate sheets to prevent "short-circuiting" during capillary transfer.
11. Paper towels.
12. Approximately 500 g weight.
13. Indian ink.

2.2. Probe Preparation

1. Low-gelling-temperature (LGT) agarose: Sigma (St. Louis, MO) type VII.
2. 5X random prime labeling buffer: 250 mM Tris-HCl, pH 8.0, 25 mM MgCl$_2$, 10 mM dithiothreitol (DTT), 1 mM HEPES (pH 6.6), and 26 A$_{260}$ U/mL of random hexadeoxyribonucleotides (store at $-20°$C).
3. dNTPS mixture: Mix equal volumes of 1.5 mM stock solutions of each of the unlabeled dNTPs (i.e., dATP, dTTP, and dGTP) (store at $-20°$C).
4. BSA: 10 mM acetylated (nuclease-free bovine serum albumin [BSA]) (store at $-20°$C).
5. [^{32}P] dCTP: Amersham (Arlington Heights, IL).

6. Klenow fragment of DNA polymerase I, usually 1 U/µL.
7. 25 mM EDTA, pH 8.0.
8. TE: 10 mM Tris-HCl, pH 8.0, 1 mM EDTA.
9. Sephadex G50: Prepare by slowly adding 30 g of Sephadex G50 (medium) to 250 mL TE, autoclave for 15 min at 15 lb/in^2, and allow to cool. Decant the supernatant and add a volume of fresh TE equal to the volume of the Sephadex sediment (store at 4°C).
10. Siliconized glass wool: Siliconize glass wool using Sigmacote (Sigma) siliconizing solution.
11. Spun column: Prepare the spun column in a 1-mL syringe: Using the plunger, pack a small amount of siliconized glass wool into the bottom of the syringe, then fill the syringe with a suspension of Sephadex G50. Centrifuge the syringe in a 50-mL centrifuge tube at 1000 rpm for 1 min. The Sephadex G50 packs down, giving ideally, a bed volume of 1 mL. Load 100 µL of 25 mM EDTA (pH 8.0), recentrifuge as before, and measure the volume of the solution coming out of the column: Repeat this step until the volume eluting from the column is equal to the volume loaded.
12. Boiling water bath, ice, and screw-capped plastic microfuge tubes (Sarstedt).

2.3. Hybridization and Autoradiography

1. 20X SSC: 3M NaCl, 0.3M sodium citrate, pH 7.0 (store at room temperature).
2. 6X SSC: Prepared by diluting 20X SSC stock.
3. 20% w/v sodium dodecyl sulfate (SDS) (store at room temperature).
4. 50X Denhardt's solution: 5 g Ficoll (mol wt 400,000), 5 g polyvinylpyrrolidone (360,000 average mol wt), and 5 g BSA in 500 mL dH$_2$O (store at –20°C).
5. Denatured salmon sperm DNA: Dissolve salmon sperm DNA in water to a concentration of 10 mg/mL, shear the DNA by drawing repeatedly through an 18-gage hypodermic needle (six to eight times) bent at 45°C or by sonication, then boil for 10 min (store at –20°C).
6. Fine mesh nylon filters.
7. Hybridization bottles and oven containing rotating support.
8. 2X SSC, 0.1% w/v SDS solution: 100 mL prepared from stock solutions. Preheated to hybridization temperature.
9. 1X SSC, 0.1% w/v SDS solution: 100 mL prepared from stock solutions, preheated.
10. 0.1X SSC, 0.1% w/v SDS solution: 100 mL prepared from stock solutions, preheated.
11. Whatman 3MM filter paper.
12. Saran wrap.
13. X-ray film: Kodak X-OMAT AR5.
14. Autoradiography cassette (if possible with intensifying screens).

3. Methods
3.1. Determination of Optimum Conditions for Separation

Prepare the blocks containing chromosomal-sized DNA and use electrophoretic conditions as described by Johnston (*see* Chapter 8). Optimize the electrophoretic conditions (*see* Note 1) to obtain the best possible separation of the chromosomes for each particular yeast strain (*see* Fig. 2A).

3.2. Preparation of Karyotypic Southern Blots

This method, based on that described by Southern *(10)*, includes depurination of the DNA to break down large DNA fragments >10 kb, which otherwise would not be efficiently transferred *(11)*.

1. Place the gel in depurination solution for 20 min with slight agitation, rinse in distilled water, and repeat.
2. Rinse the gel in distilled water and place in denaturation solution, leave for 30 min (with constant agitation), rinse the gel, and repeat.
3. Rinse the gel and place in neutralizing solution for 30 min (again with agitation), rinse the gel, and repeat. Transfer the gel to distilled water.
4. Set up the capillary transfer as described in steps 5–10, as shown in Fig. 1.
5. Using a square glass or perspex plate, slightly larger than the gel, cut four strips of Whatman 3MM paper to the same width but ≈8 cm longer than the plate.
6. Place the plate on the supports (rubber bungs ≈4 cm tall), in a larger flat-based glass or plastic dish. Place the first of the Whatman sheets onto the plate with ≈4 cm overlap at each end, wet the filter paper with 20X SSC, removing any trapped air bubbles using a flat edge, then fold the overhanging filter paper down to the base of the dish. Place the second sheet of filter paper at 90° to the first and wet as described earlier; repeat this step using all four sheets of Whatman paper, being careful to remove any trapped air bubbles. Fill the bottom of the dish with 20X SSC (≈3cm deep).
7. Slide the gel on to the bed of Whatman paper, being careful to remove/prevent any trapped air bubbles (this is easier if some 20X SSC is first poured on the filter paper).
8. Slide acetate sheets under the edges of the gel, projecting over the sides of the dish, covering the rest of the Whatman paper.
9. Wearing gloves, cut a sheet of nitrocellulose paper (*see* Note 2) together with three pieces of Whatman 3MM to the size of the gel. Wet the nitrocellulose paper by floating on 2X SSC, then place it on the gel, removing any air bubbles. This is made easier if the gel is covered with a little 2X SSC. Repeat with the three sheets of Whatman 3MM paper.

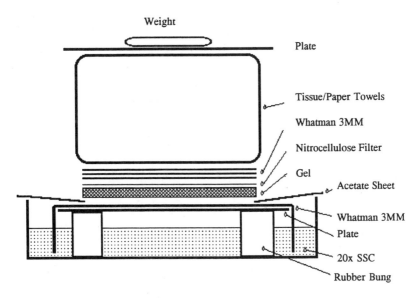

Fig. 1. Arrangement for capillary transfer of DNA (Southern blotting).

10. Place a thick layer of paper towels on the Whatman 3MM paper, completely covering the gel (*see* Note 3), then place a second glass/perspex plate on the towels and weigh them down with a 300–500-g weight.
11. Leave for 24–48 h at 4°C.
12. Remove all layers above the nitrocellulose paper, then mark the sample lanes, and any reference points (*see* Note 4) on to the nitrocellulose paper using Indian ink. Rinse the filter in 2X SSC, blot gently on Whatman paper, and then bake at 80°C for 2 h (*see* Note 5).

3.3. Probe Preparation

This method is for radiolabeling specific DNA sequences present in LGT agarose. The recombinant plasmid isolated as described in Chapter 18, must first be digested using a suitable restriction enzyme (as described in Chapter 4). The digested plasmid should then be fractionated by gel electrophoresis (*see* Chapter 18), using LGT agarose. The radiolabeling protocol is for random primer labeling using the Klenow fragment of DNA polymerase I.

1. Excise the desired DNA band from an LGT agarose gel, taking care to avoid excess gel and contamination with other DNA.
2. Transfer the gel block to a preweighed 1.5-mL microfuge tube, and find the weight of the gel block. Add distilled water in the ration 3 mL H_2O to 1 g agarose block.

3. Puncture the cap of the tube and place in a boiling water bath for 7 min, making sure the gel block is completely dissolved.
4. Aliquot into fractions containing ≈25 ng of DNA (*see* Note 6). If not used immediately, the aliquot can be frozen at –20°C, then reboiled for 1 min prior to use.
5. Transfer the tube to a 37°C water bath for 10 min.
6. To the DNA sample, in a maximum volume of 25 µL for a 50-µL reaction vol (if the DNA sample is above 25 µL, double the reaction vol), add the following solutions in order: 10 µL 5X random prime labeling buffer (20 µL for a 100-µL reaction vol), 2 µL of dNTP mix (i.e., dATP, dGTP, and dTTP), 2 µL BSA, 3–5 µL [^{32}P] dCTP (between 10 and 120 pmol), then 5 U Klenow DNA polymerase I. All solutions (except the DNA solution) should be thawed on ice prior to use, except the enzyme that should be removed from the freezer immediately before use and then returned to –20°C.
7. Gently mix the sample using a micropipet, and, if necessary, centrifuge briefly to collect the sample in the bottom of the tube.
8. Incubate the sample for 3–5 h at room temperature (*see* Note 7).
9. Dilute the sample with an equal volume of distilled water, then load onto the spun column; place the tip of the syringe in a 1.5-mL centrifuge tube and centrifuge at 1000 rpm for 1 min (in a 50-mL centrifuge tube).
10. Check incorporation using a Geiger counter; the column should give a high count toward the top of the column, reducing toward the tip. Obviously the sample in the microfuge tube should have a high activity.
11. Denature the probe DNA, in a screw-capped microfuge tube, by heating for 5 min in a boiling water bath.
12. Place the denatured probe DNA on ice, ready for addition to the hybridization solution.

3.4. Hybridization and Autoradiography

1. Soak the filters, prepared as described in Section 3.2., briefly in 6X SSC, then place onto nylon gauze and roll up and place in an appropriate hybridization bottle (*see* Note 8).
2. Add 20 mL of rehybridization solution heated to 68°C (*see* Note 9), composed of 5X Denhardt's solution, 0.5% SDS and 6X SSC: add denatured salmon sperm DNA to a final concentration of 100 µg/mL.
3. Seal the hybridization bottle and incubate at 68°C.
4. Leave the filter in prehybridization solution for a minimum of 4 h before addition of the radiolabeled probe. It is often a good idea to dilute the probe in some of the prehybridization solution before adding to the bottle,

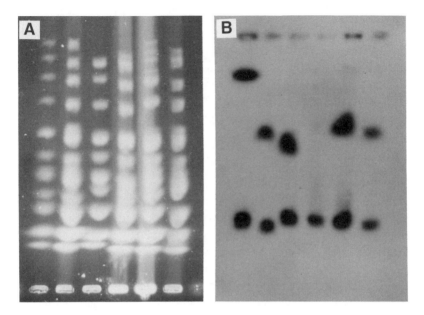

Fig. 2. **A** shows a CHEF gel separation of the chromosomes of six different strains of *Saccharomyces cerevisiae*. There is some variation in the banding pattern between the different strains—the result of chromosomal polymorphisms. The strains are A, JM1098 (*STA1*); B, JM2099 (*STA2*); C, JM2100 (*STA3*); D, DBY747 (*sta °*); E, lab[1] sp59 (*STA2*); and F, lab[1] sp5 (*STA2*). The separation was carried out using a 110-mL 1% agarose gel in 0.5X TBE maintained at a constant 10°C; the voltage was fixed at 180 V (6.4 V/cm), with the switching interval being ramped from 10–100 s over the 24-h run time. This gel was blotted and hybridized with the cloned *STA2* gene; the resultant autoradiograph can be seen in **B**. This figure shows a hybridization signal common to all strains test and represents hybridization to the internal sporulation-specific glucoamylase gene (*SGA*). One other hybridization can be seen in all the amylolytic yeast strains (i.e., all but strain D) representing hybridization to the extracellular glucoamylase genes (*STA1*, *STA2*, and *STA3*), all of which can be seen to be present on different chromosomes.

to prevent undiluted probe contacting the filter, as this lead to "hot spots" of high specific activity on the filters.
5. Incubate the filters in the hybridization solution at 68°C, usually overnight.
6. Remove the hybridization solution and wash the filter to remove nonspecific signals. The washes usually involve two 50-mL washes of 2X SSC 0.1% SDS, followed by two washes in 1X SSC 0.1% SDS, and finally two

washes in 0.1X SSC 0.1% SDS, all solutions being preheated and carried out at the hybridization temperature (68°C). Allow 20–60 min at each wash step, checking the filter between each wash, after the first two.
7. Remove excess solution from the filter by blotting on Whatman 3MM filter paper and then seal in Saran wrap.
8. Expose the filter between two sheets of X-ray film using, if possible, an autoradiography cassette with intensifying screens, at −70°C (see Fig. 2B).

4. Notes

1. It is often difficult to obtain pulsed field gel electrophoresis conditions that give good resolution of both the smaller and larger sized chromosomal DNA molecules. To obtain good resolution of all the chromosomes of a particular organism often requires using several different running conditions employing multiple or sometimes ramped switching interval cycles. For large-sized chromosomes, equal to or larger than those of *Schizosaccharomyces pombe*, it is advisable to use low voltages (≈ 1.4 V/cm), with long switching intervals: We obtained good resolution of the chromosomes of *S. pombe* using a switching interval of 4500 s over a total run time of 200 h. For smaller sized molecules, the voltage gradient can be increased up to 3.5 V/cm, or above, thereby reducing total run time; under these conditions it is often advisable to use several different switching intervals during a run, or the ramped runs mentioned earlier.
2. Schleicher and Shuell nitrocellulose BA85 paper with a pore size of 0.45 µm, is often used; however, other membranes are also available, for instance, charge-modified nylon membranes like Hybond. Obviously, these membranes require different handling and fixing techniques. Follow manufacturer's instructions.
3. The layer of paper towels acts as a sink, drawing the 20X SSC through the gel by capillary action, thereby transfering the DNA from the gel on to the nitrocellulose filter. It is important, therefore, that an excess of paper towels be used; if all towels become wet during the transfer, back-flow may reduce efficiency.
4. The marker DNA bands, together with other bands on the gel, can be marked by removal of a plug of the gel using a Pasteur pipet; these will then be visible on the nitrocellulose filter as white marks, allowing them to be marked on the filter with Indian ink. Otherwise, the gel can be traced using acetate sheet, again marking any bands, both these methods allow for quick identification of hybridization signals.
5. Nitrocellulose paper should ideally be baked under vacuum to prevent oxidation; however, this can be avoided, as long as the temperature is well regulated and does not exceed 80°C.

6. The concentration of DNA in the agarose blocks can be roughly estimated against the λ-*Hind*III marker DNA. The λ-*Hind*III digest fragments and approximate quantity of DNA per band (for 1-μg loading) are as follows: 23130 bp ≈ 480 ng, 9416 bp ≈ 200 ng, 6557 bp ≈ 140 ng, 4316 bp ≈ 90 ng, 2322 bp ≈ 50 ng, 2027 bp ≈ 40 ng, 564 bp ≈ 10 ng, and 125 bp ≈ 3 ng.
7. When labeling DNA from a gel, as in this method, it is often best to incubate the reaction overnight. The reaction may appear to solidify; this does not reduce incorporation, however. The reaction may require reheating briefly to melt the sample prior to removal of unincorporated label.
8. Using hybridization bottles in a hybridization with rotation means that a lower volume of hybridization solution is required, than in sandwich boxes, thereby allowing for a higher concentration of the probe.
9. The standard hybridization temperature is 68°C; however, when using heterologous gene probes it is often necessary to lower the temperature, thereby reducing the stringency of the hybridization. In such a case, all solutions and operations, including the washes, should be carried out at the reduced temperature.

Acknowledgments

The authors are grateful to John Hammond of the Brewing Research Foundation (UK) for supplying the *STA2* gene probe, and to Julius Marmur for supplying strains JM 1098, JM 1099, and JM 2000.

References

1. Schwartz, D. C. and Cantor, C. R. (1984) Seperation of yeast chromosome-sized DNAs by pulsed field gel electrophoresis. *Cell* **37,** 67–75.
2. Brody, H. and Carbon, J. (1989) Electrophoretic karyotype of *Aspergillus nidulans*. *Proc. Natl. Acad. Sci. USA.* **86,** 6260–6263.
3. Nagy, A., Vagvolgyi, C., and Ferenczy, L. (1994) Electrophoretic karyotype of *Mucor cirinelloides*. *Curr. Genet.* **26,** 45–48.
4. Horton, S. J. and Raper, C. A. (1991) Pulsed-field gel electrophoretic analysis of *Schizophyllum commune* chromosomal DNA. *Curr. Genet.* **19,** 77–80.
5. Smith, C. L., Matsumoto, T., Niwa, O., Klco, S., Fan, J.-B., Yanagida, M., and Cantor, C. R. (1987) An electrophoretic karyotype for *Schizosaccharomyces pombe* by pulsed field gel electrophoresis. *Nucleic Acids Res.* **15,** 4481–4489.
6. Vollrath, D. and Davis, R. W. (1987) Resolution of DNA molecules greater than 5 megabases by contour-clamped homogeneous electric fields. *Nucleic Acids Res.* **15,** 7865–7876.
7. Pretorius, I. S. and Marmur, J. (1988) Localization of the yeast glucoamylase genes by PFGE and OFAGE. *Curr. Genet.* **14,** 9–13.
8. Bignell, G. R. and Evans, I. H. (1990) Localization of glucoamylase genes of *Saccharomyces cerevisiae* by pulsed field gel electrophoresis. *Antonie van Leeuwenhoek* **58,** 49–55.

9. Adams, J., Puskas-Rotsa, S., Simlar, J., and Wilke, C. M. (1992) Adaptation and major chromosomal changes in populations of *Saccharomyces cerevisiae*. *Curr. Genet.* **22,** 13–19.
10. Southern, E. M. (1975) Detection of specific sequences among DNA fragments separated by gel electrophoresis. *J. Mol. Biol.* **98,** 503–517.
11. Wahl, G. M., Stern, M., and Stark, G. R. (1979) Efficient transfer of large DNA fragments from agarase gels to diazobenzyloxymethyl-paper and rapid hybridization using dextran sulphate, *Proc. Natl. Acad. Sci. USA* **76,** 3683–3687.

CHAPTER 10

Isolation, Purification, and Analysis of Nuclear DNA in Yeast Taxonomy

Ann Vaughan-Martini and Alessandro Martini

1. Introduction

The advent of molecular techniques has greatly influenced yeast taxonomy. Although it has become easier to identify an unknown organism, the use of these methods has caused a general reassessment of those tests previously considered important for the classification of yeasts.

Even if many traditional criteria, such as cell morphology, mode of budding, presence and type of sexual cycle, and spores are still fundamental for identification at the genus level, others tests used for classification at the species level, such as physiology, must be reconsidered as many limitations have been seen in recent years. In fact, even before molecular taxonomy became routinely employed, it was demonstrated that point mutations occur in yeast strains kept for long periods of time in collection that can result in significantly modified responses to physiological tests *(1,2)*. It was also known that even when the same methods were utilized, results could vary from one laboratory to the next. Finally, comparison of macromolecules, especially reassociation of nuclear (n) DNA (nDNA/nDNA), has shown that even when everything is rigorously controlled on the experimental level, phenotypic similarity does not necessarily correspond to synonymy. The earliest analyses of yeast macromolecules for taxonomic purposes were done by Marmur *(3)*, and Marmur and Doty *(4,5)*, and Schildkraut et al. *(6)* for the calculation of DNA base composition or mol% G + C, a value that is constant for each organism. Although this test is still routinely used in many laboratories, it is valid only as an exclusionary tool. In fact, greater than 2% difference in mol% G + C eliminates the possibility of conspecificity, whereas similar values do not necessarily mean that two strains belong to the same species.

More conclusive results can be obtained with nDNA/nDNA reassociation experiments *(8–10)*. Although extremely useful for delimitation at the species level, these methods do have some drawbacks being relatively time-consuming and requiring expensive equipment, as well as large quantities of DNA. Nevertheless, nDNA/nDNA reassociation gives very reliable results since the entire genome is used as a database, whereas all other methods, including the more recent macromolecular techniques, at best consider only a minimal part. In addition, methods such as pulsed field gel electrophoresis and restriction enzyme analysis of mtDNA, rDNA, or rRNA tend to furnish data regarding individual strains, information that is generally beyond the scope of taxonomy.

At the present time, several methods exist for the extraction of nDNA from yeast cells and they can be easily found in almost any textbook on yeast or molecular biology. The one that follows is somewhat long by current standards (about 2 wk if all goes well), but does result in clean, high molecular weight nDNA in generally good amounts (1–2 mg) applicable to a wide variety of different yeast species. Although some days are somewhat laborious, none goes over about four to five working hours and there are other days when little work is required at all. In the end, samples are of good quality and can be stored in the freezer (–18°C) for fairly long periods of time with excellent conservation. (The authors have effectively used 10-yr-old samples with good results.)

The nDNA/nDNA reassociation method described here employs the optical technique that does require a very good spectrophotometer with excellent temperature control (Gilford is used in most laboratories that use this method). Although this equipment is somewhat expensive, there are several advantages: No radioactivity is required and as a consequence no radioisotopes need to be purchased and no problem exists for waste disposal. In addition, any DNA samples available in the freezer can be used without special preparation. When a six-cuvet system is used, there can be two and even four results per working day. Finally, traditional taxonomic techniques, although apparently less expensive, require many analyses and much working time without always giving the conclusive results that nDNA/nDNA reassociation analyses generally do. This does not mean that traditional methods of yeast taxonomy are not worthwhile—on the contrary. Before time and money are spent for the extraction and analysis of nDNA, it is best to have a fairly good idea about an organism's identity. nDNA/nDNA reassociation studies can be an excellent means for the confirmation of previous results, as in the case of new species presented for publication.

The techniques for the isolation, purification, and analysis of yeast nDNA that follow have been used in the DBVPG laboratory for the last few years and usually work quite well. The general procedure is briefly outlined in the flow sheet in Fig. 1 and is then followed by a more detailed description divided into sections. The techniques used are a combination of a number of methods developed in various laboratories over the years and, if the reader needs more background, he or she is referred to the review by Kurtzman et al. *(11)*.

2. Materials

1. YEPD broth: 1.0% (all percentages w/v, unless otherwise indicated) yeast extract, 1.0% peptone, 2% glucose.
2. YEPD agar: 1.0% yeast extract, 1.0% peptone, 2% glucose, 1.5% agar.
3. Sucrose buffer: $0.02M$ Tris-HCl, $0.02M$ EDTA, 15% (w/v) sucrose, pH 7.8.
4. Glass beads (0.5 mm diameter).
5. Saline-EDTA: $0.15M$ NaCl, $0.1M$ EDTA, pH 8.0 (*see* Note 1).
6. 10% SDS: 10% Sodium dodecyl sulfate in saline EDTA (*see* Note 2).
7. Chloroform.
8. $5M$ Sodium perchlorate.
9. CIA: Chloroform and isoamyl alcohol, 24:1 (v:v).
10. 95% Ethanol (–18°C).
11. 10X SSC: $1.5M$ NaCl and $0.15M$ sodium citrate; pH 7.0.
12. 1X SSC: made from a 1:10 dilution of the 10X SSC solution.
13. α-Amylase: Type II-A (Sigma, St. Louis, MO, catalog no. A-6380) from *Bacillus* species (*see* Note 3).
14. Pancreatic ribonuclease: Type III-B (Sigma, catalog no. R-5750) from bovine pancreas, 2 mg/mL in 1X SSC heated for 10 min at 80°C in order to destroy DNase.
15. Pronase (e.g., Calbiochem): nuclease free, 2mg/mL in 1X SSC.
16. T_1 RNase (e.g., Calbiochem): 500 U/mL in 1X SSC or a less expensive type from Boehringer Mannheim (Mannheim, Germany) (catalog no. 109 193), has also been found to be effective and 1 mL of 100,000 U/mL is added to every 10 mL of sample.

CELL GROWTH & HARVEST
2-3 day YEPG slant culture inoculated into 200 ml YEPG broth incubated at 25°C, 1-2 days, 150 rpm
Preculture added to 3-6 liters YEPG broth (25°C, 1-2 days, 150 rpm)

▼

Cells harvested by centrifugation and washed twice with distilled water

CRUDE DNA EXTRACTION
(Yeast cell wall lysis)
Wet cells (30-60 grams) are treated with one of the three methods:

1) **Wet cell disintegration (Braun Homogenizer):**
cells are suspended in cold sucrose buffer with 1/3 volume of O.5 mm diameter glass beads and shaken in the homogenizer until 50% breakage

2) **Autolysis at 37°C**
A cell paste is incubated at 37°C for 16-18 hours under chloroform fumes in saline-EDTA containing 10% SDS and 1% mercaptoethanol

3) **Dry cell disintegration**
Freeze-dried cells (2-8 grams) are placed in round bottom flasks together with an equal amount of stainless steel beads (0.8 cm diameter) and kept on a rotary shaker (150 rpm) at room temperature for 3-16 hours until 50-70% breakage

SEPARATION OF CRUDE DNA
Agitation of cell lysate in saline-EDTA with 1 M sodium perchlorate and an equal volume of CIA (1-3 hours, 100 rpm)

▼

Centrifugation of the lysate for 10 minutes at 10,000g

▼

Precipitation of DNA from the top aqueous phase with cold (-20°C) 95% ethanol

▼

Second centrifugation for 15 minutes at 10,000g to recover crude DNA

ENZYMATIC PURIFICATION OF CRUDE DNA
First enzymatic treatment

Crude DNA in 20 ml IX SSC is shaken overnight (25°C, 100 rpm) together with α-amylase and Pancreatic RNase

▼

Treatment with Pronase-4 hours

▼

Agitation (150 rpm) for 30 minutes with an equal volume of CIA

▼

Centrifugation of the lysate for 15 minutes at 10,000g

▼

Precipitation of DNA from the top aqueous phase with cold (-20°C) 95% ethanol-spooling around a glass rod

▼

Resuspend DNA in 1 mM PB

Second enzymatic treatment

All day and overnight dialysis of dissolved DNA together with α-amylase, Pancreatic RNase and T_1 RNase

▼

Agitation (150 rpm) for 30 minutes with an equal volume of CIA

▼

Centrifugation of the lysate for 20 minutes at 10,000g --> make solution to O.15 M in PB

FINAL DNA PURIFICATION ON HYDROXYLAPATITE
Dilute DNA solution in O.15 M NaP is loaded onto a column of hydroxylapatite at room temperature

▼

First elution with O.15 M PB releases: Single stranded DNA / RNA / Residual carbohydrates and proteins

▼

2nd elution with O.25 M PB releases: Purified double stranded DNA

ANALYSIS OF PURIFIED YEAST NUCLEAR DNA

Determination of mol% G+C (whole or enzymatically lysed DNA)	and/or	nDNA/nDNA hybridization studies (sheared DNA)

17. 1M Sodium phosphate buffer (PB): A stock solution is prepared; 0.5M for NaH$_2$PO$_4$ and 0.5M in Na$_2$HPO$_4$. The pH is 6.8. Solutions of PB of lower molarity are made by dilution of the 1M stock solution (*see* Note 2).
18. HaP: Hydroxylapatite (Bio-Gel HTP, Bio-Rad, Richmond, CA).
19. 1 mM Ethylenediaminetetraacetate (EDTA).
20. Dimethylsulfoxide (DMSO) (*see* Note 2).
21. Mercaptoethanol.
22. Stainless steel beads (0.8 cm in diameter).
23. Dialysis tubing (*see* Note 4).

3. Methods
3.1. Growth and Harvesting of Cells

1. Transfer cultures for analysis to YEPD agar slants 2–3 d before preculture.
2. Grow a 300-mL YEPG preculture with shaking (usually at 25°C at 150 rpm) for 1 d.
3. Transfer the preculture to 3–6 L of YEPG broth and cultivated as in step 2 for an additional 1–3 d, depending on the strain studied.
4. Harvest the cells by centrifugation (*see* Note 5).
5. Wash the cells twice in distilled deionized water.

3.2. Lysis of the Yeast Cell Wall (see Note 6)
3.2.1. Wet Cell Disintegration Using a Braun Homogenizer (see *Note 7*)

1. Suspend washed cells in enough sucrose buffer to give a cream-like consistency (*see* Note 8).
2. Fill homogenizer bottle 1/3 with glass beads (0.5 mm diameter) and add cell suspension (thawed if previously frozen, and *see* Note 9).
3. Break cells in homogenizer for about 2 min with continuous CO$_2$ flow (*see* Note 10).
4. Check cells for breakage microscopically under phase contrast or regular light microscopy (*see* Note 11).
5. Remove cell suspension from the glass beads and determine volume.

3.2.2. Autolysis at 37°C

1. Wash cells twice in saline-EDTA.
2. Resuspend in the same solution to make a thick suspension (like yogurt).
3. Pour into a graduated cylinder and measure volume.
4. Place the suspension in a Fernbach flask.

Fig. 1. *(preceding page)* Major steps in the isolation and purification of high-mol-wt yeast nuclear DNA.

5. Add enough 10% SDS to give a final concentration of 2% and mercaptoethanol to give a concentration of 1%.
6. Soak the underside of a large gauze-covered cotton plug with 25 mL of chloroform. Place the plug in the neck of the flask, cover with aluminum foil.
7. Incubate at 37°C for 16–20 h. (*see* Note 12).

3.2.3. Dry Cell Disintegration (Mechanical Cell Breakage) *(According to Martini [12] and Vaughan-Martini and Martini [13])*

1. Suspend washed cells in enough saline EDTA to make a thick paste and spread it on a 20-cm diameter Petri dish.
2. Freeze at –18°C and lyophilize to complete dryness.
3. Grind the dried cells into a fine powder with a mortar and pestle and keep in a desiccator until use.
4. Put the cell powder into a heavy duty round-bottomed flask, cover completely with the stainless steel beads (0.8 cm in diameter), close the flask tightly with a rubber plug onto which a packet of filter paper containing a desiccant ($CaCl_2$) has been attached.
5. Put the flask on a rotary shaker (150 rpm) for 10–20 h until the cells are 50–75% broken (check microscopically).
6. Suspend the cells in ice cold saline EDTA to make a thick suspension (like yogurt) and measure final volume (*see* Notes 13 and 14).

3.3. Extraction of Crude DNA and Initial Purification

1. Transfer the lysed cell solution from Section 3.2. to a Fernbach or other wide-bottomed flask.
2. Add enough $5M$ $NaClO_4$ to give a final concentration of $1M$, then CIA to equal the total volume.
3. Agitate gently by hand to homogenize the two solutions.
4. Agitate the solution on a rotary shaker (100 rpm) at room temperature for 1–3 h (The two phases should not separate).
5. Centrifuge the solution for 20 min in a refrigerated centrifuge at 10,000g.
6. The centrifuged solution will be in three distinct layers: a top aqueous phase containing the crude nucleic acids, a thick central layer of cell debris, and a lower layer containing chloroform.
7. Remove the top aqueous layer using a wide-mouthed pipet so as not to shear the DNA.
8. Measure the volume of the aqueous layer and pour it into a centrifuge tube or bottle.
9. Add 1.3 vol of 95% ethanol (–18°C) and swirl gently by hand to mix. The DNA should precipitate (*see* Note 15).

10. Centrifuge the ethanol mixture in a refrigerated centrifuge at 10,000g for 10 min.
11. Pour off the ethanol and peel DNA pellet from the centrifuge bottle or tube with a metal spatula.
12. Let the crude DNA air-dry a few minutes.
13. Place the crude DNA in 20 mL 1X SSC in a 125-mL Erlenmeyer flask.
14. Add 1 mL each of α-amylase and pancreatic RNase to the DNA solution.
15. Incubate at 25°C overnight on a rotary shaker at 100 rpm or less so as to facilitate the dissolving of the DNA and to keep it from sticking to the glass of the flask.
16. After overnight enzyme treatment, add 0.5 mL Pronase solution and continue incubating for 4 h at 25°C or for 2 h at 30°C.
17. After Pronase treatment, pour the solution into a 300-mL flask.
18. Add an equal volume of CIA.
19. Shake at 150 rpm for 20 min on a rotary shaker.
20. Centrifuge at 10,000g in a refrigerated centrifuge for 15 min.
21. Remove the upper aqueous layer with a wide mouth pipet, place in a chilled graduated cylinder, and record volume.
22. Pour the crude DNA solution into a chilled beaker of appropriate size.
23. Add 1.3 vol of 95% chilled ethanol (−18°C).
24. Spool the DNA that precipitates around a glass rod that has been rinsed with 1 mM EDTA by twisting the rod between the fingers and moving in a circular motion around the beaker.
25. Compress the rod against the side of the beaker to secure the DNA against the rod and to expel excess ethanol.
26. Once all of the DNA is spooled, lay the rod across the top of the beaker for a few minutes in order to dry off the ethanol (*see* Note 16).
27. Dissolve the DNA overnight in 20 mL of 1 mM PB solution in the refrigerator.
28. The next day gently shake the DNA solution in order to thoroughly dissolve the precipitate.
29. Add the enzyme solutions: 1 mL pancreatic RNase, 1 mL α-amylase, either 0.8 mL T_1 RNase from Calbiochem or 0.1 mL of the product from Boehringer Mannheim.
30. Prepare the solution for dialysis at 25°C and do the following changes during the day:
    a. 1 h against 1 mM PB in 1 mM EDTA.
    b. 1 h against 1 mM PB in 1 mM EDTA.
    c. 2 h against 1 mM PB.
    d. Overnight against 1 mM PB.
31. The next day, carefully transfer the DNA solution from the dialysis bag into a graduated cylinder, record the volume, and then pour into a 300-mL flask.

32. Add an equal volume of CIA and shake at 150 rpm for 20 min.
33. Centrifuge the solution for 15 min at 10,000g in a refrigerated centrifuge, remove the upper aqueous layer with a wide-mouth pipet, measure volume.
34. Make the solution 0.15M in PB using the 1M PB stock solution.
35. Read the A_{260} of a 1:5 dilution of the crude DNA solution against a blank of 0.15M PB.
36. Proceed to final purification on a hydroxylapatite column (*see* Note 17).

3.4. Hydroxylapatite (HaP) Column Preparation (see Note 18)

1. Take a 50-mL plastic syringe that will act as the column.
2. Place a small piece of glass wool in the bottom of the column and cover it with a disk of Whatman (Maidstone, UK) #1 filter paper cut to fit.
3. Cover the paper disk with a 1-cm layer of 0.5-mm diameter glass beads (*see* Note 19).
4. Weigh 12.5 g of HaP (Bio-Rad, Bio-Gel HTP) for each column (*see* Note 20).
5. The HaP is placed in a flask and is suspended in about 100 mL of 0.15M PB with very gentle swirling (*see* Note 21).
6. Heat the suspension (covered with aluminum foil) for 20 min in a steam bath.
7. Let the slurry stand for 10 min in order to cool and settle.
8. Slowly decant off the buffer.
9. Gently resuspend the HaP in fresh buffer and let set 5 min.
10. Pipet off the top layer of buffer that includes the fines and add a small amount of buffer for handling as a slurry for pouring the columns.
11. Fill the columns to the 20-mL mark with 0.15M PB.
12. Pour the HaP slurry into the top of the syringe.
13. Once the bed settles, drain to the 40-mL mark and add any remainder of the HaP slurry (Flow rate = 1 drop every 2–3 s).

3.5. Purification of the DNA Solution from Section 3.3. on HaP Columns (Fig. 2)

1. Load the DNA solution, which has been dialyzed and made to 0.15M in PB (Sections 3.3. and 3.4.) onto the HaP column (which has been equilibrated with this molarity of buffer) at room temperature.
2. Wash the column with 0.15M PB (flowrate about one drop every 3 s) and monitor the A_{260} of the column eluate using a Beckman DU660 (Fullerton, CA) spectrophotometer with a flow cell where four to six columns can be monitored simultaneously (*see* Note 22).
3. There is a sharp rise in absorption as the single-stranded RNA and proteins elute. Continue washing until the A_{260} has returned to baseline.
4. Replace the 0.15M PB with 0.25M PB and elute the DNA fraction (*see* Note 23).

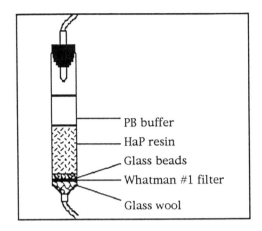

Fig. 2. Hydroxylapatite (HaP) column for DNA purification.

5. Check the purity of the DNA sample for the columns by reading the A_{230}, A_{260}, and A_{280} values of a 1:40 dilution of the DNA solution (*see* Notes 24 and 25).
6. Clean the HaP columns after use by passing 1 L of $1M$ PB per four columns slowly through the four columns (flowrate about one drop every 3 s).
7. Before starting the next series of purifications, re-equilibrate the columns by passing at least 2 L of $0.15M$ PB through the four columns (*see* Note 26).

3.6. Determination of mol% G + C by the Thermal Denaturation Method (T_m) According to Marmur and Doty (4,5) (see Note 27)

1. Dialyze an aliquot of the sample off the HaP column containing purified DNA against 1 mM EDTA and 1 mM SSC (2 mL 10X SSC/2 L of distilled deionized water) with three changes per day for 3 d in a cold room.
2. Concentrate the sample by lyophilization and resuspend in a small amount of distilled deionized water to give a concentration of at least 200 µg/mL (keep frozen at −18°C until analysis).
3. Determine T_m using 25 µg/mL nDNA in 1X SSC (*see* Note 28).

3.7. Shearing of DNA (see Note 29)

3.7.1. Shearing with a French Pressure Cell

1. Pass the dilute DNA solution from the HaP column in $0.25M$ PB twice through a French pressure cell at 10,000 lb/in.^{2}.
2. Pass the sheared sample through a nitrocellulose filter.
3. Dialyze the filtered solution against 1 mM EDTA and 1 mM SSC (2 mL 10X SSC/2 L of distilled deionized water) with three changes per day for 3 d in a cold room.

4. Concentrate the final sample by lyophilization, then resuspend it in a small amount of distilled deionized water, to give a concentration of at least 500 µg/mL, and keep frozen at −18°C until analysis.

3.7.2. Shearing DNA by Sonication

1. Dialyze the DNA sample from the HaP column against 1 mM EDTA and 1 mM SSC (2 mL 10X SSC/2 L of distilled deionized water) with three changes per day for 3 d in a cold room.
2. Concentrate the sample by lyophilization.
3. Resuspend in distilled deionized water to give a final concentration of 200 µg/mL.
4. Keep the chilled sample on ice and then treat with ultrasound (frequency 20 kc/s, with a 3/8" probe, depth 5 mm) for two 10-s periods separated by a 5-s rest period.
5. Concentrate the sample by lyophilization, resuspend in a small amount of distilled deionized water to give a concentration of at least 500 µg/mL and keep frozen at −18°C until analysis.

3.7.3. Shearing DNA by Manual Passage Through a Fine Syringe Needle (see Note 30)

1. Pass the cold DNA-water or DNA-1X SSC solution previously concentrated as in Section 3.7.2., 10–20 times through a 27-G gage syringe prior to analysis.

3.8. Determination of DNA Sequence Homology by Optical Reassociaton (see Notes 27 and 31)

1. Determine the temperature of reassociation (T_m=−25°C) prior to reassociation analysis, by performing a thermal melting reaction in 5X SSC + 20% DMSO (*see* Section 3.6.).
2. Adjust the concentration of DNA in the reassociation reaction mixture to 75 µg/mL determined spectrophotometrically with a four-cuvet system in which cuvet #1 contains a blank solution, cuvet #2 contains 75 µg/mL of sample A, cuvet #3 contains 75 µg/mL of sample B, and cuvet # 4 contains a mixture of equal amounts each of samples A and B (final concentration, 75 µg/mL).
3. Quickly heat the samples to 90°C (denaturation) and stabilize at that temperature for 10 min.
4. Lower the T_m to −25°C at a rate of 3°C/min.
5. Record the reassociation time after a T_m of −25°C has been reached.
6. Calculate the extent of DNA reassociation according to the equation of Seidler and Mandel *(7)* (*see* Note 32).

% Homology = $\{[1 - (\text{obs. Cot}_{0.5} \text{ mix} + [\text{Cot}_{0.5}\, 100 - \text{Cot}_{0.5}\, 0])] \times 100\} / \text{Cot}_{0.5}\, 100$

4. Notes

1. EDTA goes into solution only when the pH is above 7.5.
2. SDS and PB solutions as well as pure DMSO solidify under 20°C, keep in a 25°C room if necessary.
3. It is a good idea to make up extra quantities of the enzyme solutions and keep them in small aliquots at −18°C until use.
4. Dialysis tubing must be steamed for 10 min in a solution of 1 mM EDTA and then thoroughly rinsed in distilled, deionized water before use. It can be stored in distilled, deionized water at 5°C for several months.
5. A refrigerated centrifuge is best throughout; 5000g for 10 min is usually sufficient to pellet yeast cells.
6. Yeast cell walls are notoriously difficult to break. Since the amount of cells processed in this case is relatively large, it is too expensive to employ methods of enzymatic lysis. Nevertheless, at least one of the three methods listed should be possible in a normal laboratory.
7. Braun cell homogenizer (B. Braun Biotech, Melsungen, Germany) or similar.
8. Freeze if not processed immediately.
9. Keep everything on ice.
10. Exact times vary according to the strain and must be established by trial and error.
11. Approximately 50% breakage is sufficient.
12. The solution should be very viscous the next day.
13. Methods 3.2.1. and 3.2.3. using the Braun homogenizer and mechanical breakage, respectively, work for almost all yeast genera. It is very important that the final liquid solution containing the broken cell mass be kept as cold as possible until Section 3.3. in order that any nuclease activity present can be kept to a minimum.
14. Method 3.2.2. using autolysis is not recommended for those genera with known nuclease activity such as *Cryptococcus*, or for genera with particularly resistant cell walls such as *Schizosaccharomyces*. The method has worked quite well for many species of *Candida, Kluyveromyces,* and about 50% of the *Saccharomyces* species tried.
15. Some DNAs at this point will spool around a glass rod and can be removed in this manner. In most cases, however, the DNA is insufficiently pure and must be centrifuged.
16. If the DNA does not spool at this point it is probably not pure enough and requires additional enzymatic treatment or no DNA has been isolated.
17. If not processed immediately, freeze at −18°C for up to 6 mo. Thaw before further processing.

18. Hydroxylapatite (basic calcium phosphate) can be used to purify DNA and to separate DNAs of different mol % G + C. When the procedure is done at room temperature, proteins, carbohydrates, single-stranded DNAs, and RNAs usually are eluted from the HaP column at a PB molarity of less than 0.15–0.2M, whereas double-stranded DNA is generally eluted at a higher phosphate concentration. An HaP column (approx 2.5–6 cm in a 50-mL plastic syringe) is used in a single-step elution to remove non-DNA contaminants.
19. When preparing new columns, it is a good practice to prepare the empty column the first day, fill it with 0.15M PB buffer, and leave it overnight in order to check for leaks in the system. The following day proceed with HaP resin preparation.
20. Each column is best prepared separately.
21. Do not stir with a glass rod or a magnetic bar.
22. If such an apparatus with an automatic flow cell or a flow-through cuvet are not available, aliquots can be taken and read separately at A_{260}.
23. Fractions are collected as soon as the A_{260} begins to increase. The fractions containing the first 80–90% of the DNA peak are pooled.
24. The A_{230}/A_{260} and A_{260}/A_{280} values should be 0.5 and 1.86 respectively.
25. High A_{260}/A_{280} values indicate RNA contamination, whereas low A_{260}/A_{280} and high A_{230}/A_{260} values indicate protein contamination. Large deviations from ideal values suggest a need for additional column purification.
26. Regenerated HaP columns can be used up to 10 times, depending on the samples processed. Columns are discarded when the flowrate is slower than four to five drops per second.
27. A Gilford recording spectrophotometer with thermal programmer or similar is necessary.
28. When using a Gilford four-cuvet system, a blank of 1X SSC, two samples plus a control DNA (usually the type strain of *Candida parapsilosis*, T_m = 85.9) can be studied together. It is a good idea to do at least three separate determinations for each sample.

The calculation of mol% G + C is from Marmur and Doty *(5)*:
$$\text{mol\% G + C} = (T_m - 69.3) / 0.41$$

29. Shearing of the DNA is necessary for optical reassociation experiments.
30. A 27-G gage syringe is used. This method is somewhat less exact, but useful if neither a sonicator or French press is available.
31. The rate of reassociaton is increased by using a reaction solution of 5X SSC with the addition of 20% DMSO in order to depress the melting

temperature (ca. 12°C). Incubation times are approx 2.5–6 h, depending on the genus as well as the degree of relatedness between the strains studied.
32. Relationships between the strains studied is then based on the percent homology calculated. As a rule, strains are considered conspecific above 70% homology and not related below 20%. Reassociation rates between these two values have been found, and when such is the case, other factors must be considered (mating reactions when present, ecology, phenotypic similarities, or special features of the organisms) in order to evaluate actual relationships.
33. Two alternative techniques for determination of mol% C + C are:
   a. Buoyant density determination in cesium chloride. This method requires an analytical centrifuge equipped with an electronic scanner. A portion of the dilute sample (about 0.5–1.0 mL) may be saved directly from the HaP column and analyzed as such. Also, in this case, three or four separate determinations should be made and compared to an internal DNA standard can be, for example, from *Micrococcus lysodeikticus*. Further details on this method can be found in the references of Schildkraut et al. *(6)* and Kurtzman et al. *(9)*.
   b. HPLC. Some authors recently have begun to utilize high-performance liquid chromatography (HPLC) for the determination of DNA base composition of microorganisms. The method is comprised of essentially two steps: enzymatic digestion of pure DNA and subsequent determination of the individual nucleotides. Although this method has been successfully used for prokaryotes, eukaryotes can present some problems owing to the presence of more than one type of DNA of different mol% G + Cs, in the same organism (mtDNA or 2 μ plasmid DNA, for example). These must be separated out first in order that an effective measure of nuclear DNA mol% G + C can be made. The reader is referred to the review by Kaneko et al. *(14)* for further information.

References

1. Scheda, R. and Yarrow, D. (1966) The instability of physiological properties used as criteria in the taxonomy of yeasts. *Arch. Microbiol.* **55,** 209–225.
2. Rosini, G., Federici, F., Vaughan, A. E., and Martini, A. (1982) Systematics of the species of the yeast genus *Saccharomyces* associated with the fermentation industry. *Eur. J. Appl. Microbiol. Biotechnol.* **15,** 188–193.
3. Marmur, J. (1961) A procedure for the isolation of DNA from microorganisms. *J. Mol. Biol.* **3,** 208–218.
4. Marmur, J. and Doty, P. (1961) Thermal renaturation of deoxyribonucleic acids. *J. Mol. Biol.* **3,** 585–594.
5. Marmur, J. and Doty, P. (1962) Determination of the base composition of deoxyribonucleic acid from its thermal denaturation temperature. *J. Mol. Biol.* **5,** 109–118.

6. Schildkaut, C. L., Marmur, J., and Doty, P. (1962) Determination of the base composition of deoxyribonucleic acid from its buoyant density in CsCl. *J. Mol. Biol.* **4,** 430–433.
7. Seidler, R. J. and Mandel, M. (1971) Quantitative aspects of DNA renaturation: DNA base composition, state of chromosome replication and polynucleotide homologies. *J. Bacteriol.* **106,** 608–614.
8. Bernardi, G., Faures, M., Piperno, G., and Slonimski, P. P. (1970) Mitochondrial DNAs from respiratory-sufficient and cytoplasmic respiratory-deficient mutants of yeasts. *J. Mol. Biol.* **48,** 23–43.
9. Kurtzman, C. P., Smiley, M. J., and Johnson, C. J. (1980) Emendation of the genus *Issatchenkia* Kudriavzev and comparison of species by Deoxyribonucleic acid association, mating reaction, and ascospore ultrastructure. *Int. J. Sys. Bacteriol.* **30,** 503–513.
10. Price, C. W., Fuson, G. B., and Phaff, H. J. (1978) Genome comparison in yeast systematics: delimination of species within the genera *Schwanniomyces, Saccharomyces, Debaryomyces* and *Pichia. Microbiol. Rev.* **42,** 161–193.
11. Kurtzman, C. P., Phaff, U., and Meyer, S. A. (1983) Nucleic acid relatedness among yeasts, in *Yeast Genetics Fundamental and Applied Aspects* (Spencer, J. F. T., Spencer, D. M., and Smith, A. R. W., eds.), Springer-Verlag, New York, pp. 139–166.
12. Martini, A. (1973) Un nuovo metodo per la disintegrazione delle cellule di lievito. *Ann. Fac. Agr. Univ. Perugia.* **28,** 3–8.
13. Vaughan-Martini, A. and Martini, A. (1987) Three newly delimited species of *Saccharomyces sensu stricto. Antonie van Leeuwenhook.* **53,** 77–84.
14. Kaneko, T., Katoh, K., Fujimoto, M., Kumagai, M., Tamaoka, J., and Katayama-Fujimura, Y. (1986) Determination of the nucleotide composition of a deoxyribonucleic acid by high-performance liquid chromatography of its enzymatic hydrolysate: a review. *J. Microbiol. Meth.* **4,** 229–240.

Chapter 11

Isolation of Yeast DNA

Peter Piper

1. Introduction

The problems often experienced with DNA isolated from yeast are mostly owing to the impurities remaining in DNA isolated according to certain isolation protocols. Incomplete cutting of chromosomal DNA with restriction endonucleases, or poor transformation of *Escherichia coli* with plasmid DNA from yeast are the difficulties most frequently encountered. The procedures given herein have been chosen since they have used successfully by many workers to obtain DNA that does not give these problems.

It is often necessary to purify total yeast DNA in a high-mol-wt form. This is desirable irrespective of whether the DNA is to be used for the construction of genomic libraries, for the recovery ("rescue") of integrated vectors, or fragment identification by Southern hybridization. Most procedures for isolation of high-mol-wt yeast DNA incorporate spheroplast formation followed by spheroplast lysis, an approach that is widely thought to minimize the mechanical shear of the DNA. However, spheroplasting is probably not essential, as it is indeed possible (and quicker) to isolate high-mol-wt DNA after cell disruption by vortexing with glass beads *(1)*. If spheroplasting is used, it is essential that the cells have not entered the stationary phase of growth since stationary cells are refractory to spheroplast formation. Late-exponential phase YPD-grown cells of *Saccharomyces cerevisiae* or *Hansenula polymorpha* normally spheroplast very efficiently, whereas for *Schizosaccharomyces pombe* growth on minimal medium with 0.5% glucose is reported to improve the extent of spheroplast conversion *(2)*.

Section 3.1. is a straightforward means of isolating high-mol-wt DNA that can be easily scaled up or down. It is suitable for several quite diverse yeasts (including *Saccharomyces cerevisiae*, *Candida albicans*, *Schizosaccharomyces pombe*, and *H. polymorpha*). It is common practice to incorporate a sucrose gradient centrifugation step after stage 11 of this purification. This separates the high-mol-wt DNA from low-mol-wt nucleic acids (including plasmid DNAs), but is not always absolutely essential. For example, the DNA should cut well with restriction enzymes at step 11, and the sucrose gradient step can be omitted if the DNA is only to be used for purposes such as Southern blot analysis.

The extensive use of *S. cerevisiae* as a host for the expression and analysis of eukaryotic genes, often makes it necessary to recover plasmid shuttle vectors from yeast as part of the routine shuttling of plasmids between yeast and *E. coli*. These plasmids are almost always manipulated in *E. coli*. It is usually sufficient to isolate approx 1 µg plasmid DNA from the yeast transformant of interest, this being enough for an *E. coli* transformation and/or Southern analysis. Many methods for the isolation of plasmids from yeast *(4,5)* are time-consuming and give DNA that transforms *E. coli* only at low frequency. Section 3.2. *(6)*, provides a quick means of obtaining plasmid DNA from multiple small yeast cultures. This DNA can then be used to transform competent *E. coli* cells at high frequency. The procedure takes advantage of the fact that phenol/chloroform extraction of yeast in the presence of LiCl and Triton X-100 solubilizes plasmid DNA while precipitating proteins and chromosomal DNA.

2. Materials
2.1. Large-Scale Isolation of High-Mol-Wt Yeast DNA (see Notes 1 and 2)

1. TE buffer: 10 mM Tris-HCl, pH 7.5.
2. 1.2M sorbitol, 50 mM citrate-phosphate buffer pH 6.5 (autoclaved; then add 0.2% [v/v] 2-mercaptoethanol and 5 mg/mL zymolyase just before use).
3. 1.2M sorbitol, 10 mM Tris-HCl, pH 8.0.
4. 100 mM Tris-HCl, pH 8.0, 10 mM Na$_2$EDTA, 1% w/v SDS.
5. Proteinase K (stored at –20°C as use-once frozen 20 mg/mL aliquots in TE buffer, these being thawed on ice just before use).
6. Chloroform and phenol/chloroform 1:1 (w/v); the phenol should be nucleic acid grade, and saturated with 100 mM Tris-HCl, pH 8.0.
7. Absolute ethanol.
8. 2M sodium acetate, pH 5.2.
9. Enzymes: Zymolyase and RNase A.

2.2. Small-Scale Isolation of Shuttle Plasmid DNAs from Yeast for Transformation into E. coli (see Note 1)

1. 2.5M LiCl, 50 mM Tris-HCl, pH 8.0, 4% (v/v) Triton X-100, 62.5 mM Na$_2$EDTA.
2. Phenol/chloroform 1:1 (w/v); the phenol should be nucleic acid grade, and saturated with 100 mM Tris-HCl, pH 8.0.
3. Glass beads (BDH 0.45-μm mesh are suitable).
4. Absolute ethanol.
5. 70% v/v ethanol.

3. Methods
3.1. Large-Scale Isolation of High-Mol-Wt Yeast DNA (see Notes 1 and 2)

1. Grow an 1-L-YEPD culture to late log phase (approx 10^8 cells/mL). Harvest the cells by centrifugation (5 min, 1500g).
2. Resuspend the sedimented cells in 100 mL TE buffer, and pellet again to wash cells.
3. Resuspend the cell pellet in 20 mL of the sorbitol, citrate-phosphate buffer, 2-mercaptoethanol, zymolyase solution.
4. Incubate at 30°C with periodic gentle agitation. At intervals from 30–120 min check the degree of spheroplasting (add water to a streak of the cells on a microscope slide; in <1 min spheroplasts will be observed to lyse) (see Note 3).
5. Collect the spheroplasts by gentle centrifugation (5 min 2000 rpm Sorvall SS4 rotor) and wash twice in 1.2M sorbitol, 10 mM Tris-HCl, pH 8.0.
6. Resuspend in 20 mL 100 mM Tris-HCl, pH 8.0, 10 mM Na$_2$EDTA, 1% w/v SDS, 50 mg/L proteinase K (the latter being added from a 20 mg/mL stock just before use). Incubate at 37°C for 60 min; then 65°C for 15 min, before allowing to cool to room temperature.
7. Add an equal volume of 1:1 phenol:chloroform, then mix by inversion several times so as to form a white emulsion.
8. Centrifuge (10 min, 15,000 rpm Sorvall SS34), then transfer the upper aqueous phase to another tube and re-extract with chloroform.
9. Precipitate the upper aqueous phase of the chloroform extraction with 2 vol. ethanol. Place the tube on ice for 30 min. Collect the fibrous DNA precipitate with forceps, and redissolve in 10-mL TE buffer plus 50 mg/L RNase A. Incubate at 37°C for 30 min.
10. Extract once with phenol/chloroform and once with chloroform.
11. Reprecipitate DNA by adding 0.04 vol 2M NaOAc, and 2 vol ethanol. Leave on ice for 30 min, then spin (10 min 15,000 rpm Sorvall SS34), dry the DNA *in vacuo*, and redissolve it in a 2–5 mL TE buffer. Store the DNA frozen at –20°C.

3.2. Small-Scale Isolation of Shuttle Plasmid DNAs from Yeast for Transformation into E. coli (see Note 1)

1. Harvest 1.5 mL of each *S. cerevisiae* culture in an Eppendorf (microfuge) tube by centrifugation (15,000g for 20 sec).
2. Resuspend the cells in a 0.1-mL LiCl-Triton X-100 buffer.
3. Add 0.1-mL 1:1 phenol:chloroform and 0.2 g glass beads.
4. Vortex vigorously for 2 min, then centrifuge at 15,000g for 1 min.
5. Collect the upper phase (taking care not to include any interface) and precipitate with 2 vol of ethanol.
6. Leave on ice for 5 min, spin for 10 min at 15,000g.
7. Wash the pellet with ice cold 70% ethanol, dry it *in vacuo*, then dissolve it in 20–40 µL TE. This solution can be used in *E. coli* transformations without further purification.

4. Notes

1. All steps can be conducted at room temperature.
2. Section 3.1. provides DNA of >30–50 kb, suitable for recombinant library construction, yet the DNA is invariably sheared a little during isolation. Where intact chromosomal DNA is essential (as in OFAGE gel analysis of yeast chromosomes, or the reisolation of yeast artificial chromosome [YAC] vectors), the technique of zymolyase digestion in agarose blocks (*see* Chapter 8, by Johnston and ref. *3*) should be applied. When it is necessary to transfer high-mol-wt DNA solutions from one tube to another, precautions are necessary to minimize mechanical shearing. If the DNA is sufficiently diluted, it can sometimes be "tipped" from one tube into the next. Alternatively, wide-bore pipets should be employed. Sometimes the pellets from ethanol precipitations will not readily dissolve. Adding extra EDTA to the solution can help, as can patience also (leaving the solution on the bench overnight).
3. If cells are resistant to spheroplasting, it may be worth trying a different zymolyase preparation or, as a last resort, breaking cells by vortexing with glass beads *(1)*.

References

1. Huberman, J. A., Spotila, L. D., Nawotka, K. A., El-Assouli, S. M., and Davis, L. R. (1987) The *in vivo* replication origin of the yeast 2 µm plasmid. *Cell* **51**, 473–481.
2. Beach, D., Piper, M., and Nurse, P. (1982) Construction of a *Schizosaccaromyces pombe* gene bank in a yeast bacterial shuttle vector and its use to isolate genes by complementation. *Mol. Gen. Genet.* **187**, 326–339.
3. Sambrook, J., Fritsch, E. F., and Maniatis, T. (1989) *Molecular Cloning, A Laboratory Manual,* 2nd ed, Cold Spring Harbor Laboratory, Cold Spring Harbor, NY.

4. Hoffman, C. S. and Winston, F. (1987) A ten-minute DNA preparation from yeast efficiently releases autonomous plasmids for transformation of *E. coli*. *Gene* **57,** 267–272.
5. Chow, T. Y-K. (1989) Purification of yeast—*E. coli* shuttle plasmid for high frequency transformation in *E. coli*. *Nucleic Acids Res.* **17,** 8391.
6. Ward, A. C. (1990) Single-step purification of shuttle vectors from yeast for high frequency back-transformation in *E. coli*. *Nucleic Acids Res.* **18,** 5319.

CHAPTER 12

Isolation of Mitochondrial DNA

Graham R. Bignell, Angela R. M. Miller, and Ivor H. Evans

1. Introduction

As a key cell organelle, with its own semiautonomous genetic system, the mitochondrion and its enclosed DNA (mt DNA) have been the target of a huge amount of research using the methods of molecular biology (*see,* e.g., refs. *1,2*). *Saccharomyces cerevisiae* has, of course, been a particular focus of attention (*see,* e.g., refs. *3,4*), but the mtDNAs of other yeasts are becoming increasingly well known (*see,* e.g., ref. *5*).

Unlike the mtDNAs of, for example, vertebrates, fungal mtDNAs cannot usually be released efficiently into cell homogenates as compact supercoiled circular DNAs, which can be purified from chromosomal DNA contamination by the sort of technique that is suitable for plasmid DNAs, for example, ultracentrifugation on cesium chloride density gradients in the presence of ethidium bromide (but *see* Note 1). Indeed, it has recently been argued that the majority of fungal mtDNAs are actually linear in vivo, and that linearity is not merely an artifact of the isolation procedure *(6)*. However, it happens that certainly the great majority of fungal mtDNAs that have been investigated are extremely rich in A and T, and this fact can be exploited in order to separate mtDNA from nuclear DNA: density gradient centrifugation in the presence of dyes such as 4', 6-diamidino-2-phenylindole (DAPI) or *bis*-benzimide (Hoechst No. 332258), which bind preferentially to (A + T)-rich DNA, results in discrete banding of the two types of DNA, because of the differential buoyant densities caused by dye-binding.

Two protocols for isolating yeast mtDNA are described in this chapter. The first is a CsCl-dye ultracentrifugation technique that can be applied to any, even relatively crude, yeast DNA preparation, provided some mtDNA is present. The second, faster, protocol avoids the CsCl ultracentrifugation step by isolating mitochondria, from which DNA is then extracted: It is essentially that described by Querol and Barrio *(7)*.

2. Materials

2.1. Purification of mtDNA on a CsCl Gradient Containing bis-Benzimide

1. CsCl (*see* Note 2).
2. Preparation of yeast genomic DNA (*see* Note 3).
3. Sterile microfuge tubes and micropipet tips.
4. Suitable ultracentrifuge tubes; access to suitable ultracentrifuge and rotor (*see* Note 4).
5. 10 mg/mL *bis*-benzimide in dH$_2$O (store foil-wrapped at 4°C). *Caution: bis*-benzimide may be toxic and carcinogenic; handle and dispose of with appropriate precautions.
6. Long wavelength UV source, gloves and goggles or visor for UV protection, syringe, needle, and microfuge tubes for collecting mtDNA fraction(s).
7. CsCl-saturated isopropanol.
8. TE: 1 m*M* EDTA, 10 m*M* Tris-HCl, pH 8.0.
9. Ethanol (or isopropanol) chilled to 4° or –20°C.
10. 5*M* ammonium acetate, pH 5.2.
11. Liquid nitrogen (is an option for rapid DNA precipitation).

2.2. Isolation of mtDNA via a Mitochondrial Preparation

1. YEPD: 1% w/v yeast extract, 1% w/v bectopeptone, 2% w/v glucose.
2. 1*M* sorbitol, 0.1*M* EDTA, pH 7.5.
3. Appropriate sterile centrifuge tubes, transparent (e.g., ultraclear) centrifuge tubes and access to an ultracentrifuge and suitable swing-out rotor (e.g., Beckman [Fullerton, CA] SW41).
4. Zymolyase 20T (e.g., from Seikagaku Kogyo Co., Japan); alternative protoplasting enzymes are Lyticase or Novozym 235—*see* protoplasting protocol in Chapter 18, on genomic DNA banks by Bignell and Evans.)
5. 0.25*M* sorbitol 1 m*M* EDTA, 0.5 m*M* Tris-HCl, pH 7.5.
6. 20% w/v sucrose, 0.1 m*M* EDTA, 10 m*M* Tris-HCl, pH 7.5.
7. 60% w/v sucrose, 0.1 m*M* EDTA, 10 m*M* Tris-HCl, pH 7.5.
8. 50% w/v sucrose, 0.1 m*M* EDTA, 10 m*M* Tris-HCl, pH 7.5.
9. 44% w/v sucrose, 0.1 m*M* EDTA, 10 m*M* Tris-HCl, pH 7.5.

Isolation of Mitochondrial DNA

10. Syringes and needles for collection of mitochondrial band.
11. 50 mM NaCl, 1 mM EDTA, 10 mM Tris-HCl, pH 7.5.
12. SDS solution: Stock, e.g., 20% w/v sodium dodecyl sulfate.
13. Proteinase K.
14. 5M ammonium acetate, pH 5.2.
15. Sterile microfuge tubes and micropipet tips.
16. Isopropanol.
17. 70% v/v ethanol.
18. TE: 1 mM EDTA, 10 mM Tris-HCl, pH 8.0.

3. Methods
3.1. Purification of mtDNA on a CsCl Gradient Containing Bis-benzimide

1. Add CsCl (*see* Note 2) to the solution of yeast DNA (*see* Note 3) at 1.2 g/mL. Gargouri *(8)* recommends adjusting the refractive index to 1.397, but in our experience this is not necessary; the volume of the DNA solution is that able to be accommodated in one or more ultracentrifuge tubes (*see* step 3).
2. Centrifuge 16,000g for 15 min (it can be convenient to distribute the solution between microfuge tubes and use a microcentrifuge), and discard any proteins forming a pellicle above the centrifuged solution; transfer the clarified solution to one or more suitable ultracentrifuge tubes.
3. Add 5 µL of a 10 mg/mL aqueous solution of *bis*-benzimide to each ultracentrifuge tube and centrifuge at 200,000g (55,000 rpm in the 70 Ti Beckman angle rotor) and 15°C for 24 h (*see* Note 4). *Caution: Bis*-benzimide may be toxic and carcinogenic; handle and dispose of with appropriate precautions.
4. View the tube(s) using long wavelength UV light. *Caution:* Use UV-opaque goggles or visor. Two distinct blue-white fluorescing bands should be visible, the upper being mtDNA (*see* Fig. 1).
5. Collect into e.g., microfuge tubes, the upper mitochondrial DNA band, by piercing the side of the tube with a syringe needle just below the band.
6. Extract the DNA solution with an equal volume of CsCl-saturated isopropanol (this is conveniently done in microfuge tubes), briefly microfuge (16,000g for 30 s) to separate the phases and discard the upper organic phase containing *bis*-benzimide.
7. Repeat step 6 once or twice more, to thoroughly remove the *bis*-benzimide.
8. Dialyze the DNA solution against 1l TE, at 4°C, overnight, to remove the CsCl, and any residual TE (*see* Note 5).
9. Transfer the dialyzed solution to microfuge tubes and precipitate the DNA by adding 2 vol of cold (4 or –20°C) ethanol (or 1 vol of isopropanol) and 1/10 vol of 5M ammonium acetate. Mix by inversion, hold 30 min at –20°C (or 5 min in liquid nitrogen), and centrifuge 10 min at 16,000g in a microfuge.

Fig. 1. DNA from *Lipomyces starkeyi*, banded on a CsCl-*bis*-benzimide gradient, as described in Section 3.1., viewed and photographed using long wavelength UV light. The upper band is mitochondrial DNA and the lower, apparently double-band, is nuclear DNA.

10. Thoroughly drain dry and remove the last traces of organic solvent by air current or a few minutes in an evacuated desiccator.
11. Dissolve overnight (4°C) in 50–100 µL TE.

The mtDNA is now ready for analysis by a variety of procedures, such as physical mapping using restriction enzymes and blotting/hybridization experiments, cloning, and sequencing.

3.2. Isolation of mtDNA via a Mitochondrial Preparation (7)

1. Grow the yeast strain of interest overnight in 250 mL YEPD medium, in a 1-L flask, shaking thoroughly to ensure good aeration, at 30°C.

Isolation of Mitochondrial DNA

2. Harvest the cells by low-speed centrifugation (approx 4000g for 5 min); for *S. cerevisiae*, the yield is typically 3 g, wet wt.
3. Resuspend the cell pellet in $1M$ sorbitol, 0.1 mM EDTA (pH 7.5) and 0.5 mg/mL zymolyase 20T (the enzyme added just prior to use), and incubate at 37°C for 30–60 min, with periodic manual agitation. Check the efficiency of spheroplast formation microscopically.
4. Pellet the spheroplasts by centrifugation at approx 4000g for 5 min, and resuspend in $0.25M$ sorbitol, 1 mM EDTA, 0.5 mM Tris-HCl, pH 7.5, at 10 mL/g cells; incubate for 30 min at 4°C, i.e., until the spheroplasts are osmotically lysed.
5. Pellet nuclei and cell debris by centrifuging 2000g for 20 min, and transfer the supernatant to a new centrifuge tube.
6. Pellet the mitochondria by centrifuging at 15,000g for 20 min (*see* Note 6).
7. Resuspend the mitochondrial pellet in 5 mL of 20% w/v sucrose, 0.1 mM EDTA, 10 mM Tris-HCl, pH 7.5, and centrifuge at 15,000g for 20 min, in order to wash the mitochondria.
8. Resuspend the mitochondrial pellet in 2 mL 60% w/v sucrose, 0.1 mM EDTA, 10 mM Tris-HCl, pH 7.5, transfer to a suitable transparent ultracentrifuge tube, and overlay successively with 4 mL buffered 50% w/v sucrose followed by 4 mL buffered 44% w/v sucrose.
9. Centrifuge at approx 120,000g and for 90 min in a swing-out rotor (e.g., 40,000 rpm in a Beckman SW41); the mitochondria should form a tight band at the 44/55% interface.
10. Collect the mitochondria, e.g., by side puncture with a syringe, and dilute in 4 mL 50 mM NaCl, 1 mM EDTA, 10 mM Tris-HCl, pH 7.5.
11. Harvest the washed mitochondria by centrifuging at 15,000g for 20 min.
12. Resuspend the mitochondrial pellet in 2 mL of the solution used in step 10, adjusted to 0.5% w/v SDS and 50 µg/mL proteinase K; incubate 3 h at 37°C (statically).
13. Add 0.5 mL $5M$ ammonium acetate, hold at –20°C for 15 min, and centrifuge at approx 16,000g and 4°C for 10 min, to remove precipitated protein and SDS.
14. Transfer the supernatant to, e.g., microfuge tubes, add an equal volume of isopropanol, hold at room temperature for 10 min, and then collect the precipitated mtDNA by centrifuging at 16,000g for 10 min at room temperature.
15. Discard the supernatant(s), and carefully rinse the pellet(s) with 70% v/v ethanol, to remove residual isopropanol; drain and air dry the pellets until all traces of ethanol have been removed.
16. Dissolve the mtDNA pellet(s) in 200–300 µL TE; the yield of mtDNA is typically 80–120 µg, with an A_{260}/A_{280} of 1.8–1.9, indicating reasonable purity.

The mtDNA is now ready for further analysis.

4. Notes

1. In the case of the dimorphic zygomycete *Mucor racemosus*, lack of AT-enriched segments in the mitochondrial chromosome means that CsCl-*bis*-benzimide isopycnic centrifugation is not very successful in purifying mtDNA, but Schramke and Orlowski *(9)* found that CsCl-ethidium bromide centrifugation was useful in separating mtDNA from contaminating nuclear DNA, despite the fact that most of the isolated mtDNA was linear.
2. In an alternative procedure described by Fox and colleagues *(10)*, sodium iodide is used instead of cesium chloride, and is used at 0.79 g NaI/mL (final density 1.49 g/mL).
3. The DNA could be extracted by a technique such as that described in Chapter 18, which does not appear to cause undue loss of mtDNA; the DNA solution would be in a low-salt solvent such as 0.1X SSC or TE, pH 8.0. Alternatively, DNA can be rapidly extracted, on a fairly small scale, as described in the relevant chapter by Peter Piper (Chapter 11), or by the method described by Gargouri *(8)*, which is in fact aimed at mtDNA isolation. Gargouri's technique is given here:
   a. Inoculate 100 mL YEPD medium (1% w/v yeast extract, 1% w/v bactopeptone, 2% w/v glucose) in a 500-mL flask with 1 mL fresh overnight preculture, and incubate overnight at 28°C, with shaking.
   b. Harvest the cells by low speed centrifugation (e.g., 2000g for 5 min), wash with water, and repellet by centrifugation: typically, 2–3 g wet wt of cells are obtained.
   c. Resuspend the pellet in 10 mL solution A (1M sorbitol, 50 mM citric acid, 150 mM K_2HPO_4, 10 mM EDTA (from a pH 8.0 stock), 0.1% v/v 2-mercaptoethanol, 0.3 mg/mL zymolyase 100-T (100,000 U/g, Seikagaku Kogyo Co.), adjusted to pH 7.0 with NaOH (the solution is stable for a few months at 4°C), and incubate at 37°C for 30–60 min with gentle shaking. Check that most cells have spheroplasted by microscopy.
   d. Harvest the spheroplasts (4000g for 5 min); store the pellet at –20°C if necessary.
   e. Add 4 mL solution B (150 mM NaCl, 5 mM EDTA (from a pH 8.0 stock) 1% v/v Sarkosyl, 10 mm Tris-HCl, pH 7.0, to the pellet, and resuspend by up and down pipeting (or vortexing for 1 min) in order to achieve complete and drastic lysis of the cells and their mitochondria: The suspension should be very viscous and resemble an *E. coli* lysate.
   f. Clarify the lysate by centrifuging at 15,000g for 30 min (e.g., in a microcentrifuge). Retain the supernatant and discard the pellet. The solution is now ready for addition of CsCl.
4. In the case of the small Beckman vertical rotor (65 VTi), 5 h of centrifugation at 55,000 rpm at 15°C are sufficient to separate mtDNA from

Isolation of Mitochondrial DNA

nuclear DNA *(8)*. When using NaI gradients, Fox and colleagues *(9)* centrifuge for 48 h at 45,000 rpm, 20°C, in a Beckman VTi 50 rotor. We have found that centrifugation at 55,000 rpm, 15°C, for 24 h in a Beckman angle 70 Ti rotor gave a good separation.

5. This step can be replaced by diluting the DNA solution three- to fivefold with TE *(8)*.
6. In a method for preparing yeast mtDNA published by Defontaine and colleagues *(11)*, the steps followed are essentially similar (i.e., spheroplasting, lysing spheroplasts, and isolating [intact] mitochondria); at this point in the Defontaine method the mitochondria, after careful washing, are immediately lysed and extracted for DNA, so greatly abbreviating the isolation procedure. The full Defontaine protocol is given here:
   a. Grow an overnight 20-mL yeast culture in 1% w/v yeast extract, 2% w/v bactopeptone, 2% w/v glucose, at 28°C, with shaking.
   b. Harvest the cells by centrifuging 500*g* for 5 min: The yield is typically, 0.3–0.4 g, wet wt.
   c. Wash the cells centrifugally (as in b), twice with sterile dH$_2$O and once with washing solution (1.2*M* sorbitol, 2% v/v 2-mercaptoethanol, 50 m*M* EDTA, 50 m*M* Tris-HCl, pH 7.5, freshly adjusted to 2% v/v 2-mercaptoethanol and 0.2–1 mg/mL zymolyase 20T; incubate for 45 min at 37°C, with gentle agitation.

   During this step, most of the spheroplasts created by zymolyase action are osmotically lysed. Residual spheroplasts can be disrupted by sonication (e.g., using a bioruptor UCD 130, Toshi Denki Co. Ltd; at 19.3–19.5 kHz for 1–5 min, until the sample is clarified); this step is not absolutely essential, but does improve the final result.
   d. Transfer the lysate into microfuge tubes, and centrifuge at 1000*g* for 10 min to pellet large particles.
   e. Transfer the supernatant(s), which contain mitochondria, to clean microfuge tubes, and centrifuge at 15,000*g* for 15 min, to harvest the mitochondria.
   f. Wash the mitochondrial pellet centrifugally (as in step e) three or four times with solution A; the step is crucial, as it removes contaminating genomic DNA from the mitochondrial fraction.
   g. Resuspend the washed mitochondrial pellet in 0.2–0.4 mL of solution B (100 m*M* NaCl, 1% v/v Sarkosyl, 10 m*M* EDTA, 50 m*M* Tris-HCl, pH 7.8) and allow the mitochondria to lyse for 30 min at room temperature.
   h. Extract the lysate with an equal volume of phenol-chloroform (1:1) and precipitate the DNA from the aqueous phase with ethanol.
   i. Dissolve the mtDNA pellet in TE (1 m*M* EDTA, 10 m*M* Tris-HCl, pH 8.0).

Typical mtDNA yields are 10–20 µg per 0.3–0.4 g net wt, with an A_{260}/A_{280} of 1.8–1.9; if RNA contamination is a problem, it can be removed by RNase A treatment (see, e.g., gene bank Chapter 18, by Bignell and Evans).

References

1. Attardi, G. and Schatz, G. (1988) Biogenesis of mitochondria. *Annu. Rev. Cell Biol.* **4,** 289–333.
2. Lane, M. D., Pedersen, P. L., and Mildvan, A. S. (1986) The mitochondrion updated. *Science* **234,** 526–527.
3. Evans, I. H. (1983) Molecular genetic aspects of yeast mitochondria, in *Yeast Genetics: Fundamental and Applied Aspects* (Spencer, J. F. T., Spencer, D. M., and Smith, A. R. W., eds.), Springer-Verlag, Berlin, pp. 269–370.
4. Tzagaloff, A. and Myers, A. M. (1986) Genetics of mitochondrial biogenesis. *Annu. Rev. Biochem.* **55,** 249–285.
5. Griac, P. and Nosek, J. (1993) Mitochondrial DNA of *Endomyces (Dipodascus) magnusii. Curr. Genet.* **23,** 549–552.
6. Bendich, A. J. (1993) Reaching for the ring: the study of mitochondrial genome structure. *Curr. Genet.* **24,** 279–290.
7. Querol, A. and Barrio, E. (1990) A simple rapid and simple method for the preparation of yeast mitochondrial DNA. *Nucleic Acids Res.* **18,** 1657.
8. Gargouri, A. (1989) A rapid and simple method for extracting yeast mitochondrial DNA. *Curr. Genet.* **15,** 235–237.
9. Schramke, M. L. and Orlowski, M. (1993) Mitochondrial genome of the dimorphic zygomycete *Mucor racemosus. Curr. Genet.* **24,** 337–343.
10. Fox, T. D, Folley, L. S, Mulero, J. J, McMullin, T. W., Thorsness, P. E., Hedin, L. O., and Costanzo, M. C. (1991) Analysis and manipulation of yeast mitochondrial genes, in *Methods in Enzymology, vol. 194, Guide to yeast genetics and molecular biology* (Guthrie, C. and Fink, G. R., eds.), Academic, New York, pp. 149–165.
11. Defontaine, A., Lecocq, F. M., and Hallet, J. N. (1991) A rapid miniprep method for the preparation of yeast mitochondrial DNA. *Nucleic Acids Res.* **19,** 185.

CHAPTER 13

Isolation of Yeast Plasma Membranes

Barry Panaretou and Peter Piper

1. Introduction

Plasma membranes are often prepared from yeast by initially spheroplasting the cells *(1–3)*. Procedure 1 following uses spheroplasting, gives high yields, and is ideally suited to the large-scale isolation of plasma membranes. Modified after Schmidt et al. *(2)* and Cartwright et al. *(4)*, it coats the negatively charged surface of the spheroplasts with dense cationic silica beads, so as to make the plasma membrane much denser than any other membranous organelle of the cell. A washing procedure then removes the excess cationic beads, followed by addition of polyacrylic acid to block the free cationic groups on these beads. The coated spheroplasts are subsequently lysed by hand homogenization in an EGTA-containing lysis buffer (the EGTA binding divalent cations, thereby preventing aggregation of membrane components). Centrifugation of the spheroplast lysate then pellets the heavy plasma membrane-microbead assemblies, leaving intracellular membranous organelles in the supernatant.

An extended incubation (at least 30–45 min at 30°C) with zymolyase is required in order to spheroplast cells. This will almost certainly cause physiological changes, which may be reflected in alterations to plasma membrane components. Also, cells cannot be spheroplasted in certain physiological states (e.g., at stationary phase). Spheroplasting was therefore considered unsuitable for our studies on the effects of stress on the proteins of the *Saccharomyces cerevisiae* plasma membrane *(5)*. Procedure 2 following, in which cells are rapidly disrupted by vortexing with glass beads, circumvents this problem. It is our slightly modified version of the plasma membrane isolation of Serrano *(6–8)*, in which membranes are banded on sucrose density gradients. Yields are probably less than with procedure 1, yet the plasma membranes obtained are of high purity, and the procedure is ideally suited to comparative studies of the plasma membranes from cells of different physiological states *(5)*.

2. Materials

2.1. Isolation of Plasma Membranes from Spheroplasts Using Cationic Silica Beads

1. Nalco (Cheshire, UK) 1060 colloidal silica (50-nm diameter).
2. 50% (w/v) aluminum chlorohydroxide.
3. $1M$ sodium hydroxide.
4. Sorbitol buffer: $1.5M$ sorbitol, 20 mM Tris-HCl, pH 7.2, 10 mM MgCl$_2$.
5. Coating buffer: $1.5M$ sorbitol, 25 mM sodium acetate, pH 6.0, $0.1M$ KCl.
6. Lysis buffer: 5 mM Tris-HCl, pH 8.0, 1 mM EGTA.
7. 0.1 mM EDTA, 25 mM imidazole-HCl, pH 7.0, 50% (v/v) glycerol.
8. Zymolyase-20T from *Arthrobacter luteus* (ICN Immunobiologicals, High Wycombe, UK).
9. Polyacrylic acid (M^r 90,000).

2.2. Isolation of Plasma Membranes Without Spheroplasting

1. Buffer A: 2 mM EDTA, 25 mM imidazole (adjusted to pH 7.0 with HCl). This buffer is stored at 4°C, the following protease inhibitors (all available from Sigma, St. Louis, MO) being added just before use: 1 mM phenylmethylsufonyl fluoride, 1 mM *N*-tosyl-L-phenylalanine chloromethyl ketone (from 100 mM stocks in ethanol, stored at –20°C), and 2 µg/mL pepstatin A (from a 2.5 mg/mL stock in methanol, stored at –20°C). *These inhibitors are toxic and care should be taken to ensure that they do not come into skin contact.* The buffer must be kept at 4°C, not 0°C, to prevent the protease inhibitors precipitating out of solution.
2. Solutions of 0.4, 1.1, 1.65, and $2.25M$ sucrose in buffer A (containing the protease inhibitors as in step 1).
3. Discontinuous sucrose gradients, prepared by overlaying three 4-mL layers of 2.25, 1.65, and $1.1M$ sucrose in a 14 × 89 mm Beckman (Fullerton, CA) ultraclear tube.
4. 0.1 mM EDTA, 25 mM imidazole-HCl, pH 7.0, 50% (v/v) glycerol.
5. Glass beads (BDH 0.45-µm mesh are suitable).

3. Methods

3.1 Isolation of Plasma Membranes from Spheroplasts Using Cationic Silica Beads

3.1.1. Bead Preparation (see Note 1)

1. Add 35 g of aluminium chlorohydroxide (50% w/w) to 300 mL distilled water in a Waring blender.
2. Dilute 450 g Nalco 1060 colloidal silica with 90 mL water and add this to the chlorohydrol solution with the blender running at low speed.
3. Blend 2 min at high speed.
4. Incubate at 80°C for 30 min with manual stirring, followed by 16 h at room temperature without agitation.
5. Adjust pH to 5.0 with $1M$ NaOH.
6. Just before use, dilute the beads to 6–8% by weight using sorbitol buffer.

3.1.2. Isolation of Plasma Membranes from Spheroplasts Using Cationic Beads (see Note 2)

1. Harvest yeast by centrifugation (5 min, 5000g), wash once in distilled water, and twice in sorbitol buffer.
2. Resuspend cell pellet in sorbitol buffer (1.5×10^8 cells/mL), and allow to equilibrate at 30°C for 5 min.
3. Add zymolyase –20T (0.1 mg/mg dry wt cells) and incubate with gentle shaking at 30°C. After 1 h, check that most cells have spheroplasted (add water to a streak of the cells on a microscope slide; in <1 min, spheroplasts will be observed to lyse).
4. Collect spheroplasts by centrifugation (5 min, 1500g), and wash two times in coating buffer.
5. Take the cationic silica beads from Section 3.1. above, resuspend in 10 vol of coating buffer, and then remove aggregated beads by centrifugation (10 min, 1500g). Use only the supernatant in step 6.
6. Resuspend spheroplasts in coating buffer (1.5×10^8 cells/mL), and mix with bead suspension in the ratio of spheroplasts to beads 2:1.
7. After 3 min at 4°C, centrifuge at 500g for 5 min.
8. Wash pellet once in coating buffer, then resuspend in coating buffer (1.5×10^8 cells/mL).
9. Mix spheroplasts and polyacrylic acid (the latter 2 mg/mL in coating buffer) in the ratio 1:1.
10. Centrifuge 500g for 5 min, and wash pellet once in coating buffer.
11. Resuspend pellet in lysing buffer (1×10^8 cells/mL) and homogenize in a Teflon glass hand homogenizer for 5 min.
12. Centrifuge 1000g for 5 min, and wash pellet twice in lysing buffer.
13. Resuspend the plasma membrane pellet in 0.1 mM EDTA, 25 mM imidazole-HCl, pH 7.0, 50% glycerol, and store at –20°C until required.

3.2. Isolation of Plasma Membranes Avoiding the Spheroplasting of Cells (see Note 2)

1. Harvest a 1-L culture by centrifugation (5 min, 5000g), then resuspend the cells in 80 mL 0.4M sucrose in buffer A.
2. Divide the cell suspension between two 50-mL polycarbonate centrifuge tubes. Centrifuge at 5000g for 10 min.
3. Add to the cell pellet two times the pellet volume of glass beads followed by just sufficient 0.4M sucrose in buffer A to cover the cells and glass beads.
4. Vortex 5 min on a whirlimixer.
5. Dilute three times in 0.4M sucrose in buffer A.
6. Centrifuge at 530g (2500 rpm in Sorvall SS34 rotor) for 20 min.
7. Recentrifuge the supernatant from step 6 at 22,000g (16,000 rpm in Sorvall SS34 rotor) for 30 min to obtain a pellet that includes the plasma membranes and mitochondria.
8. Resuspend the pellet from step 7 in 2-mL buffer A by gentle vortexing (30 s) (*see* Note 3).
9. Load 1 mL aliquots of resuspended membranes onto discontinuous sucrose gradients (*see* Section 2.2.) and centrifuge either overnight (14 h) at 80,000g (22,000 rpm) or 6 h at 284,000g (40,000 rpm) in the Beckmann SW41 or SW40Ti rotor.
10. Membranes banding at the 2.25/1.65 interface are essentially pure plasma membranes *(7,8)*, although a smaller proportion of the plasma membranes corresponding to about one-third of the plasma membrane ATPase activity bands together with mitochondria at the 1.65/1.10 interface *(7,8)*. Collect the membranes at these interfaces from the top of the gradient with a Pasteur pipet, dilute four times with buffer A, pellet at 30,000g (18,000 rpm, 40 min, Beckman 50 Ti rotor), resuspend in 0.1 mM EDTA, 25 mM imidazole-HCl, pH 7.0., 50% glycerol, and store at –20°C.

4. Notes

1. It is necessary to prepare the cationic silica beads in advance, since they are not yet available commercially. They are most conveniently made as a large batch for use in several membrane preparations. Prepared beads can be stored (at 30% silica by weight) for several months at 4°C.
2. At 4°C, except where stated.
3. Resuspension in smaller volumes is not advisable, since it will cause the membranes subsequently to band as solid discs that are difficult to remove from the tubes during steps 9 and 10.
4. One of the best ways to assess the purity of yeast plasma membranes is to assay the fraction of the ATPase activity subject to orthovanadate inhibition *(2,7)*. The plasma membrane ATPase is inhibited by orthovanadate,

whereas the ATPase of mitochondria is not. Also, SDS PAGE analysis should show the plasma membrane ATPase (M^r 100,000) as the most abundant protein of plasma membrane preparations *(5)*.

We have found no evidence for appreciable degradation of plasma membrane proteins isolated by the method in Section 3.2. (except when heat-stressed cells were used), even though this isolation involves a 14-h 4°C centrifugation step, during which slight contamination with proteases might cause problems. We nevertheless use protease inhibitors in buffer A as a precaution against proteolysis. Protein degradation was observed during the isolation of plasma membranes from heat-shocked cells, necessitating the use of a strain substantially deficient in vacuolar proteases (BJ2168: a *leu2-3,112, trp1, ura3-52, prb1-1122, pep4-3, prc407, gal2*) in our study of the effects of heat shock on plasma membrane proteins *(5)*.

5. A recent study of an important yeast cell-surface gene product, apparently located in the plasma membrane, is that by Watan and colleagues *(9)* on the FL01 gene-mediating flocculation.

References

1. Henschke, P. A., Thomas, D. S., Rose, A. H., and Veazey, F. J. (1983) Association of intracellular low density vesicles with plasma membranes from *Saccharomyces cerevisiae* NCYC 366. *J. Gen. Microbiol.* **129**, 2927–2938.
2. Schmidt, R., Ackermann, R., Kratky, Z., Wasserman, B., and Jacobsen, B. (1983) Fast and efficient purification of yeast plasma membranes using cationic silica microbeads. *Biochem. Biophys. Acta.* **732**, 421–427.
3. Cartwright, C. P., Veazey, F. J., and Rose, A. H. (1987) Effect of ethanol on activity of the plasma membrane ATPase in, and accumulation of glycine by, *Saccharomyces cerevisiae. J. Gen. Microbiol.* **133**, 857–865.
4. Chaney, L. K. and Jacobsen, B. S. (1983) Coating cells with colloidal silica for high yield isolation of plasma membrane sheets and identification of transmembrane proteins. *J. Biol. Chem.* **258**, 10,062–10,072.
5. Panaretou, B. and Piper, P. W. (1992) The plasma membrane of yeast acquires a novel heat shock protein (hsp30) and displays a decline in proton-pumping ATPase levels in response to both heat shock and the entry to stationary phase. *Eur. J. Biochem.* **206**, 635–640.
6. Serrano, R. (1978) Characterisation of the plasma membrane ATPase of *Saccharomyces cerevisiae. Mol. Cell. Biochem.* **22**, 51–63.
7. Serrano, R. (1988) H$^+$ATPase from plasma membranes of *Saccharomyces cerevisiae* and *Avena sativa* roots: Purification and reconstitution. *Meth. Enzymol.* **157**, 533–544.
8. Serrano, R., Montesinos, C., Roldan, M., Garrido, G., Ferguson, C., Leonard, K., Monk, B. C., Perlin, D. S., and Weilar, E. W. (1991) Domains of yeast plasma membrane and ATPase-associated glycoprotein. *Biochim. Biophys. Acta.* **1062**, 157–164.
9. Watari, J., Takata, Y., Ogawa, M., Sahara, H., Koshino, S., Onnela, M. L., Airaksinen, U., Jaatinen, R., Penttila, M., and Keranen, S. (1994) Molecular cloning and analyis of the yeast flocculation gene FL01. *Yeast* **10**, 211–226.

CHAPTER 14

The Extraction and Analysis of Sterols from Yeast

Michael A. Quail and Steven L. Kelly

1. Introduction

The study of sterols is of both practical and theoretical interest, being important in the analysis of the action of antifungal compounds, in vitamin D production and investigating the role of sterols in cell division, mating *(1)*, and ethanol tolerance. The extraction of sterols from yeast is both rapid and simple, the effectiveness of the basic method being reflected by the fact that it has not been significantly altered since being performed by Woods *(2)* in 1971. The method involves saponification of harvested cell material followed by solvent extraction to recover the nonsaponifiable fraction, which contains the cellular sterols. The principal sterol, in yeast, is ergosterol (Fig. 1). Other sterols identified include intermediates in the pathway of ergosterol biosynthesis or those appearing upon inhibition of biosynthesis. These may differ from each other only in the stereochemical orientation of one of their methyl substituents or in the position of unsaturation. For instance, episterol and fecosterol, which both lie on the ergosterol biosynthesis pathway, differ only in that episterol has a Δ^7 double bond, whereas fecosterol is Δ^8 sterol.

The fundamental problem, in sterol identification, is to achieve separation of quite large molecules that differ only slightly from each other. Using current technology, no single chromatographic technique is powerful enough to separate and identify all the constituents of a cellular sterol extract.

To screen a number of yeasts for the presence of ergosterol, UV spectrophotometry scanning between 220–300 nm can be used to analyze

From: *Methods in Molecular Biology, Vol. 53: Yeast Protocols*
Edited by: I. Evans Humana Press Inc., Totowa, NJ

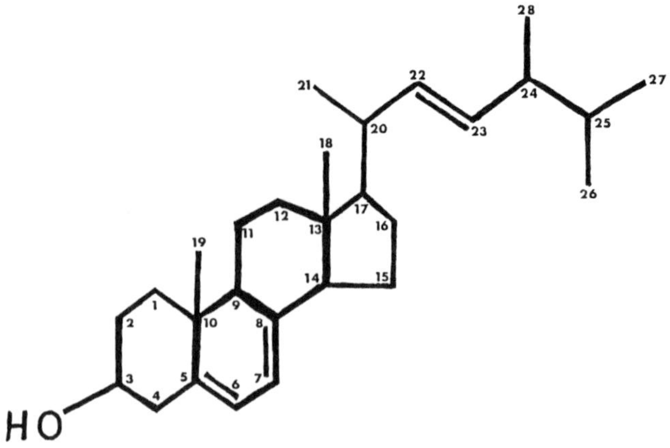

Fig. 1. The structure of the major sterol of yeast, ergosterol.

nonsaponifiable sterol extracts, prepared as in Section 2.1. Ergosterol has a conjugated Δ^5,Δ^7 double-bond system that gives a characteristic triple peak at 293, 281.5, and 271 nm *(3)*. (*see* Fig. 2).

Preparative HPLC *(4)* followed by ^{13}C NMR is required for absolute identification of sterols (^{13}C NMR is more useful than ^{1}HNMR for the structural elucidation of sterols since ^{13}C chemical shifts are more sensitive to structural changes). However, most sterols that one is likely to encounter in yeast have already been characterized and their mass spectral fragmentation patterns published (Tables 1 and 2 show the mass spectral fragmentation patterns of some yeast sterols), allowing gas chromatography mass spectrometry (GCMS) to establish identification *(5)*. Furthermore, gas chromatography (GC) has an advantage over HPLC since it allows better separation of sterols with a 14α-methyl group *(6)*.

A key step in the ergosterol biosynthesis pathway is the cytochrome P-450 catalyzed 14α-demethylation of lanosterol. This enzyme is quite labile in vitro and is the site of action of many commercial azole antifungal compounds *(7)*. Thus, treatment by azoles or inactivation of the lanosterol demethylase, results in an accumulation of 14α-methylsterols, which can be effectively separated by GC. In addition, the development of radioactivity detectors for use with GC systems may become useful, as it enables one to examine *de novo* synthesized sterols after the yeast of interest has been supplied with a labeled sterol precursor, such as ^{14}C-mevolonic acid. Previously, thin layer chromatography had been used for the separation and detection of labeled sterols. However, TLC cannot

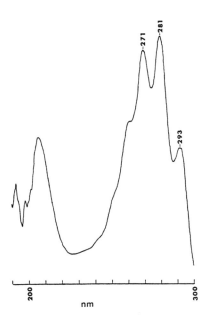

Fig. 2. UV absorption spectrum of ergosterol.

separate on the basis of 14 methyl substitution; instead; major bands comprising 4-desmethyl, 4 methyl, and 4,4-dimethylsterols are obtained.

2. Materials

1. Flasks containing medium for growing yeast (e.g., YEPO : 1% w/v yeast extract, 1% w/v bactopeptone, 2% w/v glucose).
2. The solvents used, i.e., chloroform, heptane, and methanol, are all of analytical grade (e.g., Analar grade from BDH Ltd., Leicestershire, UK). All solvent manipulations are to be carried out in a fume cupboard, while wearing protective clothing since chloroform is listed as harmful, methanol is toxic, and heptane is believed to be a possible carcinogen.
3. 0.5% w/v pyrogallol is made up in methanol immediately before use. *Caution:* Pyrogallol is toxic.
4. A 60% w/v solution of KOH is made up in water: KOH is corrosive.
5. Pyridine (e.g., Aldrich, Milwaukee, WI) and *N'O'-bis*-(trimethylsilyl)-trifluoroacetamide (e.g., Fluka, Buchs, Switzerland). Both reagents are toxic, necessitating this step to be carried out in the fume cupboard while wearing protective clothing.
6. Screwcapped glass Universal bottles.
7. A cylinder of nitrogen gas (for heptane evaporation).
8. A GC instrument with an appropriate column (*see* Note 4).

3. Methods

3.1. Isolation of Nonsaponifiable Sterol Extract

1. Grow yeast in suitable medium (e.g., YEPD) so as to obtain 50–200 mg dry cell weight; cell densities between 5×10^6 and 2×10^8 cells per mL, in 100 mL medium, should be adequate.
2. Harvest cells by centrifugation at 300g for 10 min. Pour off supernatant and discard.
3. Transfer cellular material to a 20-mL glass Universal bottle with screwcap and add 3 mL methanol, 2 mL 0.5% pyrogallol, 2 mL 60% KOH, and 50 µL 1 mg/mL cholesterol (*see* Note 1).
4. Vortex gently to mix.
5. Place tube(s) in a water bath equilibrated to 90°C and leave for 2 h.
6. Remove sample(s) from water bath and allow to cool. Extract with 3×5 mL *n*-heptane, centrifuging at 500g for 5 min to facilitate phase separation (*see* Note 2).

3.2. GC Analysis

1. Evaporate heptane in a water bath set at 55°C under a stream of nitrogen. If desired, the sample(s) can be stored at –20°C at this stage.
2. Redissolve in 50 µL pyridine and then add 50 µL N'O'-*bis*-(trimethylsilyl)-trifluoroacetamide, pipeting up and down to mix. Do this only if you are prepared to analyze the sample within 48 h (*see* Note 3).
3. Equilibrate GC column to 260°C and injector and detector to 300°C. (*see* Note 4).
4. Inject a 1-µL aliquot of the sample (with split ratio set at 1:30) onto a Durabond DB5 fused silica capillary column (30 m × 0.32 mm in., film thickness 0.25 µm) with hydrogen carrier gas adjusted to 4 mL/min. Initialize the integrator and run under isothermal conditions at 260°C for 40 min.

Figure 3 shows the sterol extracts from *Saccharomyces cerevisiae* cultures grown in YEPD and YEPD supplemented with $10^{-4}M$ fluconazole (a typical lanosterol demethylase inhibitor), respectively. Untreated cells contain mostly ergosterol, whereas treated cells contain a number of 14α methyl sterols.

4. Notes

1. Ensure the bottles used for saponification seal properly.
2. UV analysis of the heptane extract can give preliminary information on the sterol composition. Using 0.5-mL UV cuvets, dilute 50 µL extract to 500 µL with heptane, and scan absorbance between 220 and 300 nm against

Table 1
Mass Fragmentation Patterns of Some Common Fungal Sterols, Derivatized

Fragmentation[a]	1	2	3	4	5	6	7	8
[M]$^+$	468 (32)[b]	484 (38)	470 (2)	424 (49)	498 (66)	575 (55)	512 (50)	498 (41)
[M-Me]$^+$		469 (100)		409 (83)	483 (67)	557 (45)	497 (38)	483 (31)
[M-16]$^+$								
[M-42]$^+$								
[M-43]$^+$				381 (19)				
[M-84]$^+$			386 (33)					
[M-TMS]$^+$	396 (2)							
[M-TMSOH]$^+$	378 (32)							
[M-TMS-Me]$^+$								393 (100)
[M-TMSOH-Me]$^+$	363 (100)	379 (78)			393 (100)	467 (51)	407 (100)	
[M-TMS-43]$^+$			343 (100)	297 (11)				
[M-SC]$^+$								
[M-TMSOH-41]$^+$	337 (56)	318 (3)	253 (21)	325 (22)	409 (3)	393 (5)	331 (3)	255 (6)
	280 (2)	303 (27)	213 (11)	257 (15)	317 (20)	377 (90)	323 (8)	241 (20)
	253 (32)	291 (14)	211 (5)	243 (100)	241 (14)	301 (6)	255 (11)	197 (7)
		239 (12)		231 (67)	227 (35)	285 (80)	241 (40)	
		227 (15)			216 (34)	253 (4)	229 (28)	
						211 (11)	187 (17)	
						197 (100)		

[a] Me, loss of methyl; TMS, loss of trimethyl silane; TMSOH, loss of trimethyl silane hydroxide; SC, loss of side chain; 42, loss of C_3H_6 from C_{15} to C_{17}; 43, loss of C_3H_7 from C_{25} to C_{27}; 84, loss of C_{23} to C_{28} in 24-methylene side chain. Peaks are identified as: 1, Ergosterol; 2, 14α-methylfecosterol; 3, Episterol; 4, Obtusifolione; 5, Obtusifoliol; 6, 14α-methylergosta-3,6 diol; 7, 24-methylene-24(25)-dihydrolanosterol; 8, Lanosterol.
[b] Figures in parentheses represent ion intensities as a percentage of the base peak value.

Table 2
Mass Fragmentation Patterns of Some Common Fungal Sterols, Underivatized

Fragmentation[a]	1	2	3	4	5	6	7	8
[M]+								
[M-Me]+	384 (18)[b]	396 (20)	394 (37)	396 (24)	396 (30)	400 (37)	396 (6)	400 (30)
[M-H$_2$O]+	369 (15)		379 (2)	383 (15)	383 (18)	385 (12)	383 (4)	385 (9)
[M-Me-H$_2$O]+		378 (2)	376 (2)					
[M--43]+	351 (4)	363 (28)	361 (36)	365 (4)	365 (5)	367 (3)		367 (3)
[M-59]+								
[M-M-43-H$_2$O]+		337 (10)						
[M-84]+			335 (4)					
[M-84-Me]+				314 (4)			314 (17)	
[M-SC]+		271 (7)		399 (5)				
[M-SC-2H]+		269 (5)	271 (28)	273 (10)	271 (32)	273 (14)	271 (95)	273 (8)
[M-SC-H$_2$O]+	271 (10)	253 (10)	269 (12)	271 (32)				
[M-SC-2H-H$_2$O]+		251 (5)		255 (5)		255 (6)	253 (10)	255 (24)
[M-SC-27]+			251 (17)					
[M-SC-42]+	246 (8)			246 (7)		246 (4)		246 (6)
[M-SC-27-H$_2$O]+	231 (7)			231 (12)		231 (10)		231 (12)
[M-SC-56]+	228 (8)			228 (17)				
[M-SC-42-H$_2$O]+	213 (18)	211 (16)	211 (17)	213 (20)		213 (18)	213 (8)	213 (22)
[M-SC-56-H$_2$O]+			143 (30)					
[M-251]+		143 (38)						
[M-253]+								

Table 2 (continued)
Mass Fragmentation Patterns of Some Common Fungal Sterols, Underivatized

Fragmentation[a]	9	10	11	12	13	14	15	16
[M]+	428 (10)[b]	426 (35)	412 (18)	410 (65)	412 (24)	398 (20)	398 (10)	398 (55)
[M-Me]+	413 (37)	411 (49)	397 (9)	395 (20)	397 (10)	383 (10)	383 (6)	381 (28)
[M-H$_2$O]+		406 (30)		392 (11)		380 (2)	380 (2)	
[M-Me-H$_2$O]+	395 (18)	393 (42)	379 (3)	377 (18)	379 (4)	365 (2)	365 (2)	363 (7)
[M-43]+						355 (3)		353 (17)
[M-59]+								
[M-M-43-H$_2$O]+								
[M-84]+								335 (8)
[M-84-Me]+								
[M-SC]+								
[M-SC-2H]+	315 (4)		299 (4)	297 (5)	299 (3)	273 (28)	273 (14)	271 (27)
[M-SC-H$_2$O]+		313 (5)				271 (32)	271 (36)	269 (12)
[M-SC-2H-H$_2$O]+		295 (6)				255 (16)	255 (12)	
[M-SC-27]+								
[M-SC-42]+					274 (2)	248 (38)	248 (15)	
[M-SC-27-H$_2$O]+	273 (5)	273 (22)			259 (8)	231 (14)	231 (5)	
[M-SC-56]+								
[M-SC-42-H$_2$O]+	259 (6)	259 (25)						
[M-SC-56-H$_2$O]+	255 (3)	255 (26)						
[M-251]+	241 (8)	241 (36)			241 (8)	213 (18)	213 (18)	
[M-253]+								

[a] Me, loss of methyl; SC, loss of sterol side chain; 27, loss of C_2H_3 from C16 and C17; 42, loss of C_3H_6 from C15 to C17; 43, loss of C_3H_7 from C_{25} to C_{27} (in a ▲^{22} sterol); 56, loss of C_4H_3 from C15 to C17 and C32; 59, loss of C_3H_7O from C1 to C3 (in a ▲^{5,7} sterol); 84, loss of C_6H_{12} from C_{23} to C_{28} in 24-methylene side chain.

[b] Figures in parentheses represent ion intensities as a percentage of the base peak value. Peaks are identified as: 1, Zymosterol; 2, Ergosterol; 3, ergosta-5,7,22,24(28)-tetraenol; 4, Fecosterol; 5, Ignosterol; 6, Ergost-8-enol; 7, Episterol; 8, Ergost-7-enol; 9, 24,24-dihydrolansterol; 10, Lanosterol; 11, 4,4-dimethylcholesta-8,14-dienol; 12, 4,4-dimethylcholesta-8,14,24-trienol; 13, 4,4-dimethylcholesta-8,24-dienol; 14, Ergosta-8,22-dienol; 15, Ergosta-7,22-dienol; 16, Ergosta-8,14-22-trienol.

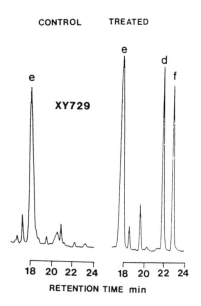

Fig. 3. Gas chromatograms of derivatized sterols from *S. cerevisiae* strain XY-729 5a; control and fluconazole treated: e = ergosterol, d = lanosterol, f = 14α-methyl-ergosta-8,24(28)-diene-3β, 6α-diol.

a heptane blank. An "ergosterol-like" conjugated Δ^5,Δ^7 double bond system gives a characteristic triple peak (Fig. 2). Other sterols with conjugated Δ^{14} systems give a characteristic peak at 250 nm. These latter sterols are generated after treatment with morpholine antifungal compounds that inhibit sterol Δ^{14} reductase *(4)*.
3. TMS derivatization enables more efficient volatilization, hence more sensitive analysis. However, certain sterol TMS derivatives are less stable than others, and once derivatized, samples should be analyzed as soon as possible. Prior to analysis, they should be stored at –20°C. Samples should not be stored in plastic tubes, since samples become contaminated with substances leached from the plastic by the solvents used.
4. The GC column used here is of intermediate polarity and as such is a compromise. More popular columns, such as SP-1000, enable better separation between molecules that differ in unsaturation pattern, whereas less polar columns (E-30 or OV-1) enable better separations based on alkyl differences.

References

1. Tomeo, M. E., Fenner, G., Tove, S. R., and Parks, L. W. (1992) Effect of sterol alterations on conjugation in *Saccharomyces cerevisiae Yeast* **8,** 1015–1024.

2. Woods, R. A. (1971) Nystatin-resistant mutants of yeast; alterations in sterol content. *J. Bact.* **108,** 69–73.
3. Breivik, O. N. and Owades, J. L. (1957) Spectrophotometric semi microdetermination of ergosterol in yeast. *Agric. Food. Chem.* **5,** 360–363.
4. Rodriguez, R. J. and Parks, L. W. (1985) High performance liquid chromatography of sterols; Yeast sterols. *Methods Enzymol.* **111,** 37–51.
5. Baloch, R. I., Mercer, E. I., Wiggins, T. E., and Baldwin, B. C. (1984) Inhibition of ergosterol biosythesis in *Saccharomyces cerevisiae* and *Ustilago maydis* by tridemorph, fenpropimorph and fenpropidin. *Phytochemistry* **23,** 2219–2226.
6. Nes, W. R. (1985) A comparison of methods for the identification of sterols. *Methods Enzymol.* **111,** 1–37.
7. Van den Bossche, H., Marichal, P., Gorrens, J., Bellens, D., Verhoevan, H., Coene, M. C., Coene, M. C., Lauwers, W., and Janssen, P. A. J. (1987) Interaction of azole derivatives with cytochrome P-450 isozymes in yeast, fungi, plants and mammalian cells. *Pest. Sci.* **21,** 289–306.

CHAPTER 15

Peroxisome Isolation

Ben Distel, Inge van der Leij, and Wilko Kos

1. Introduction

The group of De Duve first described the isolation of peroxisomes from rat liver tissue. These organelles can be separated from other subcellular organelles, such as lysosomes and mitochondria, because of their relatively high equilibrium density in sucrose (approx 1.24 g/cm^3). The isolation of peroxisomes from the yeast *Saccharomyces cerevisiae*, however, has been hampered by the fact that under standard growth conditions peroxisomes are present in low numbers. Also, the lability of peroxisomes in general has complicated their purification.

Two observations have greatly facilitated the isolation of peroxisomes from yeast: First, peroxisomes are induced by growth on a fatty acid *(1)*, and second, peroxisomes are more stable at low pH (approx 5.5). The method making use of these two findings (adapted from a procedure described by Goodman *[2]*), involves osmotic lysis of yeast spheroplasts at low pH, followed by differential centrifugation to obtain an organelle pellet. After resuspension, the pellet is layered on a discontinuous sucrose gradient to separate mitochondria and peroxisomes. Peroxisomes purified in this way are relatively stable and show only minor contamination with mitochondria.

2. Materials

Prepare all solutions in distilled water, unless otherwise indicated.

1. 2.4*M* sorbitol (sterilized in an autoclave, stable at 4°C for up to 1 mo).

2. 0.5 M MES (2 [N-morpholino] ethanesulfonic acid)/KOH, pH 5.5 (stable at 20°C for up to 1 mo).
3. 0.5 M potassium phosphate buffer (pH 7.5) (mix 83 mL 0.5 M K_2HPO_4 with 17 mL 0.5 M K_2HPO_4).
4. 0.5 M EDTA, pH 8.0.
5. 0.1 M Tris/H_2SO_4, pH 9.4.
6. Buffer A: 50 mM potassium phosphate, pH 7.5, 1 mM EDTA.
7. Buffer B: 5 mM MES, pH 5.5, 1 mM EDTA, 1 mM KCl.
8. Buffer A plus 1.2 M sorbitol.
9. Buffer B plus 1.2 M sorbitol.
10. Buffer B plus 0.25 M sorbitol.
11. 1 M DTT (dithiothreitol) prepared in H_2O, store at –20°C.
12. 0.2 M PMSF (phenylmethylsulfonyl fluoride) in ethanol (store at –20°C, handle with care, toxic). PMSF is unstable in water and must be added just prior to use.
13. Zymolyase (100,000 U/g) is obtained from ICN Immunobiologicals (High Wycombe, UK).
14. WOYglu medium: Contains 0.67% (w/v) yeast nitrogen base without amino acids (WO) (Difco, Surrey, UK) (add separately from a 6.7% [w/v] stock solution sterilized by filtration), 0.1% (w/v) yeast extract, 0.3% (w/v) glucose, and amino acids (20 µg/mL, add separately from a 100X stock solution sterilized by filtration) as needed.
15. Induction medium: Contains 0.5% (w/v) bactopeptone, 0.3% (w/v) yeast extract, 0.12% (v/v) oleic acid, 0.2% (v/v) Tween-40, and 0.5% (w/v) KH_2PO_4 (adjusted to pH 6.0 with NaOH) sterilized by autoclaving.
16. Discontinuous sucrose gradients are prepared in quick-seal tubes (polyallomer, 25 × 89 mm, Beckman Instruments, Fullerton, CA). Each gradient consists of 5 mL 60% (w/w) sucrose, 12 mL 46% (w/w), 12 mL 44% (w/w), 5 mL 40% (w/w), all in buffer B (*see* Note 6).
17. Sucrose gradient overlay consists of 20% (w/w) sucrose in buffer B.

3. Method

1. Grow a starter culture of a wild-type *S. cerevisiae* strain (e.g., D273-10B—*see* Note 1, American type culture collection, Rockville, MD) overnight in 10 mL of WOYglu medium in a 100-mL conical flask with vigorous aeration at 28°C.
2. Next morning measure the cell density at 600 nm and inoculate a flask of WOYglu at an OD_{600} of 0.15 and grow to an OD_{600} of 1.0–1.5. Dilute the culture 1:100 in fresh WOYglu and incubate at 28°C overnight.
3. Repeat step 2 (*see* Note 2). Next morning measure the cell density and inoculate 300 mL WOYglu (in 2-L flask) at an OD_{600} of 0.15. Grow to an

OD_{600} of 1.0. Pellet the cells and resuspend in 10 mL of induction medium; use 5 mL to inoculate 1 L of induction medium. Incubate overnight at 28°C with vigorous aeration.
4. Harvest the cells at room temperature by centrifugation (4000g, 5 min) and wash twice with distilled water.
5. Determine the wet weight of the cell pellet and resuspend cells at 0.125 g/mL in 0.1M Tris/H_2SO_4, pH 9.4 with 10 mM DTT (add fresh from 1M stock solution). Incubate 15 min at 30°C with gentle shaking.
6. Harvest cells (as described in step 4) and resuspend in buffer A plus 1.2M sorbitol. Centrifuge (4000g, 5 min).
7. Resuspend cells to 0.125 g/mL in buffer A plus 1.2M sorbitol. Add zymolyase (100,000 U/g) to a final concentration of 1 mg enzyme/g of cells. Incubate 30–60 min at 30°C with gentle agitation to convert the cells to spheroplasts.
8. Check osmotic fragility as follows: Gently mix one drop of cell suspension with an equal volume H_2O and observe under the microscope at 1000X magnification. Look for swollen spheroplasts. Add an extra drop of H_2O to lyse the spheroplasts and look for remnants of cell wall material. When most of the cell wall material has disappeared, the conversion to spheroplasts is complete.
9. All subsequent steps are performed at 4°C. Harvest spheroplasts (4000g, 5 min) and carefully resuspend the pellet in buffer B (*see* Note 3) plus 1.2M sorbitol. Avoid shearing of spheroplasts (*see* Note 4). Recentrifuge and wash twice in the same buffer.
10. Resuspend the spheroplasts (the same volume as step 7) in buffer B plus 1.2M sorbitol and add PMSF to a final concentration of 1 mM. To lyse the spheroplasts, slowly add buffer B plus 0.25M sorbitol and 1 mM PMSF to a final concentration of 0.65M sorbitol. Monitor lysis microscopically. We generally observe lysis of >80% of the spheroplasts. If lysis is not complete, gently shear spheroplasts using a Dounce homogenizer.
11. Centrifuge the homogenate (2000g, 10 min), save the supernatant, and resuspend the pellet in buffer B plus 0.65M sorbitol. Spin (2000g, 10 min), pool the supernatant with that of the previous centrifugation step and centrifuge (2000g, 10 min).
12. Take the pooled "low-speed" supernatant and centrifuge 30 min at 20,000g.
13. Discard supernatant and carefully resuspend organelle pellet in 30% (w/w) sucrose in buffer B to a concentration of about 5 mg of protein/mL. Avoid shearing of the organelles.
14. Apply 2 mL to each 39 mL discontinuous sucrose gradient in a quick-seal tube. Fill up the tube with 20% (w/w) sucrose in buffer B and seal.
15. Centrifuge 2.5 h in a vertical rotor (Beckman VTi 50 or equivalent) at 34,500g_{av} (*see* Note 7).

16. Two bands will be visible in the gradient, a broad band at the 46–44% sucrose interphase containing the mitochondria and a smaller band at the 60–46% sucrose interphase containing the peroxisomes (*see* Note 8).

    Gradients can be analyzed by collecting 2-mL fractions and measuring catalase *(3)* (a peroxisomal marker) and cytochrome *c* oxidase *(4)* (a mitochondrial marker). A typical enzyme profile is given in Fig. 1 (*see* Note 9).

4. Notes

1. We routinely use the *S. cerevisiae* strain D273-10B for isolation of peroxisomes; however, other strains may work as well. Before using other strains, check for peroxisome induction on oleate containing medium by measuring β-oxidation enzymes *(5)* or by inspection of thin sections under the electron microscope.
2. To get optimal induction of peroxisomes, cells must rapidly divide prior to the shift to oleate medium. This is accomplished by extensive preculturing of the cells on glucose for 3 d. However, in induction medium the glucose concentration must be kept low since peroxisome proliferation in yeast is repressed by glucose. The best results are obtained when cells are harvested 12–18 hours after the shift to induction medium.
3. Digestion of the yeast cell wall with zymolyase is optimal at pH 7.5. However, yeast peroxisomes are unstable at this pH and must be isolated at pH 5.5–6.0. Therefore, all steps following the zymolyase treatment are performed at pH 5.5.
4. Try to avoid shearing of spheroplasts (or peroxisomes) while resuspending pellets. Never use pipets with a narrow tip; use a paint brush or glass rod to resuspend pellets.
5. In the first low-speed pellet, a considerable proportion of the organelles is trapped in aggregated structures. Although resuspension of the pellet releases some of the organelles, a large part cannot be further fractionated. We (and others) have not been able to solve this problem.
6. Note that the sucrose solutions are w/w and *not* w/v.
7. The advantage of using a vertical rotor is that separation of the organelles is achieved within 3 h. This makes it possible to complete the isolation procedure within one day. A longer run in a swingout rotor is in principle also possible. However, we do not recommend this since peroxisomes are relatively labile organelles.
8. Peroxisomes can be harvested from the sucrose gradients by puncturing the tubes at the bottom with a wide gage needle and collecting the 60–46% sucrose interphase. Sucrose fractions can be stored at −80°C after rapid freezing in liquid nitrogen without severe damage of the organelles. However, they should be thawed only once and used immediately.

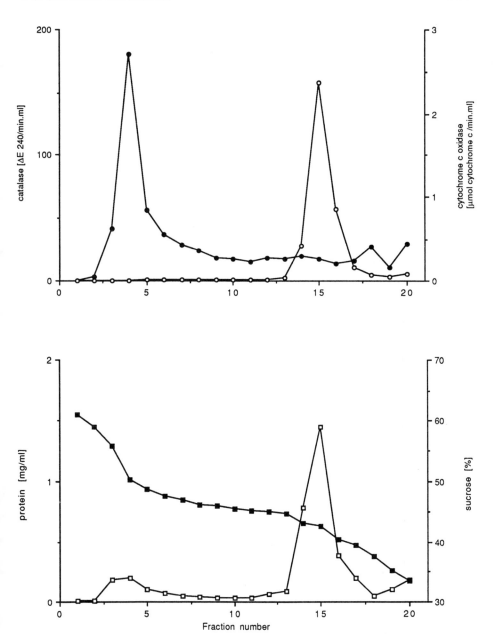

Fig. 1. Distribution of mitochondrial and peroxisomal enzymes after discontinuous sucrose gradient centrifugation of a 20,000g subcellular pellet. (■) sucrose; (□) protein; (●) catalase; (○) cytochrome c oxidase. Fractionation is from bottom (fraction 1) to top (fraction 20) of the gradient.

9. Only a fraction of the peroxisomes is recovered in the final sucrose density gradient owing to the high losses in the low-speed pellet (*see* Note 5). Contamination of the peroxisomal peak fractions with mitochondrial protein is low, however. Therefore, this fractionation procedure is suitable for the assignment of enzymatic activities or proteins to peroxisomes.
10. Alternatively, peroxisomes can be purified on continuous 15–36% Nycodenz gradients with a cushion of 42% Nycodenz, dissolved in 5 mM MES (pH 6.0), 1 mM KCl, 0.1% ethanol, and 8.5% sucrose *(6)*. Note that when applying a Nycodenz gradient, the organellar pellet of step 12 should be resuspended in 0.65M sorbitol in buffer B.
11. A recent example of a study of yeast peroxisome biology is that by Elgersma et al. *(7)* on enzyme-targeting to peroxisomes.

References

1. Veenhuis, M., Mateblowski, M., Kunau, W.-H., and Harder W. (1987) Proliferation of microbodies in *Saccharomyces cerevisiae*. *Yeast* **3**, 77–81.
2. Goodman, J. M. (1985) Dihydroxyactone synthase is an abundant constituent of the methonol-induced peroxisome of *Candida boidinii*. *Proc. Natl. Acad. Sci. USA* **260**, 7108–7113.
3. Lücke, H. (1963) Catalase, in Methods of enzymatic analysis (Bergmeijer, H. K., ed.), Academic, New York, pp. 885–894.
4. Douma, A. C., Veenhuis, M., De Koning, W., Evers, M., and Harder, W. (1985) Dihydroxyacetone synthase is localized in the peroxisomal matrix of methanol grown *Hansenula polymorpha*. *Arch. Microbiol.* **143**, 237–243.
5. Kionka, C and Kunau, W.-H. (1985) Inducible β-oxidation pathway in *Neuraspora crassa*. *J. Bacteriol.* **161**, 153–157.
6. Kunau, W. H., Beyer, A., Franken, T., Götte, K., Marzioch, M., Saidowsky, J., Skaletz-Rorowski, A., and Wiebel, F. F. (1993) Two complementary approaches to study peroxisome biogenesis in *Saccharomyces cerevisiae:* forward and reversed genetics. *Biochimie* **75**, 209–224.
7. Elgersma, Y., van Roermund, C. W. T., Wanders, R. J. A., and Tabak, H. F. (1995) Peroxisomal and mitochondrial carnitine acetyltransferases of *Saccharomyces cerevisiae* are encoded by a single gene. *EMBO J.*, **14**, 3472–3479.

CHAPTER 16

Transformation of Lithium-Treated Yeast Cells and the Selection of Auxotrophic and Dominant Markers

Robert C. Mount, Bernadette E. Jordan, and Christopher Hadfield

1. Introduction

Transformation of yeast cells can be achieved using lithium-treated or spheroplasted cells. Spheroplast transformation *(1,2)* is a high efficiency method yielding up to 10^4–10^5 transformants per µg of DNA (but more typically, in practice approximately 10^3 per µg). The preparation of spheroplasts, however, is somewhat laborious and tedious. The transformed spheroplasts require plating in a soft agar overlay to allow cell wall regeneration; additionally, the generation of undesired polyploids through protoplast fusion is also a possibility. In contrast, the lithium acetate method is relatively quick and simple to perform. It has the advantage that the cells remain intact, and so do not fuse, and can be plated directly onto the surface of selective agar plates. The transformation efficiency generally is similar to the spheroplast method.

Auxotrophic markers have long been employed for yeast transformation, but they require the strain to be genetically defined to ensure that auxotrophic complementation is possible. Amino acid and nucleotide biosynthetic genes are commonly used as auxotrophic markers; *LEU2 (2)*, *TRP1 (3)*, *HIS3 (4)*, and *URA3 (5)*, for example. Dominant markers usually confer resistance to a specific growth inhibitor and are not constrained by the recipient strain genotype. They are extremely useful when transforming industrial strains or strains not comprehensively defined, the only requirement being the relative sensitivity to the respective inhibitor compound. Markers have been described that provide resistance to antibiotics, such as G418 *(6–8)*, chloramphenicol (Cm) *(9,10)* and hygromycin B *(11,12)*; the antimetabolite methotrexate *(13)*; the herbicide sulfometuron methyl *(14)*; and heavy metals, like copper *(15,16)*.

Transformation of lithium treated cells using auxotrophic markers and the G418- and chloramphenicol-resistance dominant markers will be considered here. The method used is derived from that originally described by Ito et al. *(17)*. Initially, the yeast cells are grown to midlogarithmic phase, then washed and rendered competent by incubation in a lithium acetate solution. For transformation, DNA is added to an aliquot of competent cells, followed by polyethylene glycol (which may act to restrict the aqueous phase and thereby force the DNA into the cells). After an incubation period, the transformation is completed with a heat shock (which may transiently alter membrane dynamics and so facilitate entry of DNA into the cells).

Subsequent treatment for auxotrophic and dominant markers is different. Auxotrophic markers are biosynthetic genes and, consequently, they are generally switched on by a period of starvation. Since this occurs during the competent cell preparation and transformation procedure, such transformants can be selected directly. Dominant markers, however, require a period of expression prior to selection to enable the accumulation of sufficient marker gene product to provide resistance to the selective agent. Since the posttransformation cells are in a starved stationary phase state, an incubation in nonselective, rich, glucose-containing medium is necessary prior to plating on the selective agar.

2. Materials

All solutions and equipment must be detergent-free and sterile. New, previously unused glassware, sterile plastic pipets, Gilson (Paris) tips and 25-mL sterile plastic universal bottles (Nunc, Roskilde, Denmark) are used to reduce these problems.

1. Water for solutions and media should be distilled and then deionized (e.g., by a Millipore [Bedford, MA] milliQ system) or double distilled. Water quality can be tested by measuring the conductivity (ideally it should be <10 µS).

2. YEPD broth: Stock YEP broth, lacking a carbon source, is composed of 1% w/v yeast extract and 2% w/v bacto-peptone, dissolved in water, and autoclaved at 15 psi for 15 min. This can be stored at room temperature for up to 2–3 mo. For YEPD broth, add glucose to 2% w/v from a sterile stock solution (40% w/v, sterilized by autoclaving at 10 psi for 20 min).
3. Transformation TE: 10 mM Tris-HCl, 0.1 mM EDTA, pH 7.6 (sterile).
4. LA: 0.1M lithium acetate in transformation TE (sterile).
5. LAG: LA containing 15% v/v glycerol (sterile).
6. PEG: A 50% w/v solution of polyethylene glycol 4000 in water (sterile). To prepare, place 10g PEG in a plastic universal and add water up to the 20-mL mark; warm at 65°C until the PEG dissolves (10–15 min), then filter sterilize. This solution is better if freshly made and must be less than 1 wk old, since it degenerates.
7. Auxotrophic marker agar plates: Stock semidefined (SD) medium is composed of 0.67% w/v yeast nitrogen base (without amino acids) and 2% w/v agar, dissolved in water and autoclaved at 15 psi for 15 min. This stock can be stored at room temperature for 2–3 mo. For use, melt the solidified agar medium in a microwave oven and add glucose to 2% w/v from a concentrated stock (40% w/v), plus any supplements required to compensate for genetic deficiencies in the recipient host yeast strain that are not to be used for the selection of transformants. Amino acids should be added to 40 µg/mL (from sterile 100X stock solutions); bases should be added to 20 µg/mL (from sterile 20X stock solutions); vitamins should be added to 2 µg/mL (from sterile 1000X stock solutions).
8. Dominant marker selective plates: Stock YEP agar is prepared with YEP broth containing 2% w/v agar; autoclaved at 15 psi for 15 min. This can be stored at room temperature for 2-3 mo. For use, melt the solidified agar medium in a microwave oven and add the carbon source and antibiotic. For G418^R selection, glucose (to 2% w/v) is added from a sterile stock solution (40% w/v), and G418 is added directly to the molten agar medium as a powder, and allowed to dissolve prior to pouring the plates. For Cm^R selection, glycerol is added to the molten stock YEP agar to 1.5% w/v (from a sterile 60% w/v stock) and ethanol to 2% w/v (from an absolute ethanol stock, which is usually sterile); to avoid solubility problems, the chloramphenicol is dissolved in the ethanol before it is added to the medium.

The amount of antibiotic required in the selective plates will vary from strain to strain. Typically, the minimum amount needed to give a plating efficiency of <10^{-8} is used, since this is sufficient to preclude nontransformed, plasmid-free, resistant colonies occurring on the plates *(4,8)*. Commonly used laboratory yeast strains are typically inhibited by 0.5 mg/mL G418 or 2 mg/mL Cm in the respective medium.

3. Methods (see Notes 7–9)

3.1. Preparation of Competent Cells

1. Grow a fresh overnight culture of the recipient strain in 10 mL of YEPD broth in a 25-mL universal bottle, at 30°C with shaking.
2. Next day, use this overnight culture to inoculate 20 mL of YEPD broth in a 25-mL universal bottle at approx 2.5×10^6 cells/mL. Use a hemocytometer to determine the cell density (see Note 1).
3. Grow to a density of 1×10^7 cells/mL at 30°C with shaking (see Notes 2 and 3). This requires two cell doublings and will take 3–4.5 h.
4. Pellet the culture, in the universal bottle, using an Hereaus Christ microbiological string-out centrifuge (or equivalent model by another manufacturer) by spinning at 4000 rpm (2300g) for 5 min at 20°C.
5. Pour off the supernatant and gently resuspend the pellet in 10 mL transformation TE. Repellet as before.
6. Repeat step 5.
7. Pour off the supernatant and gently resuspend in 10 mL LA.
8. Incubate at 30°C with shaking for 1 h.
9. Repellet and gently resuspend in 2 mL LA. This is sufficient for 6 transformations (see Note 2): For storage of competent cells, see Note 5.

3.2. Transformation

The transformation procedure is the same for both types of markers.

1. Add plasmid DNA (0.1–10 µg) to each 1.5-mL microfuge tube containing 300 µL competent cells, followed by 700 µL PEG; mix the contents by several gentle inversions.
2. Incubate in a 30°C water bath, without shaking, for 60 min.
3. Heat shock by placing in a 42°C water bath for 5 min.

3.3. Post-Transformational Treatment and Plating

This differs depending on the type of marker employed.

3.3.1. Auxotrophic Markers

1. 100–200-µL aliquots can be plated out directly onto the selective agar. Alternatively, if a more concentrated plating is required, pellet the cells for 5 min at 4000 rpm (1100g) in a microfuge. Discard the supernatant and gently resuspend the pellet in 100 µL of sterile water, using the end of a micropipet tip to agitate the pellet, followed by a few *gentle* pumps of the pipet. The cells can then be plated on one or two selective agar plates.
2. Incubate at 30°C; colonies usually appear within 2–4 d.

3.3.2. $G418^R$ and Cm^R Dominant Markers

1. Dilute the cells in 2 vol of YEPD broth and incubate at 30°C to allow for expression of the marker gene. The expression time is dependent on the marker being used: $G418^R$ requires 90–120 min *(5)*; Cm^R 60 min *(8)*.
2. Pellet the cells for 5 min at 4000 rpm (2300g).
3. Gently resuspend the pellet in 200 µL YEPD broth as described earlier.
4. Divide the transformation mix between two plates with the appropriate selection (*see* Materials). $G418^R$ transformants appear within 2 d at 30°C; Cm^R transformants take 4–7 d (*see* Note 6).

4. Notes

1. Yeast cells can vary greatly in size, shape, and extent of clumping between strains. This tends to make judging cell numbers per mL by optical density somewhat precarious. A sufficiently accurate determination of culture cell density can be obtained using a hemocytometer. An improved Neubauer hemocytometer, depth 0.1 mm, 1/400 mm^2 is suitable and easy to use. A drop of culture (ca. 50 µL) can simply be placed on the counting grid of the slide and the coverslip then placed over the top. The liquid will spread under the coverslip; after a few minutes the cells will stop moving and form a single stationary layer that can be counted. When viewed under the light microscope at 200 or 400X magnification, both the cells and the grid are easily visible. The grid is composed of 25 large squares, defined by triple lines, which in turn are divided into 16 small squares. Each cell present in a large square is equivalent to 2.5×10^5 cells/mL; each cell per small square is equivalent to 4×10^6 cells/mL. Allowances can be made for bud size: Thus, a cell with a large attached daughter cell may be counted as two cells, whereas a small bud may not be counted. Highly clumped cells may be separated by a sonication or addition of EDTA to facilitate counting of individual cells. A calibration curve of counted cell numbers against optical density at 600 nm may be established for future use for strains routinely used.
2. Cultures grown in universals generally produce cells of greater competence than those grown in flasks. At midlog, cells are growing anaerobically on the available glucose, so aeration is not necessary.
3. Exponentially growing cells in the range 5×10^6 to 5×10^7 cells/mL can be used for transformation; however, cells at 1×10^7/mL tend to be optimal for the process in terms of both condition and numbers.
4. Multiple cultures can be grown for a greater number of transformations.
5. Competent cells may be stored at –70°C for up to 2 mo, if at step 6 of the competent cell protocol the cells are resuspended in LAG, rather than LA, and "flash-frozen" in 300-µL aliquots (i.e., microfuge tubes, containing the

cells, are placed in a dry ice/ethanol bath to quickly freeze the cells). The transformation efficiency is, however, reduced by storage in such a manner.
6. DNA quality affects the transformation efficiency; CsCl gradient-prepared plasmid DNA ensures optimum efficiency.
7. Gentle handling of the cells at all stages is paramount to ensure successful, efficient transformation.
8. The method described is applicable to all types of vectors (replicating plasmid or integrative).
9. Include a DNA-free transformation control to check for a clear background with both auxotrophic and dominant marker selections.
10. The lithium acetate transformation procedure has been optimized for maximum yield of transformants by Geitz et al. *(19)*.
11. Biolistics (*see* Chapter 17) and electroporation *(18)* are alternative transformation procedures.

References

1. Beggs, J. D. (1978) Transformation of yeast by a replicating plasmid. *Nature* **275,** 104–109.
2. Hinnen, A., Hicks, J. B., and Fink, G. R. (1978) Transformation of yeast. *Proc. Natl. Acad. Sci. USA* **75,** 1929–1933.
3. Hitzeman, R. A., Clarke, L., and Carbon, J. (1980) Isolation and characterization of the yeast 3-phosphoglycerate kinase gene (PGK) by an immunological screening technique. *J. Biol. Chem.* **255,** 12073–12080.
4. Struhl, K. (1982) Regulatory sites for *HIS3* gene expression in yeast. *Nature* **300,** 284–287.
5. Struhl, K., Stinchcomb, D. T., Scherer, S., and Davis, R. W. (1979) High-frequency transformation of yeast: Autonomous replication of hybrid DNA molecules. *Proc. Natl. Acad. Sci. USA* **76,** 1035–1039.
6. Hadfield, C., Jordan, B. E., Mount, R. C., Pretorius, G. H. J., and Burak, E. (1990) G418-resistance as a dominant marker and reporter for gene expression in *Saccharomyces cerevisiae*. *Curr. Genet.* **18,** 303–313.
7. Webster, T. D. and Dickson, R. C. (1983) Direct selection of *Saccharomyces cerevisiae* resistant to the antibiotic G418 following transformation with a DNA vector carrying the kanamycin resistance gene of Tn903. *Gene* **26,** 243–252.
8. Jimenez, A. and Davies, J. (1980) Expression of a transposable antibiotic resistance element in *Saccharomyces*. *Nature* **287,** 869–871.
9. Hadfield, C., Cashmore, A. M., and Meacock, P. A. (1986) An efficient chloramphenicol-resistance marker for *Saccharomyces cerevisiae* and *Eschericha coli*. *Gene* **45,** 149–158.
10. Hadfield, C., Cashmore, A. M., and Meacock, P. A. (1987) Sequence and expression characteristics of a shuttle chloramphenicol-resistance marker for *Saccharomyces cerevisiae* and *Escherichia coli*. *Gene* **52,** 59–79.
11. Kaster, K. R., Burgett, S. G., and Ingolia, T. D. (1984) Hygromycin B resistance as dominant selectable marker in yeast. *Curr. Genet.* **8,** 353–358.

12. Gritz, L. and Davies, J. (1983) Plasmid-encoded hygromycin B resistance: the sequence of hygromycin B phosphotransferase gene and its expression in *Eschericha coli* and *Saccharomyces cerevisiae*. *Gene* **25,** 179–188.
13. Zhu, J., Contreras, R., Gheysen, D., Erst, J., and Fiers, W. (1985) A system for dominant transformation and plasmid amplification in *Saccharomyces cerevisiae*. *Biotechnology* **3,** 451–456.
14. Casey, G. P., Xiao, W., and Rank, G. H. (1988) A convenient dominant selection marker for gene transfer in industrial strains of *Saccharomyces* yeast: *SMR1* encoded resistance to the herbicide sulfometuron methyl. *J. Inst. Brew.* **94,** 93–97.
15. Henderson, R. C. A., Cox, B. S., and Tubb, R. (1978) The transformation of brewing yeast with a plasmid containing the gene for copper resistance. *Curr. Genet.* **9,** 133–138.
16. Fogel, S. and Welch, J. (1982) Tandem gene amplification mediates copper resistance in yeast. *Proc. Natl. Acad. Sci. USA* **79,** 5342–5346.
17. Ito, H., Fukuda, Y., Murata, K., and Kimra, A. (1983) Transformation of intact yeast cells treated with alkali cations. *J. Bacteriol.* **153,** 163–168.
18. Faber, K. N., Haima, P., Harder, W., Veenhuis, M. and Ab, G. (1994) Highly efficient electrotransformation of the yeast *Hansenula polymorpha*. *Curr. Genet.* **25,** 305–310.
19. Geitz, D., St. Jean, A., Woods, R. A., and Schiestl, R. H. (1992) Improved method for high efficiency transformation of intact yeast cells. *Nucleic Acids Res.* **6,** 1425.

Chapter 17

Biolistic Transformation of Yeasts

Stephen A. Johnston and Michael J. DeVit

1. Introduction

Biolistic transformation is a unique process in which DNA or RNA is introduced into cells on micron-sized particles. These microparticles are accelerated to supersonic speeds utilizing forces generated by a gunpowder discharge or cold gas explosion. This technique was first developed for transformation of plant cells *(1)*. It has since been used successfully with bacteria *(2)*, fungi *(3)*, and mammalian cells, both in vitro *(4)* and in vivo *(5)*.

The biolistic protocol is effective in transforming yeasts. The efficiency is comparable to conventional protocols: One bombardment with transforming DNA of a yeast strain spread on a selective plate will produce a confluent lawn of transformants. Most strains are transformed at 10^4–10^5 transformants/µg DNA. Typically, 1–3 × 10^8 cells are spread on each plate and bombarded with 0.2–1 µg DNA. Linear DNA transforms at rates comparable to circular *(3)*. An experienced operator can bombard 30 plates/h.

For standard transformation, the technique has some potential advantages. First, stationary cells transform with the highest efficiency *(3)*. Even 3–4-d-old cultures produce transformants, with 1–2-d-old cultures optimal. Also, since 10–30 biologically active plasmids are carried-in on each microprojectile, it is easy to introduce multiple plasmids into a single cell *(3)*. A unique feature is the ability to transform many strains or isolates at the same time with one plasmid. Ten to fifty strains are spotted onto a single plate and it is bombarded with the plasmid of interest, producing one or more transformants in each spot.

The biolistic process also has some unique applications with respect to yeast transformation. One is its ability to transform hard-to-transform strains or species. The authors have found that some strains that respond little or not at all to conventional methods, will transform biolistically *(3)*. Recently, the authors have demonstrated that the technique is much more efficient in transforming *Crytococcus neoformans* than conventional methods *(6)*, and allows transformation of the pathogenic form of this yeast, which has not been accomplished by standard protocols *(6a)*.

The second unique feature is that this is the only technique for effecting mitochondrial transformation *(7,8)* (and plant chloroplast *[9]*). Presumably, the microprojectiles can penetrate both the cell wall and the mitochondrial membrane. The efficiency of this transformation is only 10^{-3}–10^{-4} of nuclear transformants. Direct selection of mitochondrial transformants (respiratory competent) has not been possible, their isolation requiring secondary screening *(7,10)*.

The original biolistic device was gunpowder-driven *(1)*. Recently, the helium-driven device has been shown to be much more effective for all applications, including yeast transformation *(11)*. Since the commercial devices now available (Bio-Rad, Hercules, CA) (Fig. 1) are of the helium-type, the authors have described the protocol as applying to this device. The specific details for construction of the helium-driven device have been published *(11)*.

The basic principles of operation are the following. DNA is coated onto microprojectiles. These are usually gold or tungsten, although there has been limited success using glass, phage and bacteria (Sanford, personal communication). For yeast, the optimal diameter of the microprojectiles is ~0.5 µm. The DNA-coated microprojectiles are spread onto a macrocarrier made of KaptonTM (Dupont, Wilmington, DE), which is 2 mil (2/1000 in.) thick. The macrocarrier is placed into the biolistic device just below a tube containing high pressure (1200 psi) helium. The helium gas is released by bursting the restraining Kapton disks, either spontaneously or with an electrically controlled lance. This sudden release of gas creates a supersonic shockwave that propels the macrocarrier against a stopping screen. The macrocarrier is retained, and the microprojectiles keep going toward the target plate. The macrocarrier also serves to block-off the opening through the stopping screen and deflects the escaping gas away from the target cells. The bombardment takes place in a partial vacuum (29 in. mercury) in order to decrease the impedance of the microprojectiles *(10)* (Fig. 2). A general review of biolistic applications has recently been published *(12)*.

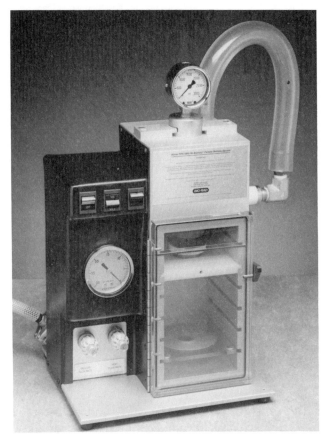

Fig. 1. Biolistic device (PDS-1000) commercially available from Bio-Rad.

2. Materials

1. Microparticles (Bio-Rad or Alfa, Ward Hill, MA): M10 tungsten or 0.8 µm gold, 60 mg/mL H_2O. Prepare by washing one time with 95% ethanol, then two times with sterile H_2O. These 1-mL aliquots of the microprojectile-H_2O slurry can be frozen and stored for several weeks.
2. Expendables: 2-mil Kapton disks are used as the rupture membrane. Each 2-mil disk retains ~350–400 psi. Disks (2 mil) are also used as the macrocarrier. Both the rupture disks and macrocarriers are used once. Steel screens are the stopping plates. They can be used for several bombardments. The disks and screens can be purchased from Bio-Rad or punched out from sheets by the user.
3. $2.5M$ $CaCl_2$ and $1M$ spermidine-free base. Filter sterilize both and store frozen in 1-mL aliquots.

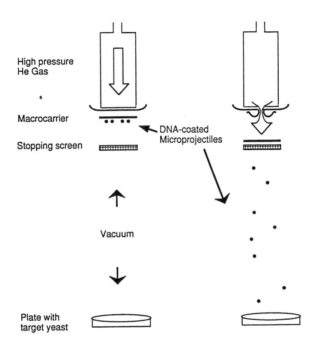

Fig. 2. Principles of operation of the biolistic device. Bombardment takes place within a chamber as shown in Fig. 1.

4. Ethanol: 70% v/v and 100% (must be stored water-free over desiccant).
5. DNA (1 µg/µL) plasmids can be linear or circular with linear giving ~0.3 the number of transformants. Miniprep DNA can be used but the cleaner the DNA the higher the efficiency.
6. Selection plates: Either with $1M$ sorbitol or $0.75M$ sorbitol + $0.75M$ mannitol. The mannitol-containing plates give somewhat higher frequencies.
7. Yeast cells: Stationary cells (12–48 h poststationary) give the highest efficiency. Spin down, take off 2/3 of media and spread cells on selection plates at $1-3 \times 10^8$ cells/plate. Let them dry. The plates can be stored at 4°C overnight, though with some loss of efficiency.
8. Helium gas: 2500 psi tank, high purity.

3. Methods

3.1. Preparation of Microprojectiles (for Three to Four Shots)

1. Pipet 25 µL of microprojectile slurry (the 60 mg/mL aliquots described earlier) into a microfuge tube.

Biolistic Transformation of Yeasts

2. Add 2.5 µL DNA (1–5 µg) while mixing well.
3. While gently vortexing, rapidly add 25 µL 2.5M CaCl$_2$ and then 5 µL 1M spermidine. The CaCl$_2$ must be added first.
4. Let sit for 5 min.
5. Centrifuge briefly (pulse in microfuge) to pellet microprojectiles and then remove supernatant.
6. Resuspend in 200 µL 70% ethanol, pellet, and remove supernatant.
7. Repeat using 100% ethanol (*see* Note 1).
8. Resuspend microprojectiles in 25 µL 100% ethanol (*see* Note 2).
9. Sonicate microparticles, 1–2 s, or vortex briefly if clumping is a problem. The requirement for this step varies from batch to batch of microprojectiles.
10. While vigorously stirring or agitating the slurry, withdraw 6 µL and place it as a droplet onto the center of the Kapton macrocarrier. Because of the inhibitory effects of water (*see* Note 2), this step is carried out over a desiccant in a small container or glove box.
11. When the ethanol has evaporated, the disk can be fired. The disks should be kept under desiccation until ready to be fired. Tungsten-coated particles should be used within a few hours as the DNA will eventually degrade.

3.2. Operation of the Biolistic Device

1. Detailed instructions based on our protocols are provided with the commercial device available from Bio-Rad. The details of construction and principles of operation of the helium-driven device recently have been published *(10)*. In the following, the authors comment on some particulars of operation for yeast transformation.
2. Pressure settings: The pressure setting for the system is set at the regulator. The authors have found that 1200 psi is optimal, i.e., producing maximal velocity with least cell damage owing to the shockwave. In the commercial system the burst pressure is determined by the number of restraining 2-mil Kapton disks screwed onto the restraining cap. The authors are using a system in which the membranes are burst at a particular pressure with a solenoid-driven, sharpened lance. Though these two methods of gas release produce qualitatively different shockwaves, the authors have noticed no reproducible differences in yeast transformation rates.
3. The macrocarrier coated with microprojectiles is placed on a "launch pad" approx 1.5 cm below the rupture disks. The disk should be held with minimal adhesion. A tiny amount of vacuum grease applied to the back of the macrocarrier works well.
4. The stopping screen is fixed approx 1.5 cm below the macrocarrier. The screens can be reused several times and may actually give a better spray pattern of the microprojectiles after several uses.

5. The sample plate should be placed near the bottom of the chamber, 12–14 cm from the stopping screen.
6. A vacuum is applied to the chamber. The highest vacuum (32" Hg) gives the highest rate. The chamber should be sealed just before firing to avoid air disturbances.
7. After bombardment, the plates are treated as usual. Unlike the spheroplasting method, the cells can be transferred to a nonsorbitol medium within a few hours of bombardment.
8. To avoid crosscontamination, the screens and the inside of the chamber should be ethanol cleaned between bombardments.

4. Notes

1. The ethanol must be completely free of water. It should be stored over desiccant. If water is present in the ethanol, the microprojectiles tend to adhere to the macrocarrier, decreasing the launch velocity and the transformation frequency.
2. Recently, the authors have started using an alternative protocol that is not sensitive to water. After the last ethanol rinse, the ethanol is removed and the microprojectiles completely dried. The microparticles are then sifted through a 50–100 μ screen onto the Kapton macrocarrier. The particles are gently rubbed through the screen with the blunted end of a plastic microtip. The disadvantage of this protocol is that some of the microprojectiles fall off when the macrocarrier is inverted, leading to two- to fivefold fewer transformants/plate compared to the standard protocol.

References

1. Sanford, J. (1990) Biolistic plant transformation: a critical assessment. *Physiologia Plantarum* **79**, 206–209.
2. Smith, F. D., Harpending, P. R., and Sanford, J. C. (1992) Biolistic transformation of prokaryotes-factors that effect biolistic transformation of very small cells. *J. Gen. Microbiol.* **36**, 239–248.
3. Armaleo, D., Ye, G.-N., Klein, T. M., Shark, K. B., Sanford, J. C., and Johnston, S. A. (1990) Biolistic nuclear transformation of *Saccharomyces cerevisiae* and other fungi. *Curr. Genet.* **17**, 97–103.
4. Zelenin, A. V., Titomirov, A. V., and Kolesnikov, V. A. (1989) Genetic transformation of mouse cultured cells with the help of high velocity mechanical DNA injection. *FEBS Lett.* **244**, 65–67.
5. Williams, R. S., Johnston, S. A., Riedy, M., DeVit, M. J., McElligott, S. G., and Sanford, J. C. (1991) Introduction of foreign genes into tissues of living mice by DNA-coated microprojectiles. *Proc. Natl. Acad. Sci. USA* **88**, 2726–2730.
6. Toffaletti, D. L., Rude, T. H., Johnston, S. A., Durack, D. T., and Perfect, J. R. (1993) Gene transfer in *Cryptococcus neoformans* using biolistic delivery of DNA. *J. Bacteriol.* **175**, 1405–1411.

6a. Cox, G. M., Toffaletti, D. L., and Perfect, J. R. (1995) A dominant selection for use in *Cryptococcus neoformans* (submitted).
7. Johnston, S. A., Anziano, P. Q., Shark, K., Sanford, J. C., and Butow, R. A. (1988) Mitochondrial transformation in yeast by bombardment with microprojectiles. *Science* **240,** 1538–1541.
8. Belcher, S. M., Perlman, P. S., and Butow, R-A. (1994) Biolistic transformation of mitochondria in *Saccharomyces cerevisiae*, in *Particle Bombardment Technology for Gene Transfer* (Yang, N.-S. and Christou, P., eds.), W. H. Freeman and Co., New York.
9. Svab, Z., Hajdukiewicz, P., and Maliga, P. (1990) Stable transformation of plastids in higher plants. *Proc. Natl. Acad. Sci. USA* **87,** 8526–8530.
10. Fox, T. D., Sanford, J. C., and McMullin, T. W. (1988) Plasmids can stably transform yeast mitochondria totally lacking endogenous mt DNA. *Proc. Natl. Acad. Sci. USA* **85,** 7288–7292.
11. Sanford, J. C., DeVit, M. J., Russell, J. A., Smith, F. D., Harpending, P. R., Roy, M. K., and Johnston, S. A. (1991) An improved, helium-driven biolistic device. *Technique* **3,** 3–16.
12. Sanford, J. C., Smith, F. D., and Russell, J. A. (1993) Optimizing the biolistic process for different biological applications, in *Methods in Enzymology* (Wu, R., ed.), Academic, New York, pp. 483–509.

CHAPTER 18

Genomic Yeast DNA Clone Banks

Construction and Gene Isolation

Graham R. Bignell and Ivor H. Evans

1. Introduction

Techniques for constructing genomic yeast DNA libraries and for isolating yeast genes from them, mainly by complementation, have been known for over 15 yr. The first method *(1)* involved making the library in an *Escherichia coli* plasmid vector, but as soon as "shuttle" vectors—capable of stable propagation in both *E. coli* and *Saccharomyces cerevisiae*—became available, these were used in preference. The key advantage of shuttle vectors is that they permit selection for expression of yeast genes in a yeast *S. cerevisiae* host, but the vector containing the desired yeast gene can then be transferred ("shuttled") back to allow easier plasmid DNA isolations and manipulations. In essence, the usual approach is to isolate pure genomic DNA from the yeast of interest, partially digest the DNA with an appropriate restriction enzyme (so ensuring any particular sequence of interest will occur intact, on a reasonably sized DNA fragment), size-fractionate the cut DNA (to eliminate small, under gene-sized fragments), insert the DNA in an appropriate shuttle vector, and transform it into *E. coli*, to make an initial, bacterial, clone library. *E. coli* transformation is very efficient, compared with yeast transformation, and so allows a good-sized clone library to be generated from even relatively small quantities of ligated genomic-vector DNA. Library DNA is then extracted from the pooled *E. coli* clones and transformed into an appropriate host strain of *S. cerevisiae*, thus creating a clone library in yeast, which is then used to screen or select for a desired gene.

From: *Methods in Molecular Biology, Vol. 53: Yeast Protocols*
Edited by: I. Evans Humana Press Inc., Totowa, NJ

All the experimental manipulations needed to make and use a yeast genomic library in the way just described, are detailed in the following, in the expected sequence. This is not the place to review yeast vectors (an extensive survey of yeast vectors has been published by Parent and colleagues *[2]*), but choice of an appropriate vector is a particularly important decision for the experimenter.

YRp vectors (which contain a yeast chromosomal DNA replicator sequence), although an early type of construction, mitotically unstable, and less efficient in yeast transformation than later 2-µ plasmid-based (YEp) vectors, have been used to make a very widely used S. cerevisiae genomic library *(3)* and, more recently, a *Yarrowia lipolytica* genomic library *(4)*. YRp vectors arguably have the advantage that their mitotic instability means that transformants are easily distinguished from revertants. YEp vectors are more stably inherited in *S. cerevisiae* and transform the yeast very efficiently; they have also been used in constructing useful gene libraries (for example, for *Candida tsukubaensis [5]* and *Schwanniomyces occidentalis [6]*). YCp vectors are a more recent type of construction, being circular plasmids with an *S. cerevisiae* centrometric sequence that confers exceptional mitotic stability and low copy number (one or two per haploid genome). The low copy number means that artifacts caused by artificially high levels of expression of a cloned gene—reflecting high copy number—cannot arise, although YCp plasmids are more difficult to recover from yeast than high copy number vectors. Nevertheless, YCp vectors have been used to make yeast genomic DNA banks *(7,8)*. Yeast artificial chromosome vectors (YACs) tend to be most useful in analysis of large, complex, eukaryotic genomes and in linkage and sequencing studies *(9)*. pRS vectors are a fairly recent useful group of *S. cerevisiae* vectors introduced by Sikorski and Hiefer *(10)*. They have pBLUESCRIPT phagemid attributes and most standard DNA manipulations can be performed in the same plasmid that is introduced into yeast.

Many experimenters wish to clone genes from non-*Saccharomyces* yeasts for which homologous transformation systems and suitable vectors have not been developed. In a number of cases, genes from these yeasts do express in *Saccharomyces* and so have been successfully cloned in what is a heterologous expression system (examples are found in references cited earlier).

Indeed, it is worth recalling that a significant number of *S. cerevisiae* genes are expressed—and were initially cloned—in *E. coli*, through fortuitous occurrence of appropriate transcription signals. However, especially when intending to clone genes from yeasts taxonomically remote from *Saccharomyces*, it would be worth considering using a *Saccharomyces* expression vector in which, typically, the cloned fragment is inserted 3' to a powerful *Saccharomyces* promoter—often derived from glycolytic genes. The unpredictable and quirky aspects of heterologous gene expression are illustrated by our recent experience with the TRP1 (*N*-[5'-phosphoribosyl] anthranilate isomerase-encoding) gene of the yeast *Lipomyces starkeyi*: When cloned in the vector YCp50, this gene expresses in *E. coli*, but does not do so in *S. cerevisiae* (Bignell et al., in preparation). Expression of heterologous genes in yeasts has recently been reviewed by Romanos and colleagues *(11)*, and a succinct overview of genetic engineering in yeast has been published by Brearley and Kelly *(12)*.

2. Materials
2.1. Isolation of Genomic DNA
2.1.1. Preparation of Protoplasts

1. YEPD: 1% w/v yeast extract, 1 w/v bactopeptone, 2% w/v glucose.
2. 0.05M EDTA, pH 7.5.
3. 0.5M EDTA, pH 9.0.
4. β-mercaptoethanol.
5. Protoplasting solution: 1M sorbitol, 0.1M EDTA, pH 7.5.
6. Novozyme 234 (Novo, Copenhagen, Denmark: A yeast cell-wall degrading enzyme preparation with β-glucanase activity).

2.1.2. Lysis of Protoplasts

1. Saline-EDTA: 0.15M NaCl, 0.1M EDTA, pH 8.0.
2. Proteinase K: 20 mg/mL stock solution in saline EDTA, freshly made.
3. 20% SDS (w/v).

2.1.3. Isolation of DNA from Lysate

1. Chloroform, 4% v/v for isoamyl alcohol.
2. Absolute ethanol, ice cold or –20°C (freezer).
3. 20X SSC buffer: 3M NaCl, 0.3M trisodium citrate. Adjust to pH 7.0 with sodium hydroxide and sterilize by autoclaving. Store in aliquots at 4°C for maximum shelf life.
4. 0.1X SSC, prepare just before use by dilution of the 20X stock with water.
5. Pancreatic RNase A: Dissolve pancreatic RNase (RNase A) at a concentration of 10 mg/mL in 0.1M sodium acetate (pH 5.2); heat to 100°C for 15 min, and allow to cool slowly to room temperature (RNase precipitates when concentrated solutions are heated to 100°C at neutral pH). Neutralize by adding 0.1 vol Tris-HCl, pH 7.4. Distribute in tubes as small aliquots and store frozen at –20°C.

6. T1 RNase: This is typically available at 300,000–600,000 U/mg. Make a approx 10,000 U/mL stock solution in 10 mM sodium acetate (pH 5.0), heat to 80°C for 10 min, allow to cool to room temperature, then adjust the pH to 7.0 with NaOH. Aliquot and store frozen at −20°.

2.2. Digestion of Genomic DNA

2.2.1. Pilot Digestion

1. Restriction enzyme buffer: Follow the restriction enzyme manufacturer's recommendations or use the concentrated buffer that is usually supplied with the enzyme. (Optimal buffer conditions vary between the enzymes, and the manufacturer will supply the optimal buffer for their particular enzyme product, *see* Note 9.)
2. Appropriate restriction enzyme (*see* Note 8).
3. Sterile distilled water, carefully prepared to avoid traces of detergent in bottles.
4. Sterile 1.5-mL Eppendorf tubes and micropipet tips.
5. 0.5M EDTA (pH 8.0).
6. Stock (X6) gel-loading buffer: 40% w/v sucrose, 0.25% bromophenol blue store at 4°C (*see* Note 11).
7. Agarose (with a low coefficient of electroendosmosis ($-M_r$), e.g., Sigma (St. Louis, MO) type 1.
8. Gel electrophoresis apparatus for running horizontal submerged DNA gels, and power pack.
9. Gel running buffer (TBE): X10 stock, 0.9M Tris, 0.9M boric acid, 25mM Na$_2$ EDTA, the whole being adjusted to pH 8.2 with HCl. Can be stored indefinitely at 4°C.
10. DNA size-markers (*see* Note 12).
11. Ethidium bromide stock: X20,000, 10 mg/mL, store in foil-wrapped bottle at room temperature. *Caution:* Ethidium bromide is toxic and mutagenic/potentially carcinogenic. Handle using disposable plastic gloves and dispose of used ethidium bromide-containing solutions safely (*see* Note 15).
12. Transilluminator for viewing DNA gels (i.e., emitting UV near 300 nm) and camera or scanning device for recording the positions of DNA fragments in the gel (*see* Note 16).

2.2.2. Scaled up Digestion of Genomic DNA

1. Restriction enzyme.
2. Suitable restriction enzyme buffer.
3. Sterile distilled water.
4. Sterile 1.5-mL Eppendorf tubes and micropipet tips.
5. 0.5M EDTA pH 8.0 (*see* Note 35).

2.3. Size-Fractionation of DNA

1. Sucrose solutions in TE (10 mM Tris-HCl, pH 8.0, 1mM EDTA, sterile (*see* Note 12) suitable for preparing a linear 10–40% w/v density gradient (total vol approx 20 mL). If using the diffusion method for gradient preparation, the solutions needed are 40, 30, 20, and 10%, all w/v (*see* Note 18). It is best to store the solutions at 4°C and use within a few days to avoid any risk of (usually fungal) contamination.
2. Suitable centrifuge tubes, rotor, and ultracentrifuge for centrifuging gradient(s) in swing-out mode at approx 100,000g (*see* Note 18).
3. Suitable device for fractionating density gradients (commercially available, or homemade) (*see* Note 20).
4. Sterile 1.5-mL Eppendorf tubes for collecting gradient fractions.
5. Materials and equipment for analyzing DNA samples by agarose gel electrophoresis (*see* Section 2.2.1.).
6. Dialysis tubing (10,000 cutoff): Pretreat by boiling 10 min in a large volume of 2% w/v sodium bicarbonate and 1 mM EDTA (pH 8.0), rinse thoroughly in distilled water, boil an additional 10 min in 1 mM EDTA (pH 8.0), cool, and store in the 1 mM EDTA at 4°C. Subsequently, only handle the tubing with gloves.
7. TE: 10 mM Tris HCl, pH 8.0, 1 mM EDTA (*see* Note 35).
8. Absolute ethanol (ice cold, or –20°).
9. 5M ammonium acetate, pH 5.2.

2.4. Isolation of Purified Vector DNA

See Sections 2.11., 2.12., 3.11., and 3.12.

2.5. Treatment of Vector DNA Prior to End-Filling

1. Vector DNA.
2. Appropriate restriction enzyme and buffer for cutting vector DNA at cloning site (*see* Sections 2.2.1. and 3.2.1.).
3. Equipment and materials for analyzing DNA by agarose gel electrophoresis (*see* Sections 2.2.1. and 3.2.1.).
4. Sterile 1.5-mL Eppendorf tubes and micropipet tips.
5. Absolute ethanol: ice cold or –20°C.
6. Phenol (buffer equilibrated), (*see* Note 34).
7. TE, pH 8.0: 10 mM Tris-HCl, 1 mM EDTA, pH 8.0. (*see* Note 35).
8. Phenol/chloroform reagent: 25 parts of buffer-equilibrated phenol to 24 parts chloroform to 1 part isoamyl alcohol (reduces foaming); store under 100 mM Tris-HCl, pH 8.0, in a light-excluding bottle at 4°C (stable for 1 mo).
9. Chloroform.

2.6. End-Filling of Vector DNA

1. Cut vector DNA (approx 5 µg in a small [<10 µL] vol of 10 mM Tris-HCl, pH 8.0, 1 mM EDTA or water, and *see* Notes 8 and 24).
2. NTSB buffer (X10): 0.5M Tris, HCl, pH 7.8, 90 mM MgCl$_2$, 500 mg/mL BSA.
3. Stock solutions (10 mM) of the required deoxynucleoside triphosphates (store frozen).
4. 2-Mercaptoethanol.
5. Klenow fragment of DNA polymerase 1.
6. DEPC: 10% v/v diethylpyrocarbonate in ethanol.
7. Absolute ethanol (ice cold or −20°C).
8. 5M ammonium acetate, pH 5.2.
9. TE (pH 8.0): 10 mM Tris-HCl, pH 8.0, 1 mM EDTA.
10. Sterile 1.5-mL microfuge tubes and micropipet tips.

2.7. End-Filling of Genomic DNA

See Section 2.6., except different nucleotides are required (*see* Note 24).

2.8. Ligation of Vector and Genomic DNA

1. Approximately equal quantities (e.g., 5 µg) of end-filled vector DNA and end-filled genomic DNA (in TE, pH 8.0).
2. Sterile 1.5-mL Eppendorf tubes and micropipet tips.
3. Phenol (buffer-equilibrated, *see* Note 34).
4. Phenol:chloroform:isoamyl alcohol (25:24:1).
5. Chloroform.
6. Absolute ethanol (ice cold or −20°C).
7. 5M ammonium acetate, pH 5.2.
8. Sterile water.
9. X10 Ligase buffer: 200 mM Tris-HCl, pH 7.6, 50 mM MgCl$_2$, 50 mM dithiothreitol, 500 µg/mL BSA (Fraction V, Sigma); store in small aliquots at −20°C. Suitable ligase buffer may be supplied with the ligase by the manufacturer.
10. T4 DNA ligase: Usually supplied at 1–5 U/µL.

2.9. Transformation of E. coli

1. LB broth: 0.5% w/v yeast extract, 1% w/v NaCl, 1% w/v tryptone. Adjust the pH to 7.0 with 5M NaOH (approx 0.2 mL, per I medium) before autoclaving.
2. LB agar: As described earlier, plus 2% w/v agar added prior to autoclaving. According to requirements, antibiotics are added to the described media after autoclaving. Ampicillin is prepared as a stock solution of 50 mg/mL in water (prepare in sterile water, or filter-sterilize) and stored (light-protected) at −20°C. Tetracycline stock solution is 15 mg/mL in ethanol (need not be sterilized) and is also stored in a light-proof bottle at −20°C. Typical

Genomic Yeast DNA Clone Banks

final working concentrations are 50 µg/mL for ampicillin and 15 µg/mL for tetracycline. Antibiotic-containing medium should be used within 1–2 wk.
3. Sterile 50-mL plastic centrifuge tubes.
4. $0.1M$ $CaCl_2$, sterile, ice cold.
5. Sterile 1.5-mL microfuge tubes and micropipet tips.
6. Sterile saline (1% w/v NaCl).
7. Sterile glycerol.

2.10. Assessment of the Clone Bank
2.10.1. Alkaline Lysis Extraction of Plasmid DNA
1. LB broth containing an appropriate antibiotic for plasmid selection (*see* Section 2.9.1.).
2. Sterile 1.5-mL microfuge tubes and micropipet tips.
3. AL I: 50 mM glucose, 25 mM Tris-HCl, pH 8.0, 10 mM EDTA, pH 8.0. Sterilize by autoclaving and store at 4°C. Prior to plasmid extraction, put on ice.
4. AL II: $0.2M$ NaOH (freshly diluted from a $10M$ stock), 1% w/v SDS. Prepare immediately before use; do not chill.
5. AT III: Prepare by mixing 60 mL of 5-mL potassium acetate, 11.5 mL glacial acetic acid, and 28.5 mL water (yielding a solution $3M$ for potassium and $5M$ for acetate). Prechill on ice.
6. Phenol:chloroform:isoamylalcohol (25:24:1).
7. Absolute ethanol (room temperature).
8. 70% v/v ethanol (4°C).
9. TE, pH 8.0: 10 mM Tris-HCl, pH 8.0, 1 mM EDTA.

2.10.2. Analysis of Sample of the Clone Bank
Materials as in Section 2.2.1.

2.11. Bulk Isolation of Plasmid DNA from Clone Library
1. 500 LB broth containing an appropriate antibiotic for plasmid selection (*see* Section 2.9.1.).
2. Centrifuge bottles (250 mL) and tubes (50 mL).
3. STE: $0.1M$ NaCl, 10 mM Tris-HCl, pH 8.0, 1 mM EDTA, pH 8.0, ice cold.
4. AL I: *See* Section 2.10.1., ice cold. (Note that the pH of AL I must not be less than 8.0, as lysozyme is not effective then.)
5. Lysozyme.
6. AL II: *See* Section 2.10.1.
7. AL III: *See* Section 2.10.1., ice cold.
8. Sufficient muslin to form four layers for filtration.
9. Isopropanol.
10. 70% ethanol.
11. TE, pH 8.0: *See* Section 2.10.1., item 9.

2.12. Isopycnic Centrifugation of DNA

1. CsCl (solid).
2. 10 mg/mL ethidium bromide (stored in a light-excluding bottle at room temperature). *Caution:* Ethidium bromide is a potential carcinogen, wear disposable plastic gloves, and *see* Note 15.
3. Tubes (plus cap assemblies and paraffin oil for eliminating air spaces, if required), rotor, and access to ultracentrifuge to permit centrifugation at approx 200,000g (*see* Note 30).
4. Disposable plastic syringes and needles (e.g., gage 19) for DNA band extraction.
5. Sterile 1.5-microfuge tubes for collecting gradient fractions and for DNA precipitation, and micropipet tips.
6. Long wavelength UV source and UV-protective goggles or visor, in case the DNA bands are not readily visible in ordinary light.
7. 1-Butanol (water-saturated).
8. Dialysis tubing (10,000M wt cutoff), pretreated by boiling in 10 mM EDTA for 15 min, followed by two 15-min treatments in boiling distilled water. Store tubing, e.g., in sterile 1 mM EDTA (pH 8.0) at 4°C, and handle with gloves.
9. TE, pH 8.0:10 mM Tris-HCl, pH 8.0, 1 mM EDTA.
10. Absolute ethanol (ice cold or −20°C).
11. 5M ammonium acetate, pH 5.2.
12. 70% v/v ethanol (ice cold).

2.13. Transformation of Yeast

1. Suitable yeast strain (*see* Note 32).
2. Growth media
   a. YEpD: 1% w/v yeast extract, 1% w/v bactopeptone, 2% w/v glucose.
   b. SM: Selective medium, i.e., 0.67% w/v Difco (Detroit, MI) yeast nitrogen base *without* amino acids, 2% glucose, supplements—as required to permit only the growth of host yeast cells expressing the vector marker gene (for concentrations of the various supplements used in yeast "genetic" media, *see* Chapter 16). Solidify with 2% w/v agar.
3. Hemacytometer, or cuvets, if estimating cell density spectrophotometrically.
4. Sterile plastic 50-mL centrifuge tubes.
5. TE, pH 8.0: 10 mM Tris, pH 8.0, 1 mM EDTA, sterile.
6. 0.2M lithium acetate, sterile.
7. Sterile 1.5-mL microfuge tubes and tips.
8. 70% v/v PEG 4000, sterile.
9. Distilled water, sterile.

2.14. Isolation of Desired Clones

The precise materials required will be specific to the particular gene of interest, e.g., appropriate supplemented minimal plates that will select for the expression of the wild-type allele of a metabolite biosynthetic gene, or plates containing as sole carbon source a substrate that can be utilized by the genome donor, but not the yeast host strain.

2.15. Isolation of Plasmid DNA from S. cerevisiae

1. Extraction medium: $2.5M$ LiCl, 50 mM Tris-HCl, pH 8.0, 4% v/v Triton X-100, 62.5 mM EDTA.
2. Phenol:chloroform:isoamyl alcohol (25:24:1) (*see* Section 2.5.8.).
3. Glass beads (0.45–0.50 mm).
4. Sterile 1.5-mL microfuge tubes and tips.
5. Absolute ethanol (ice cold or –20°C).
6. 5M ammonium acetate, pH 5.2.
7. TE, pH 8.0.

3. Methods

Unless otherwise stated, all operations are at room temperature. In general, it is best (although not always absolutely essential) for those undertaking DNA experiments to work aseptically, using sterile reagents and materials, and disposable plastic gloves: This minimizes contamination with both living microorganisms and particles containing DNase activity.

3.1. Isolation of Genomic DNA (see Note 1)

3.1.1. Preparation of Protoplasts

1. Grow a culture of the chosen yeast strain to early to midstationary phase (*see* Note 2) in 1.5-YEpD medium.
2. Harvest by centrifugation in preweighed centrifuge bottles in a prechilled (4°C) rotor, at 3500g for 5 min. (*see* Note 3).
3. Discard the spent culture medium supernatants and reweigh the bottles in order to determine the cell pellet weights.
4. Resuspend the cell pellets in prechilled (4°C) 0.05M EDTA, pH 7.5, using 30 mL per 10 g (wet weight) of cell pellet. At this stage, the cell suspensions are most conveniently handled in (sterile) 50-mL plastic screwcapped centrifuge tubes.
5. Harvest each resuspended 10 g cell pellet as in step 2.
6. Resuspend each cell pellet with 20 mL dH$_2$O plus 2 mL 0.5M EDTA, pH 9.0, plus 0.5 mL β-mercaptoethanol (all reagents at room temperature), and allow to stand 15 min at room temperature, with occasional resuspension/agitation by hand.

7. Again harvest as in step 2 and resuspend in 2 mL protoplasting solution (at room temperature) per gram (wet weight) of cells.
8. Add a suitable cell-wall-digesting (β-glucanase-containing) enzyme preparation to an appropriate final concentration, e.g., 0.5 mg Novozyme 234 per mL cell suspension (*see* Note 4).
9. Incubate the cell suspension at 37°C until protoplasting is complete, as judged by microscopy; this usually takes a few (2–6) hours (*see* Note 5).
10. Harvest the protoplasts by centrifuging in a prechilled (4°C) rotor at 2800g for 10 min at 4°C (the g force is reduced to minimize possible damage to the protoplasts, which are more fragile than cells). Take care when decanting the supernatant, as the pellet is sloppy.
11. Wash the protoplasts by resuspending the pellet in approx 30 mL chilled (4°C) protoplasting solution and harvesting as in step 10. If desired, the procedure can be interrupted at this point, and the washed protoplast pellet can be stored for several days at –20°C *(13)*.

3.1.2. Lysis of Protoplasts

1. Resuspend the pelleted protoplasts from Section 3.1.1., step 11, in saline-EDTA (at room temperature), using 2 mL/g *initial* wet weight of cells, in a 50-mL plastic screw-capped centrifuge tube.
2. Add proteinase K to a final concentration of 50–100 g/mL from a stock solution (20 mg/mL, freshly made up in saline-EDTA).
3. Add SDS to a final concentration of 1% (w/v), from a stock solution (20% w/v).
4. Incubate at 37°C (statically), with occasional (approx 10-min intervals) manual inversion for 3–4 h, until all protoplasts have lysed, checking by microscopy.
5. Incubate the lysate in a water bath at 60°C for 30 min statically (mixing as in step 4) and then cool to room temperature.
6. Dilute the lysate with an equal volume of saline-EDTA. This step reduces the concentration of solutes in the lysate and so reduces interfacial precipitation and possible DNA-trapping during chloroform extraction in Section 3.1.3.; also, a less viscous lysate is easier to handle.

3.1.3. Isolation of DNA from Lysate

1. Carefully mix the lysate with an equal volume of chloroform (containing 4% v/v isoamyl alcohol) until a homogeneous white emulsion is formed. (It is best to use glass-stoppered glass vessels for this.) Mix by gentle swirling or inversion, but *not too violently,* as this will shear the DNA and make spooling of the ethanol-precipitated DNA (*see* step 5) more difficult.
2. Centrifuge the emulsion (using chloroform-resistant centrifuge bottles or tubes at 12,000g) for 20 min at room temperature.

3. Aspirate the upper aqueous phase into a glass beaker capable of receiving another 2 vol of liquid. Take care not to transfer precipitated material at the interface. Make an approximate estimate of the volume of the aqueous layer. Discard the lower, chloroform, layer (*see* Note 6).
4. Carefully layer 2 vol of ice-cold (or −20°C) ethanol onto the aqueous phase, so that a discrete upper phase of ethanol is formed (this can be done by gently dribbling in the first 20 mL or so of ethanol from a pipet with the tip close to the aqueous meniscus, but directed toward the beaker's wall. Subsequent ethanol additions can be made more rapidly).
5. Using a glass rod, stir the interfacial region:macromolecular material, including DNA, precipitates as sticky fibers that wind ("spool") around the base of the rod. Continue stirring until both phases are completely mixed (*see* Note 7).
6. Dissolve the ethanolic precipitate in 15–20 mL of 0.1X SSC, assisting with occasional manual agitation. This can be a slow process, and it may be convenient to leave the preparation overnight, at 4°C, at this stage.
7. Adjust the dissolved precipitate to 1X SSC, using a 10X SSC stock solution.
8. Add pancreatic RNase A and T1 RNase to final concentrations of 40–50 µg/mL and 50–100 U/mL, respectively.
9. Incubate the preparation statically for 1 h at 37°C.
10. Repeat the chloroform extraction and DNA precipitation, i.e., steps 1–5.
11. Dissolve the ethanolic precipitate in 5 mL of 0.1X SSC. We find that DNA at this stage is suitable for genomic library construction, but, if deemed necessary, repeat cycles of RNase treatment and deproteinization by chloroform extraction can be carried out; to eliminate possible polysaccharide (usually glycogen) contamination, α-amylase (DNase-free!) can be used at approx 0.03 U/mL during an RNase digestion step *(13)*. Density gradient centrifugation in CsCl (for example, using Section 3.12., but omitting ethidium bromide) can also further purify DNA, as it should band separately from contaminating protein, RNA, or polysaccharide components.

3.2. Partial Restriction Enzyme Digestion of Genomic DNA

3.2.1. Pilot Digestion

1. Statically incubate 10 µg of genomic DNA with 1 U or less of a chosen restriction enzyme (*see* Note 8) in a total volume of 100 µL of appropriate buffer (*see* Note 9) in a stoppered 1.5-mL conical plastic microcentrifuge ("Eppendorf") tube at 37°C.
2. At zero time and at short (e.g., 2–5 min) intervals after starting the incubation, remove a 10-µL sample and stop digestion by adding 1 µL 0.5*M* EDTA, pH 8.0, and heating to 70°C for 15 min.

3. Prepare a suitable agarose gel for analyzing all the samples and corunning DNA size markers (*see* Note 10).
4. Place the gel, on its tray, in the electrophoresis tank, add sufficient electrophoresis buffer (TBE) to cover the gel to a depth of 1–2 mm, and gently remove the slot-former, taking care to avoid damaging the wells.
5. Add stock loading buffer to each sample, so that the loading buffer is correctly diluted, the maximum total volume of the sample is somewhat less than the slot volume, and the DNA content is approx 1 µg (*see* Note 11). Having thoroughly mixed the sample (in the microfuge tube) by, e.g., up and down micropipeting, carefully introduce the sample into a well, avoiding, in particular, contamination of neighboring wells: It is helpful to hold the pipet tip just above the well, as this minimizes the risk of damaging the gel with the tip, and the dense blue sample is seen to stream down into the well.
6. Load samples of DNA size markers, approx 1 µg per sample, into wells on each side of the samples to be analyzed (*see* Note 12).
7. Having ensured the tank is safely connected to a power pack with the cathode at the well end, i.e., samples are migrating toward the anode (*see* Note 13), start the electrophoretic "run" using a constant voltage of 3–5 V/cm gel tank length (*see* Note 14).
8. At the end of the run (marker dye at the end of the gel), transfer the gel to a suitable dish containing TBE (e.g., spent electrophoresis buffer) adjusted to 0.5 µg/mL ethidium bromide, using ethidium bromide stock solution. If ethidium bromide is added after gel immersion, ensure the stock solution is mixed in away from the gel or there may be localized high-background staining of the gel. It is sensible to minimize the volume of gel-staining solution as it will be discarded after use. *Caution:* Ethidium bromide is a potential carcinogen; use disposable plastic gloves. After use, ethidium bromide solutions should be disposed of safely (*see* Note 15).
9. Allow the gel to stain for a minimum of 15 min at room temperature. If the gel is overstained (it should not be obviously pink!) it can be destained by immersion in TBE or water for 15–30 min.
10. Transfer the gel onto a transilluminator and record the band patterns photographically (*see* Note 16).
11. Using a calibration curve constructed from the marker DNA fragment data ($\log_{10}$ molecular weight [kbp] vs distance migrated in cm), estimate the size-range of the bulk of fragments from the partial digest of the genomic DNA sample (*see* Note 17). If appropriately sized genomic DNA fragments are not adequately represented in any of the samples, the pilot digestion should be repeated using a different enzyme concentration and/or incubation times.

3.2.2. Scaled-up Digestion of Genomic Insert DNA

1. Scale up from the pilot digestion conditions (*see* Section 2.1.) that give genomic fragments of the desired size and quantity; ensuring that the DNA and restriction enzyme concentrations, in particular, are *exactly* the same as the test digestion. A quantity of DNA, such as 250 µg, should yield sufficient DNA of the correct size for library construction.
2. Stop the digestion at the chosen time by adding 0.5M EDTA, pH 8.0, to a final concentration of 50 mM, and incubating at 70°C for 15 min.
3. Store the DNA on ice or in the refrigerator at 4°C.

3.3. Size Fractionation of DNA on a Sucrose Gradient (see Note 18)

1. Prepare a linear 10–40% w/v sucrose density gradient in an appropriate centrifuge tube (*see* Note 19).
2. Carefully layer the DNA sample on to the sucrose gradient and place the tube in the swing-out rotor, ensuring that the tube/rotor bucket remains vertical and there is no sample spillage.
3. Centrifuge for 24 h at 110,000g (middle of tube) and 20°C, and at the end of the run, allow the rotor to decelerate without braking.
4. Collect the gradient in suitably sized fractions, e.g., 0.5 mL for a 20-mL gradient (*see* Note 20).
5. Take a 10-µL sample from every third fraction, dilute with an equal volume of water, and analyze by agarose gel electrophoresis, with appropriate DNA size markers (*see* Section 3.2.1.). If the midpoint of the genomic fragment smear is at, say, 25% w/v sucrose (i.e., 12.5% after dilution), adjust the marker DNA sample(s) to the same sucrose concentration, to make an approximate adjustment for the retarding effect of the sucrose.
6. Combine all fractions in the correct size range (say, 5–15 kbp) and dialyze against 2 TE, at 4°C, changing the buffer twice.
7. Precipitate the DNA in plastic 1.5-mL microcentrifuge tubes by adding 2.5 vol of ice-cold (or −20°C) ethanol and 0.1 vol of 5M ammonium acetate, pH 5.2, and hold at −20°C for 30 min (or −70°C for 10 min).
8. Recover all precipitated DNA by centrifuging at 16,000g (i.e., 13,000 rpm max speed, in a typical microcentrifuge) at 4°C; carefully pour off supernatant and drain dry by inversion over tissue paper, then complete drying of the pellets using an air current (or evacuated desiccator) for several minutes (but beware if overdried, the pellets can be difficult to dissolve).
9. Dissolve the DNA in a minimal volume of TE and check the concentration by an A_{260} spectrophotometry, measurement store at −20°C (in a freezer without a freeze–thaw cycle).

3.4. Isolation of Purified Vector DNA

Vector (plasmid) DNA should be isolated from a relevant *E. coli* strain following Sections 3.11. and 3.12 (*see* Note 21).

3.5. Treatment of Purified Vector DNA Prior to End-Filling (see Note 22)

1. Completely digest a useful quantity (e.g., 5 µg) of purified vector DNA with the appropriate restriction enzyme (e.g., for YCp50, Sal l, which cuts the vector once, within the tetracycline-resistance gene); check for apparent completeness of digestion by analyzing a small sample using agarose gel electrophoresis (*see* Note 23 and Note 36).
2. Precipitate the DNA with 2.5 vol of ice-cold ethanol and 0.1 vol of $5M$ ammonium acetate, pH 5.2, leaving for 30 min at ice-temperature (or 10 min at $-70°C$) before centrifuging at $16,000g$ for 10 min at $4°C$ (microfuge).
3. Carefully drain-dry and then air-dry the pellet (barely visible) and redissolve the pellet in a precise, suitable (300–400 µL) vol of TE, pH 8.0.
4. Add an equal volume of buffer-saturated phenol to the DNA solution, vortex thoroughly, and centrifuge for 1 min at $16,000g$ to break the emulsion. Carefully aspirate the upper aqueous DNA-containing layer (micropipet) avoiding any phenol, and transfer to a fresh microfuge tube.
5. Re-extract the DNA solution firstly with phenol/chloroform (1:1) and second with chloroform, exactly as in Section 3.5.4.
6. Precipitate the extracted DNA as in Section 3.5.2.
7. Dissolve the DNA in TE, pH 8.0 to a nominal (i.e., assuming there have been no DNA losses) DNA concentration of 1 µg/mL.

3.6. End-Filling of Vector DNA (see Note 24)

1. To the DNA solution, add 5 µL of 10X NTSB, the appropriate nucleotides (e.g., dCTP and dTTP for Sal l-cut vector) to final concentrations of 0.1 mM, and 2-mercaptoethanol, to a final concentration of 0.1 mM; then make up to 49 µL with dH$_2$O.
2. Initiate the reaction by addition of 1 µL of Klenow fragment of DNA polymerase 1 and mixing.
3. Incubate for 10 min at $16°C$ (refrigerated water bath or water bath in cold room).
4. Terminate reaction by adding 0.02 vol of DEPC solution, mixing, and incubating for 10 min at $65°C$.
5. Precipitate DNA by adding 2.5 vol of ice-cold ethanol and 0.1 vol of $5M$ ammonium acetate solution, pH 5.2, and holding 30 min at ice temperature.
6. Centrifuge for 15 min at $16,000g$.
7. Discard the supernatant, drain and air-dry the pellet thoroughly, then redissolve the pellet in TE, pH 8.0, to a concentration of 0.2 µg/µL. The preparation can be stored frozen at this point.

Genomic Yeast DNA Clone Banks

3.7. End-Filling of (Partially Digested Fractionated) Genomic DNA

The procedure followed is exactly as for the vector DNA, with the exception that the two nucleotides used for end-filling are dATP and dGTP (*see* Note 24).

3.8. Ligation of Vector and Genomic DNA

The success of any particular ligation (i.e., in terms of frequency of recombinant molecules) is very difficult to predict, being dependent on the quality of the vector and insert DNAs, the particular method used (if any) to minimize self-ligation of cut vector, and the particular ligation conditions. It is therefore sensible to undertake one or more test ligations, using sub-1 µg quantities (e.g., 50 ng of each DNA) to check ligation efficiency: This is done by transforming the ligated DNA into *E. coli* and examining a sample of plasmids extracted from clones for the frequency and size of inserts (Sections 3.9. and 3.10.). It is obviously also useful as a control to compare the transformation frequencies of uncut, and cut-treated-religated (in the absence of insert DNA) vector: The latter should yield <1% of the transformant frequency of the former. The parameters to alter, in improving ligation efficiency are, essentially, DNA and enzyme concentrations—usually increasing them.

Typically, equal quantities of (treated) vector DNA and insert DNA are used, and approx 5 µg of each should be adequate for a library.

1. Mix equal quantities of end-filled vector DNA and end-filled genomic DNA in a microfuge tube.
2. Extract once with phenol by vortexing vigorously, in the microfuge tube with an equal volume of buffer-saturated phenol, briefly microfuge ($16,000g \times 1$ min) to break the emulsion, and discard the phenol layer.
3. Extract as in step 2, but with phenol/chloroform (1:1) and then with chloroform alone.
4. Precipitate the DNA by adding 2.5 vol of ice-cold ethanol and 0.1 vol $5M$ ammonium acetate, pH 5.2, and holding 30 min at ice temperature (or 10 min at $-70°C$). Microfuge at $16,000g$ for 10 min.
5. Discard the ethanol supernatant, allow the (usually invisible) DNA pellet to air-dry, and dissolve in 8 µL sterile dH_2O.
6. Add 1 µL × 10 ligase buffer, followed by 1 µL T4 DNA ligase. Briefly vortex (do not overvortex, as this may denature the ligase), microfuge

(16,000g × 1 min) to recover the contents to the bottom of the tube, and incubate overnight at 12–16°C (higher temperatures, up to 20°C may be used, although the stability of the hydrogen bonding between the cohesive ends of vector and insert DNAs will be reduced).

There is no need to terminate the ligation reaction.

3.9. Preparation of Bacterial Clone Library by Transformation of E. coli with Ligated DNA

(Note that this procedure requires *thorough* asepsis throughout.)

1. Grow the selected strain of *E. coli* (*see* Note 25) as a small, overnight culture at 37°C, in LB broth.
2. Use 0.5 mL of the overnight culture to inoculate a fresh 50-mL batch of LB; grow with rapid shaking/thorough aeration, at 37°C until an A_{600} of 0.2 has been reached, which is usually within 1.5–2 h.
3. Cool the culture for 10 min on ice, then pellet the cells by centrifugation in a 50-mL sterile plastic centrifuge tube (1600g × 10 min at 4°C).
4. Discard the supernatant, aseptically resuspend the pellet in 20 mL (for 50 mL original culture) ice-cold sterile 0.1M $CaCl_2$, and incubate in ice water for 20 min.
5. Repellet the cells as in step 3, resuspend the pellet in 1 mL (per 50 mL original culture) ice-cold sterile 0.1M $CaCl_2$, and incubate in ice water for a minimum of 3 h (*see* Note 26).
6. Dispense as 100-μL aliquots into sterile plastic 1.5-mL microfuge tubes on ice, for the required number of transformations.
7. Add approx 50 ng of ligated DNA, in a volume of 1–15 μL per tube, and leave on ice for 30 min.
8. "Heat-shock" by incubating at 42°C (temperature control is important at this point) for 45 s and return to ice for 2 min.
9. Add 0.9 mL of LB per tube, transfer to a larger tube or vessel (for better aeration), and incubate in an orbital shaker (200 rpm) at 37°C for 60–90 min.
10. Plate as 100- to 200-μL aliquots onto solid LB medium containing the relevant antibiotic for selection of the transforming DNA, and incubate overnight at 37°C.

3.10. Assessment of the Clone Bank

This is done by analyzing the plasmid DNAs in a sample of transformants (*see* Note 27). Alkaline lysis is a convenient method for extracting plasmids from *E. coli* (*see* Note 29 and ref. *14*).

3.10.1. Alkaline Lysis Extraction of Plasmid DNA

1. Inoculate 5 mL LB medium (in a vessel permitting good aeration, e.g., a McCartney bottle), containing the appropriate antibiotic for plasmid selection, from single colony, and incubate overnight at 37°C with vigorous shaking.
2. Transfer 1.5 mL of the overnight culture to a microcentrifuge tube, centrifuge 16,000g for 1 min at room temperature, and thoroughly resuspend the cell pellet in 100 µL of ice-cold solution ALI by vigorous vortexing.
3. Add 200 µL of freshly prepared solution ALII and thoroughly mix by manual shaking (inversion—do not vortex). Leave at room temperature for 5 min.
4. Add 150 µL of ice-cold solution ALIII, mix by rapid manual inversion, and leave on ice for 5 min.
5. Pellet bacterial debris and chromosomal DNA by centrifugation in a microfuge (16,000g) for 5 min at 4°C, and transfer the supernatant (carefully avoiding any contamination with pellet material) to a fresh microfuge tube.
6. Add an equal volume of phenol/chloroform (1:1) to the supernatant, vortex thoroughly, and separate the phases by centrifuging at 16,000g for 2 min, at room temperature. Carefully transfer the upper aqueous phase to a fresh microfuge tube, avoiding contamination with the interfacial and chloroform layers.
7. Add 2 vol of room-temperature ethanol, vortex, leave 2 min at room temperature, then pellet the precipitated plasmid DNA by (micro)centrifugation at 16,000g for 5 min at room temperature.
8. Gently rinse the surface of the pellet with 70% v/v ethanol (4°C), drain off, and allow the pellet to air dry for 10 min (no visible liquid remaining).
9. Dissolve the plasmid DNA in 50 µL TE, pH 8.0, containing 20 µg/mL DNase-free pancreatic RNase and leave at room temperature for 1 h.

The DNA is now suitable for restriction enzyme digestion.

3.10.2. Analysis of Sample of Clone Bank.

Prepare plasmid DNA from a reasonable sample of, e.g., 25–50 clones from your clone bank, fully digest samples of the DNAs with a restriction enzyme cutting the vector once outside the cloning site, and analyze the DNA fragments by agarose gel electrophoresis, with appropriate size markers (*see* Section 3.2.1.). Deduce the total size of the insert in each case (in interpreting the gels remember that the cloned genomic DNA may well contain sites for the restriction enzyme that you use), then determine the average insert size and so assess your clone bank (*see* Note 28).

3.11. Bulk Isolation of Plasmid DNA from Clone Library

This bulk isolation procedure is essentially a scaled-up version of Section 3.10.1. (*see* Note 29).

1. Grow a sample of pooled *E. coli* clone bank cultures overnight at 37°C in 500 mL LB medium containing the relevant antibiotic for plasmid selection (in a 2-L flask); aerate efficiently, e.g., by shaking in an orbital shaker at 200 rpm.
2. Harvest the cells by centrifuging at 4000*g* for 10 min at 4°C and discard the supernatant.
3. Resuspend the bacterial pellet(s) using a total volume of 40 mL ice-cold STE, transfer to a 50-mL plastic centrifuge tube, and reharvest centrifugally as in step 2.
4. Discard the supernatant and thoroughly resuspend the pellet in 10 mL of ice-cold solution ALI, in which lysozyme has been freshly dissolved to a concentration of 5 mg/mL.
5. Incubate statically at room temperature for 5 min.
6. Add 20 mL freshly prepared solution ALII and mix thoroughly by inversion (this at room temperature).
7. Incubate 10 min on ice.
8. Add 15 mL ice cold solution ALIII, mix thoroughly by inversion, and incubate on ice for an additional 10 min. A flocculent white precipitate of chromosomal DNA, and SDS/protein/membrane complexes should form.
9. Centrifuge at around 50,000*g* (i.e., approx 20,000 rpm using a standard 50-mL angle rotor in a typical preparative centrifuge) to pellet bacterial debris and chromosomal DNA.
10. Filter the supernatant through three to four layers of muslin (cheesecloth) into a plastic centrifuge tube or bottle. Mix with 0.6 vol of isopropanol and leave at room temperature for 15 min.
11. Centrifuge at 12,000*g* for 30 min at room temperature, to pellet the plasmid DNA (a low temperature may cause precipitation of salt).
12. Discard the supernatant, gently rinse the pellet with 70% v/v ethanol, carefully drain dry, and allow to air dry until no ethanol remains.
13. Dissolve the pellet in a small volume (500–1000 µL TE, pH 8.0, or an appropriate larger volume, e.g., 7.5 mL, if subsequently transferring to an ultracentrifuge tube). The DNA may now be pure enough for efficient transformation, or other purposes, but it is probably advisable to purify it further, by isopycnic centrifugation (*see* Note 30).

3.12. Purification of Plasmid Library DNA by Isopycnic Centrifugation

1. Add 1.0 g CsCl/mL DNA solution (i.e., 7.5 g CsCl if following on from Section 3.11.13.), and then add ethidium bromide solution to a concentration of 0.6 mg/mL (i.e., 0.8 mL of 10 mg/mL ethidium bromide solution in 10 mL CsCl DNA solution). *Caution:* Wear disposable plastic gloves when handling ethidium bromide, as it is a potential mutagen/carcinogen. After use, ethidium bromide solutions should be disposed of safely. *See* Note 15.
2. Check the density of the solution by weighing a known volume, and adjust, if necessary (by adding solid CsCl or TE), to a final value of 1.55 g/mL.
3. Transfer the solution to an appropriate ultracentrifuge tube (short, wide tubes give the sharpest bands), capping or sealing as directed by the manufacturer, check the tube balance as required for the particular rotor/centrifuge (a blank tube for balancing will have to contain a comparable dense CsCl solution), and centrifuge at approx 195,000g (typically, approx 45,000 rpm) for 36 h at 20°C.
4. The DNA bands may be visible in ordinary light, otherwise view the tube with long wavelength UV light. *Caution:* Use UV-opaque goggles, or better, a visor, for protection from UV. Two well-defined fluorescent bands should be visible near the center of the tube, separated by about 1 cm. The lower (denser) band contains closed circular plasmid DNA.
5. Adjust the tube cap to allow air to enter, pierce the tube above the lower band using a disposable syringe fitted with a wide-bore (e.g., gage 19) needle, and withdraw the fluorescent plasmid band. (An alternative is to aspirate the upper part of the gradient, including the chromosome DNA, after opening the top of the tube, using a Pasteur pipet, then position the needle of a syringe just below the band and aspirate it.) It is useful to have a helper holding the tube, or to use a clamp. When all the plasmid has been removed, a very thin band may appear to remain—this is merely an optical effect caused by removal of some gradient, resulting in an abrupt change of refractive index.
6. Add an equal volume of water-saturated 1-butanol, mix/vortex vigorously, and separate the emulsion by centrifuging at 16,000g for 3 min. Discard the upper pink-colored organic phase containing extracted ethidium bromide.
7. Repeat the butanol-extraction (step 6) until no more pink color can be removed; to avoid increasing the volume of the DNA solution too much during extractions, it can be useful to alternate between extracting with water-saturated butanol and pure butanol.
8. Dialyze the plasmid DNA solution against two changes of 0.5–1.0 L TE at 4°C. (The dialysis tubing should be pretreated by boiling in 10 mM EDTA for 15 min, followed by two 15-min treatments in boiling distilled water.)

9. Distribute the dialyzed DNA solution as 0.4-mL aliquots in 1.5-mL microfuge tubes, add 0.1 vol 5M ammonium acetate and 2.5 vol ice-cold ethanol to each, hold 10 min on ice, and centrifuge at 16,000g for 15 min.
10. For each tube, discard the supernatant, gently rinse the pellet with 70% v/v ethanol, carefully drain dry, and dry the pellet for an additional 5 min or so under vacuum or in an air current until no traces of ethanol remain.
11. Dissolve each pellet in TE, pH 8.0, pool and estimate the DNA concentration by measuring the absorbance at 260 nm (use, e.g., restricted volume quartz cuvets). Measure also the A_{280} of the preparation. If the A_{260}:A_{280} ratio is less than 1.7, the DNA is impure, and the organic solvent extractions and ethanol precipitation should be repeated. The DNA can be stored frozen—in several aliquots, for safety—at $-70°C$ or at $-20°C$ in a nonfreeze-thaw cycle freezer.

3.13. Transformation of Yeast with Library DNA (see Note 31)

Note that this procedure requires *thorough* asepsis throughout.

1. Inoculate the recipient yeast strain (*see* Note 32) from a slope or plate into 100 mL YEPD medium in a 500-mL flask and grow aerobically, with shaking, at 28°C (assuming no temperature-sensitive mutations are present) until the cell density reaches between 5×10^7 and 1×10^8 per mL (mid-late log phase—use a hemacytometer); this usually takes 15–20 h. (An alternative to using a hemacytometer is to determine the strain's growth curve spectrophotometrically [A_{600} vs time] in YEPD in a prior experiment, then to grow the culture to be used for transformation to an A_{600} known to be in the mid-late log range.)
2. Harvest the cells *aseptically* by centrifugation (approx 1600g for 10 min at 4°C) and resuspend and combine the cell pellets to a total volume of approx 50 mL, using sterile TE, in a sterile plastic 50-mL centrifuge tube.
3. Repellet the cells centrifugally as in step 2 and again resuspend in approx 50 mL sterile TE.
4. Repellet the cells centrifugally as described, and resuspend with sterile TE, pH 8.0, to a cell density of between 1×10^9 and 4×10^9 cells/mL (i.e., total volume approx 10 mL).
5. Transfer, aseptically, 0.5 mL to a fresh sterile 50-mL plastic centrifuge tube and mix with 0.5 mL sterile 0.2M lithium acetate.
6. Incubate the suspension at 28°C in an orbital shaker at 140 rpm for 1 hr.
7. For each transformation reaction/plate (at a typical conservative efficiency of 2×10^3 transformants per µg, and with an insert size of, e.g., approx 7 kb and a yeast genome size of 14 Mbp, 10 transformations, i.e., 20,000 clones, should yield an adequately sized library in yeast), aseptically transfer 0.1 mL of the cell suspension to a sterile 1.5-mL microfuge tube.

Genomic Yeast DNA Clone Banks

8. To each transformation tube, add 1–5 µg library DNA, mixing gently by swirling/tapping.
9. Incubate statically at 28°C for 30 min, mixing gently every 10 min.
10. Aseptically add 0.1 mL sterile 70% v/v PEG 4000, mix by vortexing, and incubate for 1 h, as in step 9.
11. Heat shock by incubating statically at 42°C for 5 min, then immediately to room temperature.
12. Pellet the cells centrifugally ($16,000g \times 1$ min), resuspend in 1-mL sterile distilled water and pellet again.
13. Repeat the mentioned washing step twice, finally resuspending in 1 mL sterile dH_2O.
14. Plate the tube contents onto approx five plates of SM-medium that only permits the growth of transformants (i.e., selects for expression of the yeast marker gene on the vector).
15. Incubate at 28°C: Pin-prick colonies appear in approx 2 d and full-sized colonies should be visible in 3–5 d.
16. If the yeast clone bank size is inadequate, carry out further yeast transformation using the plasmid DNA bank until an adequate collection of yeast transformants has been obtained. The bank can be used immediately to isolate genes of interest, but the sample should be stored frozen at –70°C in glycerol (final concentration 15% v/v), so that replica libraries can be recovered when required by replating on selective medium.

3.14. Isolation of Desired Clones

1. Arrange to plate a representative yeast library sample (e.g., a few million cells in total) on medium selecting (or allowing screening) for the gene of interest.
2. Further test a small representative group of clones of interest to ensure they are likely to contain the desired gene(s) (*see* Note 33). Plasmid DNA can now be extracted from the purified yeast clones, so enabling a large range of molecular genetic studies to be undertaken on the gene(s).

3.15. Isolation of Plasmid DNA from S. cerevisiae

The rapid, small-scale method given herein is based on that of Ward *(15)*. It is intended to provide plasmid DNA adequate for immediate transformation back into *E. coli* from which successive isolations can then be made readily.

1. Suspend a single colony (2–3 mm diameter) of the yeast strain of interest in 100 µL extraction medium, in a microfuge tube.
2. Add 100 µL of 1:1 (v/v) phenol:chloroform and 0.2 g glass beads (0.45–0.50 mm).

3. Vortex vigorously for 2 min.
4. Centrifuge 16,000g (microfuge) for 1 min.
5. Remove the supernatant to a fresh microfuge tube, add 2.5 vol of ice-cold ethanol and 0.1 vol of 5M ammonium acetate (pH 5.2), hold at ice temperature for 30 min, and centrifuge at 16,000g for 15 mins.
6. Discard the supernatant and air dry the pellet until no traces of ethanol are evident.
7. Dissolve the precipitated DNA in 50 µL TE (several hours at 4°C, for safety).
8. Use 5 µL of this DNA solution to transform *E. coli* by Section 3.9: It is much easier to recover plasmid DNA in high yield from *E. coli* than from yeast.

The cloned gene will then be available for verification (*see* Note 33), subcloning, restriction-mapping, sequencing, and a variety of other types of analysis, such as in vitro mutagenesis to investigate key control or catalytic sequences, construction of highly expressed clones for possible commercial reasons, and so on.

4. Notes

1. This method is based on that developed by Cryer, Eccleshall, and Marmur *(13)* for *S. cerevisiae*. Different yeast strains/species may well present particular problems, necessitating some modifications of the method. Some yeasts may be especially difficult to protoplast, presumably because of the nature of their cell walls, though a large variety of yeasts respond quite well to commercially available lytic enzyme preparations based on β-glucanases.

   Another range of problems may be presented by particular cell components, notably storage carbohydrates and lipids. Polysaccharides can be coprecipitated with nucleic acids by ethanol, and are not easily eliminated by chemical extraction or precipitation; treatment with (DNase-free!) carbohydrase preparations should hydrolyze polysaccharides to sugars, which can then be removed from DNA by dialysis. Removal of lipid contaminants may be accomplished by repetitive extractions with chloroform (Section 1.3.1.).

2. As microbial cell walls are generally thinner and more delicate in log-phase than in stationary phase, it is important to ensure that the culture of the strain in question is still rapidly growing (ideally, mid-to-late log phase to maximize cell yield) when exposed to the cell wall-degrading enzymes. It may be necessary to define growth conditions in a prior, small-scale experiment in which the growth curve is determined by plotting cell number (A_{600} is usually adequate) vs time.

3. Chilling is important at this stage to arrest growth and prevent progression into stationary phase. Some yeasts, e.g., those with a high fat content may

need more g min for good centrifugal pelleting. Asepsis is not necessary in the DNA isolation procedure, though minimizing possible microbial and DNase contamination is obviously good practice, and use of the widely available pre-sterilized disposable plasticware is convenient.

4. Suitable cell wall-digesting preparations include lyticase (from *Arthrobacter luteus*), available from a variety of suppliers, and Novozym 235, available only from Novo-Nordisc.
5. Protoplasts may be evident as spherical cells, if the intact yeast cells normally have some other shape. The presence of protoplasts can be confirmed by diluting a small sample of the suspension (e.g., 10 µL) into (e.g., 200 µL of) 1% w/v SDS and leaving for a few minutes before viewing in the microscope (×400 magnification is adequate): Protoplasts should have burst, releasing sub-cellular debris.
6. Certain cell components may cause a problem at this stage. In the case of the lipid-accumulating soil yeast *Lipomyces starkeyi (16)*, the cell homogenate can—in later growth stages—contain an oily, yellow substance that requires repetitive chloroform extractions for removal. In this case, the lipid contaminant was also much reduced by harvesting the cells somewhat earlier in exponential phase.
7. If desired, unspooled material can be harvested by centrifugation at 2000*g* for 1 min at room temperature, but this is not worthwhile if substantial quantities of precipitate have spooled, as the unspooled material tends to be low in DNA content (and the DNA may well be of low molecular weight).
8. The choice of restriction enzyme is dictated partly by the method chosen for attachment to the vector—which usually exploits compatible sticky ends, e.g., *Sau*3A—cut genomic DNA with *Bam*Hl-cut vector, or *Sau*3A-cut partially filled-in genomic DNA with *Sal*I-cut partially end-filled vector (*see* the following). Enzymes, such as *Sau*3A, which recognize and cut at tetranucleotide sites, are usually used for restricting genomic DNA, as the sites are very frequent, so the fragments produced by partial digestion will approximate to a random selection from the genome. Restriction enzymes are stable when stored at −20°C in a buffer containing 50% v/v glycerol. When carrying out restriction enzyme digestions, reactions should be prepared so that all reagents are present except the enzyme. The enzyme stock should be taken from the freezer immediately before use and placed *on ice*, appropriate volumes (usually 1 µL or less) should be dispensed rapidly to each reaction tube, and the restriction enzyme should be returned promptly to the freezer. Fresh sterile pipet tips should be used each time enzyme is dispensed to avoid any possibility of contamination of the enzyme stock.

9. Buffers for different restriction enzymes differ mainly in the concentration of sodium chloride they contain. Restriction enzymes are adequately active in one of thee different buffers: The appropriate one will be indicated by the enzyme manufacturer (or given in, e.g., refs. *17* and *18*), and you can prepare this if you wish. Prepare as a 10X stock and store at –20°C (concentrations in the following are for the X10 stock solutions).
   a. Low salt: 100 mM Tris-HCl, pH 7.5, 100 mM MgCl$_2$, 10 mM dithiothreitol.
   b. Medium salt: 0.5M NaCl, 100 mM Tris-HCl, pH 7.5, 100 mM MgCl$_2$, 10 mM dithiothreitol.
   c. High salt: 1M NaCl, 0.5M Tris-HCl, pH 7.5, 100 mM MgCl$_2$, 10 mM dithiothreitol.

   A single buffer (KGB), which can be used for virtually all restriction enzymes, has been devised by McClelland and colleagues *(19)*. Working concentration for different enzymes are given in *(17)*.

   2X KGB is: 200 mM potassium glutamate, 50 mM Tris-acetate, pH 7.5, 20 mM magnesium acetate, 100 µg/mL BSA (Fraction *V*, Sigma), 1 mM β-mercaptoethanol.

10. This type of analysis is conveniently carried out in a submarine minigel system. To make the gel, prepare an appropriate volume of (in this case) 1% w/v agarose (DNA electrophoresis grade) in electrophoresis buffer (TBE) and boil until all agarose grains have melted, producing a transparent, homogeneous solution (this is conveniently accomplished in a microwave oven). Cool the agarose solution to approx 60°C and pour into the gel tray (in some systems, the gel is cast directly in the tank), the two short ends being temporarily sealed with tape (e.g., autoclave tape) and with the slot-former (or "comb"; typically giving 6–12 sample wells of 8–20-µL vol) in place (check that no air bubbles are trapped by the teeth of the comb). The gel is set and ready for use within about 30 min, though gels can be stored for some days at 4°C, provided they are protected from dehydration. Obviously, tapes should be removed before placing the gel in the tank.

    Approximate separation ranges for different strength agarose gels are:

Agarose concentration in gel (% w/v)	Efficient separation range for linear DNA molecules (kbp)
0.3	60–5
0.6	20–1
0.7	10–0.8
0.9	7–0.5
1.2	6–0.4
1.5	4–0.2
2.0	3–0.1

Genomic Yeast DNA Clone Banks

It should be noted that low agarose gels (0.5% w/v) are soft and difficult to handle and are prone to electrical heating, which may cause band distortion.

Tris-acetate (TAE) and Tris-phosphate (TPE) buffers may also be used for gel electrophoresis of DNA, instead of Tris-borate (TBE), but TAE, although cheaper, has poorer buffer capacity than the other two. TBE is by far the most widely used and it is now realized that a working strength of 0.5X (0.045M Tris borate, 0.00125M EDTA) provides adequate buffering capacity for almost all types of agarose gel electrophoresis of DNA *(17)*.

11. The maximum sample volume that ran be introduced into a well obviously depends on the dimensions of the "teeth" of the slot-former and the depth of the gel—as this, typically, determines the height of the sample slot/well. Check that you know the well volumes, for a particular system, before loading, and remember to allow for the addition of loading buffer to samples before application (usually this increases the sample volume by one-fifth). A typical slot (5 mm × 5 mm × 1.5 mm) will hold about 37.5 µL. Gel-loading "buffers" increase sample density and give color, to facilitate easy loading, and provide a visible marker to track the progress of the electrophoresis: In 0.5X TBE, bromophenol blue migrates at about the same rate as a linear 300 bp DNA fragment. There are several different gel-loading solutions, including dense solutes such as Ficoll, and dyes such as xylene cyanol*(17)*, and choice is largely a matter of personal preference.

    The maximum amount of DNA that can be applied to a well, without impairing resolution, depends on the number and size of the fragments in the sample. For a 3-mm wide well the maximum amount of DNA, which may be separated as a single band without smearing or trailing, is about 50 ng; such a well can therefore be used for 50–500 ng of a simple population of DNA fragments. However, if the sample consists of many fragments of different sizes (e.g., restriction digests of whole genomes), up to 20 µg per slot can be loaded without serious loss of resolution.

    The highest detection sensitivity achievable by photography of ethidium bromide stained gels is about 2 ng in a band.

12. DNA size markers covering the size range from whole yeast chromosomes (a few Mbp) to small fragments (<10 bp) are now commercially available. One of the restriction enzyme digests of phage λ DNA is suitable for assessing partial digests of genomic yeast DNA, e.g., the *Hin*dIII digest (eight fragments of sizes 23,130, 9416, 6557, 4361, 2322, 2027, 564, and 125; the two smallest fragments are likely to be quite faint).

13. The pH of running buffers for DNA electrophoresis are such that the phosphodiester backbone hydroxyls are highly dissociated, and DNA fragments have a strong, size-dependent, net negative charge.

14. The electrophoretic mobility of linear DNA fragments is only proportional to molecular weight (kbp) at low voltages, the mobility of large fragments increasing disproportionately at high voltages. Therefore, to maximize resolution of fragments greater than 2 kbp, the voltage gradient in an agarose gel should not be more than 5 V/cm.

    Remember, in calculating the voltage gradient, the linear distance *between the electrodes* must be used. TBE is sufficiently strong a buffer that recirculation between cathode and anode compartments is not necessary. If the buffer gets excessively hot (say >50°C), either the electrical conditions are incorrect or the buffer has been wrongly prepared: "hot-running," although fast, will cause band distortion and poor resolution.

15. As ethidium bromide is a potential carcinogen, solutions containing this compound should be handled with care (use disposable gloves) and solutions containing ethidium bromide to be discarded should either be incinerated (ethidium bromide decomposes at 262°C) or decontaminated prior to disposal. Dilute (approx 1 μg/1 mL) solutions can be decontaminated by the following procedure (using gloves throughout the procedure):
    a. Add 100 mg activated charcoal per 100 mL staining solution.
    b. Mix well and leave at room temperature, mixing occasionally.
    c. Filter through a Whatman (Maidstone, UK) No. 1 filter and discard the filtrate down the sink.
    d. Seal the filter and activated charcoal in a plastic bag and dispose of in the hazardous waste.

16. DNA in gels is visualized by the orange-red fluorescence emitted when DNA–ethidium bromide complexes are irradiated with UV light. The easiest way to view a gel is using a transilluminator—a UV light box with a built in UV-transmission-specific filter on which the stained gel is laid. Transilluminators are commercially available and mostly emit light at 302 nm. (Lower wavelengths cause much greater nicking of DNA and at higher wavelengths the intensity of the fluorescence is reduced.)

    Caution: Ethidium bromide-stained gels should be handled with gloves and UV-opaque goggles (or, preferably a face mask) should be worn when the transilluminator is on (not all laboratory safety goggles are UV-protective!).

    It is normal to record gel results by photography. Conventional (e.g., Ilford, Essex, UK, FP4) 35-mm film in a mounted 35-mm camera, or sheet film in an enlarger, is very flexible in terms of developing and printing and can give very good results, but Polaroid film, although more expensive, is quicker and easier. Various Polaroid cameras are available that can be mounted in an enlarger system or are mounted in a cone device (with filter already in place) that is designed for photographing gels and sits over the gel, on the transilluminator; all cameras must be used with an orange-red filter (e.g., Wratten 22A). Suitable Polaroid films are 667 (no negative), which typically requires a 0.5–1 sec exposure and 45 s for development, and 665 (yields a negative) exposure of 60–75 s and 45 s for development.

17. For the DNA markers, plotting distance migrated (cm) against $\log_{10}$ kbp should yield a largely linear calibration curve that can be used to size the genomic digest, seen as a smear. A useful size range of fragments for a yeast genomic library is, e.g., 5–15 kbp. For comments on library/clone bank size, *see* Note 28.
18. The fractionation method detailed in this protocol is rate centrifugation on a sucrose density gradient, and this has been selected because DNA recovered from sucrose gradients is readily ligatable. An alternative approach is to fractionate the DNA by agarose gel electrophoresis, recovering DNA in the appropriate size range by techniques such as electroelution or solvent extraction of the melted (low melting point agarose) gel *(20)*. This approach has the drawback that the recovered DNA is not always readily ligatable, probably because of polysaccharide contaminants extracted from the agarose or the agarose itself, but many recently developed commercial "clean-up" kits (e.g., Geneclean) are overcoming this problem.
19. Fractionation by rate centrifugation on a gradient should be conducted in a swing-out rotor, therefore allowing the centrifuge tube and contents to be perpendicular to the axis of rotation: This minimizes wall effects and so optimizes resolution. Carefully pipet into the upright centrifuge tube successive layers of ice-cold sucrose solutions (all w/v in TE) in the indicated ratios: 40% (added first, to the bottom of the tube, 1 vol), 30% (2 vol), 20% (2 vol), 10% (1 vol). Ensure that the interfaces between the sucrose solutions are initially sharp and distinct, then place the tube, secured in a vertical position, in a cold (4°C) vibration-free environment and leave overnight. Diffusion generates a linear 10–40% sucrose density gradient. In our experience, the diffusion method produces an adequate and surprisingly linear density gradient. Gradients can also be generated by commercial or homemade gradient makers. The total gradient volume should be such as to provide reasonable resolution of DNA fragments loaded onto the gradient in a particular volume. We have found that a 0.25-mL sample volume (containing 250 µg DNA) loaded onto a 20-mL gradient, in a centrifuge tube of approximate dimensions 13 cm long × 2 cm diameter (i.e., fits the bucket of the 3 × 20 mL swing-out rotor of an MSE Superspeed 50 centrifuge) worked well.
20. A convenient method for gradient collection is bottom displacement: 50% w/v sucrose can be introduced into the bottom of the centrifuge tube, e.g., via a thin steel pipe secured in a bung, which also has a short exit pipe, which fits the top of the tube: The pipe is connected to a reservoir of 50% sucrose via plastic tubing routed through a peristaltic pump, and the pipe is filled to the tip with sucrose solution. The tube, mounted vertically, e.g., on a lab-

jack, is slowly elevated over the pipe, minimizing disturbance to the gradient, until the pipe just reaches the tube base and the bung is seated in the top of the tube. Fifty percent sucrose is then introduced into the tube bottom, at a rate of about 1 mL/min or less, slowly displacing the top of the gradient through the short exit pipe and via a plastic tube to the fraction collection tubes, e.g., microfuge tubes. The fractions can be conveniently collected manually. Although the gradient could be passed through a flow-cell to monitor the gradient absorbance at 260 nm and so localize the DNA, this is not necessary as the fractions will need to be checked out, anyway, by agarose gel electrophoresis, and the UV irradiation could cause some damage to the DNA.

21. This is not the place for an extended discussion of yeast vectors, although some general points are given, and references are cited, in the Introduction. The particular joining method described here was chosen for use with the centrometric plasmid shuttle vector YCp50, but is broadly applicable.

22. The prime concern in designing a ligation strategy is to maximize the yield of recombinant vectors containing single fragments of genomic DNA and to minimize the number of clones containing vector DNA without genomic inserts. In one approach, alkaline phosphatase is used to remove 5' phosphoryl groups from the cut vector DNA, and so it prevents self-ligation *(21)*, but problems in getting reasonable recombinant yields when using dephosphorylated (vector) DNA in ligations are not uncommon. An alternative approach has therefore been suggested here, in which the cut ends of vector and genomic DNAs are partially filled in, using Klenow enzyme, so that they are incapable of self-ligation, but are capable of cohering and ligating with each other.

23. Even after prolonged digestion with excess enzyme, significant numbers of plasmid molecules may remain uncut, although invisible on an agarose gel, and contribute to "background" transformants that lack inserts, hence the usefulness of being able to select or screen for recombinants.

24. In the case of *Sal*I-cut YCp50 DNA, end-filling is with dTTP and dCTP, and the genomic insert DNA is end-filled using dATP and dGTP:

	Restriction site cohesive end		Partially filled sequence	
*Sal*I-cut vectorDNA	5'G 3'CAGCT	TCGAC 3' G 5'	5'**GTC** 3'CAGCT	TCGAC 3' **CTG** 5'
*Sau*3A-cut insert DNA	5'N 3'N'CTAG	GATCN 3' N' 5'	5'**NGA** 3'CTAG	GATCN 3' **AGN** 5'

It is evident that the partially filled-in DNAs cannot self-ligate, but the modified vector and insert DNAs can ligate with each other.

25. Many strains of *E. coli* suitable for transformation have been developed, e.g., HB101 (highly transformable and useful for plasmid production; *see* ref. *18*); they should be *rec A*$^-$ to prevent recombination between plasmid and chromosomal DNA.
26. At this stage, cells are competent to take up DNA. Competent cells can be conveniently stored by adding sterile glycerol to 25% v/v final concentration, dispensing 125-µL aliquots into sterile plastic 1.5-mL microfuge tubes and keeping frozen at −70°C. When needed for transformation, frozen competent cells should be allowed to thaw in ice water, but cannot then be refrozen and used subsequently.
27. The clone bank will inevitably contain some colonies in which the vector either lacks an insert of genomic DNA, or the insert is too small to contribute to the gene library. It is therefore necessary to assess the clone bank by estimating the average genomic DNA insert per clone. This is conveniently done by using a rapid small scale plasmid DNA isolation method—for example, alkaline lysis—to isolate plasmid DNA from a random selection of, e.g., 25–50 colonies, cutting the DNA using a restriction enzyme that cleaves the vector once, outside the cloning site, and analyzing the products by agarose gel electrophoresis with appropriate DNA size markers. If insertion of genomic DNA causes a selectable or screenable phenotype, then a sample of clones should be checked, to determine the frequency of clones lacking inserts, and only those apparently containing inserts should be used for DNA analysis. In the case of YCp50 and many other yeast-*E. coli* shuttle vectors, insertion of genomic DNA into the *Bam*H1 cloning site inactivates the tetracycline resistance gene, so "toothpick" subculturing of the ampicillin-resistant transformants onto LB-tetracycline plates identifies those that are tetracycline-sensitive and therefore likely to contain inserts.
28. Having established the average size of genomic DNA insert in the clone library, the following formula should be used to check that the library is large enough to give the experimenter a good chance of isolating any particular unique sequence (i.e., sequence occurring only once in the haploid genome) of interest:

$$N = \ln(1 - P)/\ln(1 - 1/n)$$

where N = number of clones, P = probability of library containing a single copy gene, and n = genome size divided by average insert size.

If necessary, the library can be built up to the right size by a series of transformations, or ligation plus transformation experiments. It is not necessary to eliminate clones lacking inserts (e.g., tetracycline-resistant colonies) from the library.

The library can be stored as cells (essential initially, but more problematic for long-term storage, because of loss of viability and risk of contamination) or as DNA (convenient for long-term storage and suitable for transformation/complementation experiments with a variety of different strains/organisms). For cell storage using strict asepsis (e.g., inoculation cabinet), wash transformant colonies off the plates using small volumes of sterile saline (1% w/v NaCl), pool the cell suspensions, adjust them to 15% v/v (final concentration) with sterile glycerol, and store the library as aliquots in sterile tubes at $-70°C$. Replicate tube storage will minimize subjection of the library to freeze–thaw cycles when taking cell samples for experiments.

29. In this method, based on the procedure of Birnboim and Doly *(14)*, the cell walls are initially weakened by EDTA treatment. (Lysozyme also used to digest the cell walls, but certainly at the small scale; this is not now deemed necessary *[17]*.) Lysis is completed by addition of alkaline SDS. Chromosal DNA is extracted as linear fragments that denature at the high pH, and, on rapid neutralization, the single strands form an insoluble network that can be removed by centrifugation. By contrast, intact plasmid molecules, being supercoiled, covalently closed, circular, do not denature into fully separated strands and, on neutralization, rapidly renature, and remain in solution. The high salt concentration precipitates SDS–protein complexes, and so removes much cell protein. Ethanol precipitation further purifies the DNA.

30. Plasmid DNA prepared by the cleared lysis technique will inevitably be significantly contaminated by chromosomal DNA. Further purification can be effected by the classic isopycnic centrifugation technique. In this, ethidium bromide binds to both chromosomal and plasmid DNA by intercalation, but the linear chromosomal DNA is able to accommodate much more ethidium bromide then the topologically constrained covalently closed, circular, plasmid DNA. Since ethidium bromide is less dense than DNA, the chromosomal DNA–ethidium complex has a lower density than the plasmid DNA–ethidium complex, and the two DNA species therefore band at different positions in the cesium chloride density gradient generated by centrifugation.

31. Yeast transformation is detailed in Chapter 16, but the version used in our laboratory's gene library work is also given here. The lithium protocol detailed here is simpler, although less efficient in terms of yeast transformants per μg DNA, than the procedure using protoplasts. A biolistic approach (*see* Chapter 17) may also be considered.

32. Typically, the recipient strain of *S. cerevisiae* will carry a stable recessive mutation in the gene used as a selectable marker in the shuttle vector (e.g., *leu2*, *his3*, *ura3*, *trp1*) as in recipient strain DBY747. Unless the gene(s) of interest in the yeast used for library construction are likely to be dominant or semi-dominant to the wild-type allele in the host, and confer a select-

able or screenable phenotype (e.g., production of a particular carbohydrase allowing growth of the transformant on a carbon source that cannot be utilized by the recipient stain), the recipient yeast strain obviously should also carry recessive mutations in genes of interest, allowing selection of the genes by complementation. It should be realized that genes from yeasts distantly related to *S. cerevisiae* may not express well or at all in *S. cerevisiae*, mainly because of different transcription signals (through evolutionary divergence). Assuming no homologous transformation system is available, a cDNA/expression vector approach may be necessary, and this could allow the use of a yeast host to be entirely bypassed. However, a number of genes, especially auxotrophic markers, may be directly selectable in *E. coli*. It is also possible for heterologous yeast genes to express in *E. coli* but *not* in *S. cerevisiae*!

33. First, the clone should be tested for the presence of the host markers not complemented by the vector, to exclude contaminants. The desired phenotypes should also, under selection, be convincingly and stably expressed and maintained in successive subcultures. With the less stable vectors, e.g., YRp and, to some extent, YEp, the vector is lost at a significant frequency during mitotic proliferation under nonselective conditions: Co-loss of vector marker and phenotype of interest therefore strongly indicate successful cloning (rather than, e.g., some kind of reversion event). However, the most important indication of successful cloning is the ability to retransform the host yeast strain to the desired phenotype, with high efficiency, using vector DNA originally isolated from the initial yeast transformants of interest.

34. Commercially available liquefied phenol that is clear and colorless can be used without redistillation (crystalline phenol contains DNA-damaging oxidation products, and needs to be redistilled). The phenol must first be equilibrated with buffer of a pH >7.8, because in acid conditions DNA will partition into the organic phase *(17)*. Liquefied phenol can be stored at –20°C, at which temperature it solidifies, so samples for equilibration should first be allowed to warm to room temperature and then melted in a water bath at 68°C. (These and subsequent operations to be conducted in a fume hood!) Next, add hydroxyquinoline to a final concentration of 0.1% v/v. As an antioxidant, hydroxyquinoline preserves the useful life of the phenol reagent, and its yellow color allows the phenol phase to be recognized easily during extractions of DNA samples. Add an equal volume of $0.5M$ Tris-HCl, pH 8.0, and stir thoroughly (magnetic stirrer) for 15 min at room temperature. Stop stirring, allow the phases to separate, and aspirate (safety pipet), and discard as much of the upper phase as possible. Repeat the extractions, but using $0.1M$ Tris-HCl, pH 8.0, until the pH of the phenolic phase is >7.8 (pH paper). After removal of the final aqueous phase, add

0.1 vol of 0.1M Tris-HCl, pH 8.0, containing 0.2% v/v β-mercaptoethanol (as a further antioxidant): In a light-excluding bottle at 4°C, this reagent is stable for up to 1 mo.

35. Tris-HCl-based buffers and EDTA are constantly used in DNA work. It is convenient to have stock sterile solutions available. A stock 1M Tris solution should be adjusted to the required pH by adding concentrated HCl, the pH being measured at room temperature, as Tris pH is temperature dependent. The stock should be dispensed into aliquots, sterilized by autoclaving, and can be stored at room temperature. A 0.5M EDTA, pH 8.0, stock can be made by adjusting the pH with NaOH pellets (EDTA at this molarity will not go into solution fully until the pH is approx 8.0); aliquot, sterilize by autoclaving, and store at room temperature.

36. It may not be necessary to purify the DNA after restriction, and prior to end filling, as the Klenow fragment of *E. coli* polymerase I works well in virtually all restriction enzyme buffers *(17)*, but the authors have followed the practice of thoroughly purifying the DNA, i.e., Section 3.5., steps 2–7.

37. For comparing reactant molarities, it may be useful to know that 1 kbp of double-stranded DNA at a concentration of 50 μg/1mL (i.e., $A_{260} = 1$) contains 4.74×10^{13} molecules/mL, is 78.7 nM, and the molar concentration of termini is 157.4 nM. This should help, if desired, in estimating molarities of yeast vectors (usually 5–10 kbp) or genomic DNA of a known average size.

38. The method detailed here exploits continuous CsCl density gradients that, in angle rotors, typically require 36–48 h to generate. The more expensive vertical rotors, now available from well-known manufacturers, such as Beckman (Fullerton, CA) and Sorvall-DuPont (Wilmington, DE), permit higher centrifugation speeds, and a short continuous gradient is rapidly produced (in a few hours) at right angles to the tube axis, the gradient reorienting as the rotor slows and stops, and very good resolution can be achieved. Discontinuous CsCl gradients can also be used, cutting centrifugation time to 6 h *(17)*, but take more time and care to set up.

References

1. Struhl, K., Cameron, J. R., and Davis, R. W. (1976) Functional genetic expression of eukaryotic DNA in *Escherichia coli. Proc. Natl. Acad. Sci. USA* **73,** 1471–1475.
2. Parent, S. A., Fenimore, C. M., and Bastian, K. A. (1985) Vector systems for the expression, analysis and cloning of DNA sequences in *Saccharomyces cerevisiae. Yeast* **1,** 83–138.
3. Naysmith, K. A. and Reed, S. I. (1980) Isolation of genes by complementation in yeast: molecular cloning of a cell-cycle gene. *Proc. Natl. Acad. Sci. USA* **77,** 2119–2123.
4. Tréton, B. Y., Le Dall, M-T., and Gaillardin, C. M. (1992) Complementation of acid phosphatase mutation by a genomic sequence from the yeast *Yarrowia lipolytica* identifies a new phosphatase. *Curr. Genet.* **22,** 345–355.

5. Kinsella, B. T., Larkin, A., Bolton, M., and Cantwell, B. A. (1991) Molecular cloning and characterisation of a *Candida tsukubaensis* α-glucosidase gene in *Saccharomyces cerevisiae. Curr. Genet.* **20**, 45–52.
6. Abarca, D., Fernandez-Lobato, M., and Jiminez, A. (1991) Isolation of new gene *(SWA2)* encoding an α-amylase from *Schwanniomvces occidentalis* and its expression in *Saccharomyces cerevisiae. FEBS Lett.* **279**, 41–44.
7. Stark, M. J. R. and Milner, J. F. (1989) Cloning and analysis of the *Kluyveromyces lactis TRP1* gene: a chromosomal locus flanked by genes encoding inorganic pyrophosphatase and histone H3. *Yeast* **5**, 35–50.
8. Tsay, Y. H. and Robinson, G. W. (1991) Cloning and characterisation of ERG8, an essential gene of *Saccharomyces cerevisiae* that encodes phosphomevalonate kinase. *Mol. Cell. Biol.* **11**, 620–631.
9. Burke, D. T., Carle, G. F., and Olson, M. V. (1987) Cloning large segments of exogenous DNA into yeast by means of artificial chromosome vectors. *Science* **2336**, 806–812.
10. Sikorski, R. S. and Hiefer, P. (1989) A system of shuttle vectors and host strains designed to give efficient manipulation of DNA in *Saccharomyces cerevisiae. Genetics* **122**, 19–27.
11. Romanos, M. A., Scorer, C. A., and Clare, J. J. (1992) Foreign gene expression in yeast: a review. *Yeast* **8**, 412–488.
12. Brearley, R. D. and Kelly, D. E. (1991) Genetic engineering techniques in yeast, in *Genetically-engineered proteins and enzymes from yeasts: production and control.* (Wiseman, A., ed.), Ellis Horwood, Chichester, pp. 75–95.
13. Cryer, D. R., Eccleshall, R., and Marmur, J. (1975) Isolation of yeast DNA, in *Methods in Cell Biology* (Prescott, D. M., ed.), Academic, New York and London, pp 39–44.
14. Birnboim, H. C. and Doly, J. (1979) A rapid alkaline extraction procedure for screening recombinant plasmid DNA. *Nucleic Acids Res.* **7**, 1513–1523.
15. Ward, A. C. (1990) Single-step purification of shuttle vectors from yeast for high frequency back transformation into *E. coli. Nucleic Acids Res.* **18**, 5319.
16. Bignell, G. R., Bruce, I. J., and Evans, I. H. Electrophoretic karyotype of the amylolytic yeast *Lipomyces starkeyi* and cloning and chromosomal localisation of its *TRP1* gene, *Current Genetics,* in press.
17. Sambrook, J., Fritsch, E. F. and Maniatis, T. (1989) *Molecular Cloning: A Laboratory Manual*, 2nd ed, Cold Spring Harbor Laboratory, Cold Spring Harbor, NY.
18. Gingold, E. B. (1984) The use of restriction endonucleases, in *Methods in Molecular Biology*, vol 2, *Nucleic Acids* (Walker, J. W., ed.), Humana, Clifton, NJ, pp. 217–223.
19. McClelland, M., Hanish, J., Nelson, M., and Patel, Y. (1988) KGB: a single buffer for all restriction endonucleases. *Nucleic Acids Res.* **16**, 364.
20. Gaastra, W. and Jorgensen, P. L. (1984) The extraction and isolation of DNA from gels, in *Methods in Molecular Biology*, vol 2 (Walker, J. M., ed.), Humana, Clifton, NJ, pp. 67–76.
21. Dale, J. W. and Greenaway, P. J. (1984) The use of alkaline phosphase to prevent vector regeneration, in *Methods in Molecular Biology*, vol 2, *Nucleic Acids* (Walker, J. W., ed.), Humana, Clifton, NJ, pp. 231–236.

CHAPTER 19

Yeast Colony Hybridization

Matthew John Kleinman

1. Introduction

Yeast colony hybridization is a method for determining the presence or absence of a sequence of DNA, be it of chromosomal or plasmid location, in an individual (single cell-derived) colony. It can be used, for example, to screen transformants for a nonselectable sequence or to estimate the percent of individuals in a population still containing a nonselective element like the 2-μ plasmid (when estimating plasmid copy number of a nonselectable plasmid such as the 2-μ, it is vital to know the percent of cells in a population without any plasmid at all; failure to take this into account can lead to misleading results).

The colony hybridization method used in most laboratories is based on the Grunstein-Hogness Technique *(1)* and involves transferring yeast colonies to a membrane where the cells are lysed *in situ*, washed from below with buffers designed to make the DNA single-stranded, which is then bound to the membrane. The membrane containing bound DNA can then be probed in the same way as for the Southern blot procedure. With a suitable probe, single copy genes or plasmids can be detected.

2. Materials

1. Nitrocellulose or nylon membranes, circular so as to fit Petri dishes, e.g., Gelman (Ann Arbor, MI) Biotrace Charge Mod Nylon 82 mm, part no. 664835, or Millipore (Bedford, MA) HATF 08525 HA 0.45UM Nitrocellulose filters.
2. Petri dishes.

3. Petri dishes with appropriate agar media.
4. Plating grid (e.g., made from graph paper).
5. 3MM Whatman (Maidstone, UK) filter paper; sheet and circles to fit Petri dishes.
6. $1M$ Sorbitol, 20 mM EDTA, 50 mM dithiothreitol.
7. $1M$ Sorbitol, 20 mM EDTA, 1 mg/mL zymolyase 60,000 (*see* Note 1).
8. 5% w/v SDS.
9. $0.5M$ Tris-HCl, pH 7.5, 10X SSC (1X SSC is $0.15M$ NaCl, $0.015M$ sodium citrate).
10. 2X SSC.
11. Tweezers, glass spreading rod, toothpicks (autoclaved in a glass beaker with foil lid).

3. Method

1. Using a grid to ensure even spacing, use toothpicks to transfer cells from colonies to be analyzed onto agar Petri plates (with suitable media) or pick directly onto a membrane placed on the surface of an agar plate (*see* Note 2) and allow colonies to grow about 12 h (agar plate) or 6–8 h (membrane on plate) at 30°C.
2. If colonies are not already on the membrane, lift colonies from the surface of the agar onto the membrane (*see* Note 3) by gently lowering the membrane from one side using tweezers onto the surface of the agar and gently rubbing the back of the filter with a glass spreading rod. Make identification/orientation marks with ballpoint pen or by a cut or piercing of the membrane. Gently peel off the membrane and lay (colony side up) onto 3MM paper (two to four layers, fresh pieces for each solution change) placed in a Petri plate, saturated with the following solutions (about 5 mL), in the following order:
    a. $1M$ Sorbitol, 20 mM EDTA, 50 mM dithiothreitol for 30 min at room temperature.
    b. $1M$ Sorbitol, 20 mM EDTA, 1 mg/mL Zymolyase 60,000 for 2–24 h at 37°C (*see* Note 4).
    c. $0.05M$ NaOH for 7 min at room temperature.
    d. $0.05M$ Tris-HCl, pH 7.5, 10X SSC, 5 min at room temperature.
    e. $0.05M$ Tris-HCl, pH 7.5, 10X SSC, 5 min at room temperature.
    f. 2X SSC, 5 min at room temperature.
3. Air dry on filter paper for 30 min to 1 h.
4. If using nitrocellulose membranes, bake in a vacuum oven at 80°C for 1–2 h. If using nylon membranes, depending on type, air dry, expose to UV, or bake (*see* manufacturer's recommendations). The membranes are now ready to probe in the usual way (*see,* e.g., Section 3.4., Chapter 9).

4. Notes

1. Other yeast wall digesting enzymes can be used, such as Novozyme (SP234) and glucuronidase, both at 1% concentration.
2. With bacteria, good results can be obtained by simply lifting colonies onto filters from spread plates. However, with yeast for clearest results, colonies must be picked and evenly spotted onto a fresh plate, allowed to grow, and then lifted onto a membrane. An alternative that gives very clear (and quicker) results, but is slightly more time-consuming, is to spot colonies directly onto membranes (that have been gently lowered from one side, so as to avoid air bubbles, onto agar plates) and then allowing colonies to grow for 6–8 h at 30°C before cell wall digestion. However, if viable cells need to be recovered, it is important at the same time to spot (or replica plate) a duplicate set of "master" agar plates, because the cells on the filter will be destroyed. Agar plates should not be too wet, because this will result in colonies spreading into each other.
3. Nylon membranes can usually be used straight from the pack and wet satisfactorily on the agar plate or on the first pile of filter paper; if contamination problems occur, treat as for nitrocellulose. Nitrocellulose often has to be prewetted by being placed into a dish of distilled water. The membranes must then be autoclaved: the prewetted membranes sandwiched between damp filter paper and wrapped in foil.
4. Other cell-wall digesting enzymes, such as glucuronidase or Novozyme (SP234), both can be used at 1%. Digestion time will vary depending on the yeast strain and the enzyme used. Potential lysis can be checked by picking a few cells with a toothpick onto a microscope slide, placing a coverslip over cells, and putting a drop of 5% SDS on edge of the slip. If, on viewing under a microscope (preferably phase contrast) a significant proportion of cells (say more than 20%) swell up and burst, continue. If less, allow more digestion time.
5. It is always good practice to include positive and negative controls; in this case, ideally, colonies with and without the desired sequence. If such a positive control is unavailable (i.e., the point of the experiment), a very small sample of denatured unlabeled probe applied to the membrane before baking can be used to indicate that the hybridization procedure is working.
6. Do not panic if bits of colony debris wash off during the hybridization procedure. It is the DNA bound to the membrane that is important.
7. Clearest results usually occur with colonies pieced and thinly spread into small circles (about 3 mm), and grown for a few hours on a membrane. A (just visible) colony of young, easily lysed cells is preferable to a thick colony of tough, geriatric cells.

Reference

1. Grunstein, M. and Hogness, D. S. (1975) Colony hybridization: a method for the isolation of cloned DNAs that contain a specific gene. *Proc. Natl. Acad. Sci. USA* **72,** 3961–3965.

CHAPTER 20

Determination of Plasmid Copy Number in Yeast

Bernadette E. Jordan, Robert C. Mount, and Christopher Hadfield

1. Introduction

Autonomously replicative plasmids are widely employed as vectors for the expression of cloned genes in yeast (1). A number of different kinds of such plasmid vectors exist, which fall into three classes: *ARS* (autonomously replicating sequence) based plasmids, 2-µm-based plasmids, and *ARS* plasmids containing a centromere.

ARS-based plasmids are highly unstable because of maternal bias: this is an unequal inheritance of the vector at cell division, in which the plasmid tends to stay in the mother cell rather than entering the budding daughter cell. This generates plasmid-free segregants, but can also result in very high copy numbers in the plasmid-containing cells. *ARS*-based plasmids, can, in addition, integrate and in this way become stably maintained.

Addition of a centromere (*CEN*) counteracts the instability of *ARS* plasmids, but although these *ARS-CEN* plasmids exhibit greater stability, they are present at only one to three copies per cell.

The most widely employed vectors are those based on the yeast 2-µm plasmid. This is a naturally occurring plasmid present in most strains of *Saccharomyces cerevisiae*. It is normally present in the cell at between 40–100 copies per cell and is stably maintained through mitosis and meiosis by its partitioning system.

Essentially, two types of 2-µm-based vector exist: those containing the entire 2-µm sequences, which approach the stability of the native

From: *Methods in Molecular Biology, Vol. 53: Yeast Protocols*
Edited by: I. Evans Humana Press Inc., Totowa, NJ

plasmid, and those that incorporate only those sequences required for autonomous replication and stability, *ORI* and *STB*, respectively. In this instance, the *trans*-acting factors required for 2-μm partitioning need to be provided by endogenous 2-μm plasmids within the host cell. Unequal segregation occasionally can still occur during mitosis, and such vectors are therefore very highly, but not completely, stable.

The 2-μm *ORI-STB*-based plasmid vectors are often present at around 20 copies per cell, although this can vary greatly. A vector containing a *leu2-d* marker, for example, which has a defective promoter, can drive the copy number of the plasmid up to around 100 copies per cell, when it is grown under selection.

Since most plasmid vectors are not 100% stable, any copy number determination must also involve the estimation of the stability of the vector. Two copy number values can therefore be derived: the average copy number per cell in the population and the copy number per plasmid-containing cell.

The expression level of a plasmid-borne gene cannot be interpreted meaningfully without knowing the plasmid copy number, as gene expression level often directly reflects gene dosage. Regulatory or other factors may determine whether the level of expression of a particular gene is linearly related to its copy number. Often, particularly at high copy numbers, the relationship does not hold true, indicating limiting factors affecting expression. In some cases, the gene product may have a toxic effect on the cell, and this often is manifested by decreased plasmid copy number and stability.

The method employed for plasmid copy number determination in yeast involves the use of blotting-hybridization techniques to determine the amount of a plasmid-borne yeast gene (normally a selectable marker) relative to the corresponding single copy chromosomal gene, as outlined in Fig. 1. Total DNA is isolated and then restricted with an endonuclease that cleaves the DNA on either side of both the plasmid-borne and chromosomal marker. This digestion must be arranged to generate marker fragments that are dissimilar in size from the plasmid and chromosome (usually the plasmid marker fragment is larger). After subsequent agarose gel electrophoresis, the genomic DNA gel is Southern blotted and hybridized with a [^{32}P]-labeled marker fragment. This probe will hybridize to both the single copy chromosomal fragment and the marker gene from the plasmid. After autoradiography, two differently sized bands will

Determination of Plasmid Copy Number

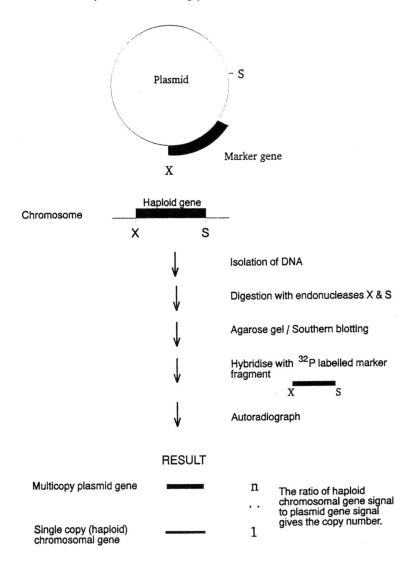

Fig. 1. A schematic representation of the method employed for copy number determinations.

be apparent on the X-ray film. The ratio of the amount of bound probe to the chromosomal and plasmid bands gives a measure of the copy number.

There are several important stages in this method. The initial DNA extraction method is perhaps the most important, since it is crucial that the isolation does not cause uneven losses of chromosomal or plasmid

DNA, or the results will be meaningless. Total cellular DNA must therefore be isolated by a method that will not incur such losses. This generally means using a serial extraction/precipitation procedure such as that described by Cashmore et al. *(2)*, Piper, or Evans and Bignell, in this volume (Chapters 11 and 18, respectively). A second highly important part of the protocol is obtaining complete restriction of the total DNA. This can be a problem as the DNA obtained by the aforementioned method is not totally pure and, consequently, digestion efficiency of the endonuclease is often significantly reduced. Other stages in the method that demand particular care are:

1. The requirement for an even and complete Southern blot, so that it will be quantitative for the two DNA fragments;
2. Setting up the hybridization to avoid nonspecific background contamination of the Southern blot filter; and
3. An accurate method of measuring the amount of bound hybridization probe to the bands on the filter.

2. Materials

1. Total DNA is required for the method, which has been isolated from the plasmid-containing cells by a procedure that does not lead to uneven loss of plasmid DNA. The method described by Cashmore et al. *(2)* involves the digestion of the yeast cell walls to produce spheroplasts, which are then lysed, followed by phenol-chloroform extraction to remove proteins, and ethanol precipitation to recover total nucleic acids. RNA is removed by RNase treatment. The DNA obtained in this way requires additional treatment to ensure it can be fully restricted. The DNA sample is normally incubated at room temperature for 20 min with 3 U of porcine α-amylase, followed by a 30-min incubation at 55°C with 1 U of amyloglucosidase, isolated from *Aspergillus niger*. These two enzymes incorporate different activities to break down contaminating carbohydrates, which are present in this type of preparation. (Volumes 2 and 4 in the Methods series can also be consulted, and *[5]*.)
2. Materials for DNA restriction and agarose-gel electrophoresis. Sample loading buffer at 5X final concentration is composed of 25% (w/v) Ficoll 400, 0.125% (w/v) bromophenol blue, and 0.25% (w/v) xylene cyanol FF, in gel buffer.
3. TE buffer: 10 mM Tris-HCl, 1 mM EDTA, pH 8.0.
4. Solution O: 1.25M Tris-HCl, 0.125M MgCl$_2$, pH 8.0 (store at 4°C).
5. dNTPs: 100 mM solutions of dATP, dTTP, and dGTP (store at –20°C); [α-^{32}P]dCTP (370 MBq/mL) (e.g., Amersham [Arlington Heights, IL] PB10475).

Determination of Plasmid Copy Number

6. Solution A: 1 mL solution O, 18 µl β-mercaptoethanol, 5 µL each of dATP, dTTP, and dGTP.
7. Solution C: Suspend 50 A_{260} U of random hexadeoxynucleotides (Pharmacia [Uppsala, Sweden] 27-2166-01) in 550 µL TE, to give a concentration of 90 A_{260} U/mL (store at –20°C).
8. OLB buffer: Mix solutions A:B:C in the ratio 10:25:15 (total 50 µL, sufficient for 10 labeling reactions) (store at –20°C).
9. Stop buffer: 20 mM NaCl, 20 mM Tris-HCl, pH 7.5, 2 mM EDTA, 0.25% SDS, 1 µM dCTP.
10. Acetylated bovine serum albumin (BSA) (Gibco [Gaithersburg, MD] BRL 540-5561UA).
11. Sephadex G50, medium grade (Pharmacia 17-0043-01).
12. Polyallomer wool (aquarium filter, available from pet shops).
13. Materials for Southern blotting and hybridization (*see* for example, Chapter 9, ref. *5,* and vols. 2 and 4 of the Methods series). A wide assortment of containers and incubators can be used for hybridizations; however, if possible, it is recommended that dedicated equipment, designed specifically for this task (e.g., Hybaid, Teddington, UK), be used, as this optimizes both effectiveness and protection from irradiation. Hybridization conditions that keep background radioactivity to a minimum are essential. The recommended hybridization solution for this purpose consists of 1% BSA (fraction V), 1 mM EDTA, 0.5M NaH_2PO_4, pH 7.2 and 7% SDS *(3)*. It is advisable to ensure that the BSA has fully dissolved before the addition of SDS, which is normally added as a 20% solution. The solution normally employed for washing is 4X SSC (35 g sodium chloride, 17.64 g sodium citrate/L in distilled water) containing 0.1% SDS.
14. The estimation of the band ratios requires a scintillation counter set for [^{32}P] and radioactive ink, which is required for markings to enable autoradiographic bands to be aligned with their filter location. The ink is made by the addition of radioactive dCTP to 500 µL writing ink to give approximately 1×10^6 cpm. (*See* Note 13 for alternative method.)

3. Method

1. Restrict approximately 5 µg of total genomic DNA with restriction endonucleases that will cleave the chromosomal and plasmid DNA on either side of the yeast marker, so that it liberates a fragment containing the same gene of different size from the chromosome (Fig. 1). Since the DNA must be fully restricted for this method, it is often beneficial to restrict the DNA in a large volume, approx 400 µL, to dilute impurities that are normally present in the DNA sample. Additionally, extra endonucleases can be added, which cut outside the marker region; these do not affect the two

marker fragment sizes, but break down the chromosomal DNA and make it more amenable to digestion by the endonucleases that do count. Furthermore, prolonged digestion, preferably overnight, usually is necessary. In some cases, subsequent addition of further enzyme and further prolonged digestion may be necessary to obtain fully restricted DNA. The extent of the digestion is determined by agarose-gel electrophoresis *(see the following)*.

2. Since the restriction is in a large volume, it is necessary to ethanol precipitate the sample to enable it to be subsequently loaded on a gel. This is done by the addition of 40 µL $3M$ sodium acetate, pH 5.6, and 800 µL ethanol; after mixing, the sample is placed in a dry ice/ethanol bath for 5 min. The DNA is pelleted by spinning at 1300 rpm (11,500g) for 15 min in a benchtop microfuge. The supernatant is discarded and the pellet washed by the addition of 400 µL of 70% ethanol. It is then repelleted by spinning at 1300 rpm for 5 min, then dried under vacuum, and resuspended in 20 µL of TE buffer.

3. For electrophoresis, to each of the samples a 25% volume of 5X loading buffer is added (resulting in a 1X concentration) and then they are loaded into the wells of a 0.8% agarose gel. Electrophoresis should be sufficient to obtain a good separation between the plasmid and chromosomal marker fragments. Figure 2 illustrates the difference between completely restricted and incompletely restricted total DNA samples following agarose gel electrophoresis.

4. The gel is Southern blotted *(see* Note 1 and Chapter 9 as well as vol. 2 in this series, and ref. *5)*. Then the blot filter is prehybridized by incubation in the hybridization buffer for 30 min at 65°C, with continuous agitation. The volume of the buffer for prehybridization is not important, provided sufficient buffer is present to keep the filter covered; usually an excessive volume is used at this step.

5. The auxotrophic marker gene present on the plasmid is employed as the hybridization probe. It is liberated from a sample of purified plasmid DNA by digestion with an appropriate restriction endonuclease and then separated on a low-gelling temperature agarose (0.8–1.0% w/v) gel. The fragment is cut out of the gel and placed in a 1.5-mL microfuge tube. Water is added at a ratio of 1.5 mL/g of gel slice. The tube is then placed in a 100°C bath for 7 min to melt the agarose and denature the DNA *(see* Note 2).

6. Afterward, the tube is placed directly into a 37°C water bath, where it can be kept for 10–60 min prior to use for the labeling reaction.

7. An aliquot of the fragment DNA is [^{32}P]-labeled to a high specific activity by the random primer method *(4)*, as follows:
    a. To a microfuge tube, at room temperature, add the following in the stated order: 16 µL DNA fragment (25 ng, made up to volume by addition of water); 5 µL OLB buffer; 1 µL acetylated BSA (10 mg/mL);

Fig. 2. Restricted genomic DNA samples run on a 0.8% agarose gel. Tracks 3 and 4 show samples that are incompletely restricted and underloaded and are not adequate for a copy number determination. Tracks 1 and 2 show samples that have been fully restricted, but are slightly overloaded.

      2.5 µL α-[^{32}P] dCTP (10 µCi/µL = 0.37 MBq/µL); 0.5 µL Klenow (large fragment) DNA polymerase I (2 U).
  b. Incubate at room temperature for at least 5 h, or overnight if more convenient.
  c. Stop reaction by adding 100 µL stop buffer.
8. The labeled DNA probe fragments are separated from the unincorporated nucleotides (*see* Note 3) by Sephadex G50 minicolumn fractionation, as follows:
  a. Remove the long, narrow end-tip of a glass Pasteur pipet by etching at the "neck" constriction where it first becomes narrow, using a diamond-tipped pen, and then breaking off the long narrow-stemmed tip.
  b. Insert a small amount of sterile polyallomer wool into the top of the residual Pasteur pipet stem and, using the detached narrow tip as a "ramrod," push the wool down to the "neck" constriction, to form a bung. It is important that this bung should not be too tightly compacted, as otherwise liquid will not flow through it; yet it must be tight enough to retain the G50 particles.

c. Vertically clamp the Pasteur pipet to form a column and place a small beaker underneath. Then load the G50 column, as follows: Shake the bottle to suspend the particles; take a new Pasteur pipet, with an attached filling device, and quickly fill the pipet with the suspension; insert the tip of the pipet into the vertical column, right down to just above the polyallomer wool bung; then discharge the G50 into the column, raising the tip of the pipet as the column fills, so that it stays at the surface. It is important that this is done quickly, before the G50 particles settle out. It is also important that no air bubbles are formed in the column material (*see* Note 4). Fill the column to ~1.5 cm from the top. Then wash the column with TE buffer by filling the small 1.5-cm-long reservoir and allowing it to drain away. Repeat twice and check that the buffer flows through the column freely; otherwise, it will not be satisfactory and will have to be redone.

d. Replace the beaker under the column with a rack containing 1.5-mL microfuge tubes, numbered 1–15. Add the stopped labeling reaction (125 µL) to the top of the column and collect the liquid displaced from the column in the first tube. Add 125 µL of TE to the top of the column and collect the displaced liquid in the second tube. Repeat the process for the remaining tubes.

e. Measure the amount of radioactivity in each fraction by Cerenkow counting. There should be two peaks: The first one (starting about fraction 5 or 6) contains the labeled fragment DNA; the second contains the unincorporated nucleotides. Pool the probe peak fractions (usually about four fractions). In total this should give rise to $>10^6$ counts/min; a good high-specific activity labeled probe will give rise to $\sim 10^7$ c.p.m.

9. Decant the prehybridization buffer from the filter and replace with fresh hybridization buffer, in a minimum volume necessary to cover the filter.

10. Denature the probe by heating at 100°C for 7 min (*see* Note 2). Add the denatured probe directly to the hybridization buffer and ensure that it is mixed in thoroughly. Hybridize at 65°C for 18 h, using an agitation system that provides continuous, even distribution of the probe solution over the surface of the blot filter. Once this is complete, the filter should be washed six times, each at 65°C for 30 min using the washing solution, which should also be kept at 65°C.

11. After hybridization and washing, the filter is blotted dry between 3MM Whatman (Maidstone, UK) filter paper. It is then sandwiched between two pieces of cling film and adhesive labels affixed around the periphery of the filter. Radioactive ink markings are then made on the labels using a toothpick and once dry, further cling film is used to cover these markings. The filter is then exposed to X-ray film.

Determination of Plasmid Copy Number

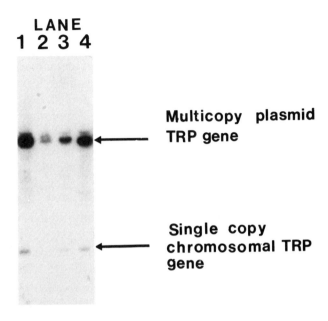

Fig. 3. An autoradiograph showing the bands obtained from a filter containing restricted genomic DNA samples probed with a *TRP* probe. Lanes 1 and 4 are overexposed, but can be employed to facilitate the excision of the single copy band from the filter. The exposure of lane 2 is that required to remove the plasmid *TRP* band.

12. On developing the film, two bands should be visible (*see* Fig. 3 and Note 5). It is usually arranged so that the upper band (larger fragment) represents the marker present on the plasmid, whereas the lower band is the single copy chromosomal band. The X-ray film image is then aligned over the cling film covered filter, using the radioactive ink markings. The edges of the band areas on the X-ray film are pierced with a syringe needle through to the filter, which enables the areas of the filter containing the two bands to be cut out, using the holes as a guide. The cling film is peeled away from both sides of each filter fragment, which are subsequently placed into a separate scintillation vial. To each vial, add 2.5 mL of nonaqueous scintillation fluid and count in a scintillation counter (*see* Notes 6–9).
13. The ratio between the counts for the upper and lower band areas can be calculated and this is the average plasmid copy number for cells in the population (*see* Notes 10 and 11). If the average copy number per plasmid-containing cell is required, then the average copy number must be adjusted to take into account the percentage of plasmid-free segregant cells in the population (*see* Note 12).

4. Notes

1. Nylon filters (such as Hybond-N [Amersham]) are recommended for the Southern blotting of the DNA, since the nylon membrane is more robust than nitrocellulose and is therefore easier to handle.
2. The 100°C bath can be boiling water, a heated polyethylene glycol bath, or dry-block heater. It is important that the caps of microfuge tubes placed at 100°C are pierced with a needle-hole prior to heating, as otherwise internal pressure will cause the cap to fly off in an explosive manner.
3. Although the probe may be used directly following the labeling reaction, this is inadvisable, as it usually results in a high background owing to unincorporated nucleotides.
4. If a trapped air bubble forms in the G50 column, fill the column with TE buffer; insert a Pasteur pipet (attached to a filling device) and gently swirl to suspend the GSO particles above the bubble; then suck the air bubble into the Pasteur pipet. The column material will settle again and surplus buffer drain away.
5. If additional bands are observed on the autoradiograph, this is usually indicative of partial restriction if 2-µm-based vectors are used. Alternatively, it could be a result of an integration, particularly if an *ARS* based vector has been used.
6. It is recommended that scintillation vials are treated with an antistatic device or that static is removed by some other means, prior to scintillation counting. Static can interfere markedly with the counting process and cause abnormally high counts, which is a common source of error. It is also advisable to count the vials several times, to be sure that static interference has been eliminated.
7. A clean radioactivity-free background is crucial, as background radioactivity can interfere with the counts obtained and hence the accuracy of the determination. Most affected will be the single-copy chromosomal band counts since these are the weakest. If the background is not clean, it may be possible to subtract the background radioactivity counts from the band counts. To estimate background, excise an appropriate piece of filter of similar area to the band, scintillation count, and subtract the scintillation counts obtained from the fragment band counts.
8. To ensure that the whole band area has been excised from the filter, the filter can be re-exposed to X-ray film. If band areas can still be observed, the data obtained clearly will not be correct.
9. To aid the excision of the filter bands, a number of autoradiograph exposures generally are required. Generally, longer exposures will be required for the single copy chromosomal band and shorter exposures for the multicopy plasmid fragment (Fig. 3).

10. Scintillation counting has been shown to be accurate for copy-number determinations of up to at least 80 plasmids per cell.
11. Densitometer scans of autoradiographs are not so easy to employ accurately for estimating plasmid copy numbers, as the signal from the multicopy plasmid fragment is usually manyfold greater than that of the single chromosomal fragment. Consequently when the single-copy band is visible, the plasmid band is overexposed on the autoradiograph, so that its density range has gone beyond the reactive range.
12. To determine the percentage of plasmid-containing cells, a microbiological plate assay can be performed. Ideally, this should be performed with cells used to make the total DNA preparation, or failing that, cells grown under identical conditions. The cells are plated out onto nonselective agar medium, to give a number of plates containing a maximum of 300 colonies per plate. The plates are subsequently replica-plated onto selective agar medium and the number of colonies growing, represent plasmid-containing cells. These are scored and the percentage calculated.
13. The recent development of phosphoimager technology now enables the amount of radiolabeled probe bound to restriction fragment bands on filter blots to be quantified directly. If available, this is easier than the scintillation counting method described.

References

1. Romanos, M. A., Scorer, C. A., and Clare, J. J. (1992) Foreign gene expression in yeast: a review. *Yeast* **8,** 423–488.
2. Cashmore, A. M., Albury, M. S., Hadfield, C., and Meacock, P. A. (1986) Genetic analysis of partitioning functions encoded by the 2-µm circle of *Saccharomyces cerevisiae*. *Mol. Gen. Genet.* **203,** 154–162.
3. Church, G. M. and Gilbert, W. (1984) Genomic sequencing. *Proc. Natl. Acad. Sci. USA* **81,** 1991–1995.
4. Feinberg, A. P. and Vogelstein, B. (1984) A technique for radiolabelling DNA restriction fragments to high specific activity. *Anal. Biochem.* **137,** 266–267.
5. Sambrook, J., Fritsch, E. F., and Maniatis, T. (1989) *Molecular Cloning: A Laboratory Manual.* 2nd ed. Cold Spring Harbor Laboratory, Cold Spring Harbor, NY.

CHAPTER 21

Determination of Chromosome Ploidy in Yeast

Christopher Hadfield, Bernadette E. Jordan, and Robert C. Mount

1. Introduction

Wild-type and industrial strains of *Saccharomyces cerevisiae* appear to be typically polyploid/aneuploid. Such strains are also essentially sterile, but will on rare occasions produce spores. The mating of such spores with reference laboratory strains can be used to give a general estimation of the overall ploidy level. Similarly, measurement of a parameter that is proportionally maintained relative to ploidy level in a euploid reference series, such as DNA content or a certain enzyme activity, can be used to estimate the general overall ploidy of wild-type or industrial strains *(1,2)*. However, this does not take account of interstrain variability and, in particular, aneuploidy.

Recently, a probe method has been developed to determine the copy number of chromosomes at specific loci *(3)*. Since such determinations are specific for small points on the chromosomes examined, their accuracy is not affected by aneuploidy or other chromosomal rearrangements. On the contrary, a second ploidy probing along the same chromosome—particularly if it is at a locus on the other side of a centromere (whose presence is essential for chromosomal maintenance)—can be used to detect differences owing to internal deletions. Alternatively, ploidy differences between two different chromosomes can indicate aneuploidy. Thus, by using a number of different ploidy probes, a picture of the overall ploidy composition of a particular strain can be built up.

From: *Methods in Molecular Biology, Vol. 53: Yeast Protocols*
Edited by: I. Evans Humana Press Inc., Totowa, NJ

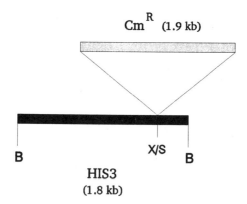

Fig. 1. Ploidy probe for chromosome XV at the *Bam*HI-*HIS3* fragment locus *(3)*. This was prepared by insertion of a *Sal*I-ended Cm^R determinant fragment into an internal *Xho*I site within the plasmid-cloned *HIS3* fragment. The resultant plasmid is simply restricted with *Bam*HI to release the ploidy probe fragment for use.

Ploidy probes are simply derived from cloned fragments of chromosomal DNAs, into which a dominant selectable marker, capable of functioning effectively in single copy, has been inserted. One such ploidy probe is shown in Fig. 1. In this case, the yeast DNA fragment comes from chromosome XV, in the form of a *HIS3*-containing *Bam*HI fragment; a chloramphenicol-resistance (Cm^R) determinant, on a *Sal*I fragment, has been inserted into the *Xho*I site within the *Bam*HI chromosomal fragment.

Following transformation of a yeast strain with the ploidy probe fragment, Cm-resistant integrant transformants are selected. The ploidy probe fragments integrate at the respective homologous chromosomal locus, via the homology of the yeast DNA, replacing the native allele, as shown in Fig. 2. This creates a new allele that is larger in size than the wild-type allele, owing to the insertion of the Cm^R determinant. This ploidy probe allele can then be used as a reference for determining the chromosomal locus ploidy.

As indicated in Fig. 2, chromosomal DNA is then isolated from the transformant and restricted with the same restriction endonuclease that was used to produce the ploidy probe fragment in the first place—in this case *Bam*HI. The next step is to separate the DNA fragments according to size by agarose-gel electrophoresis, and Southern blot. The blot filter

Determination of Chromosome Ploidy

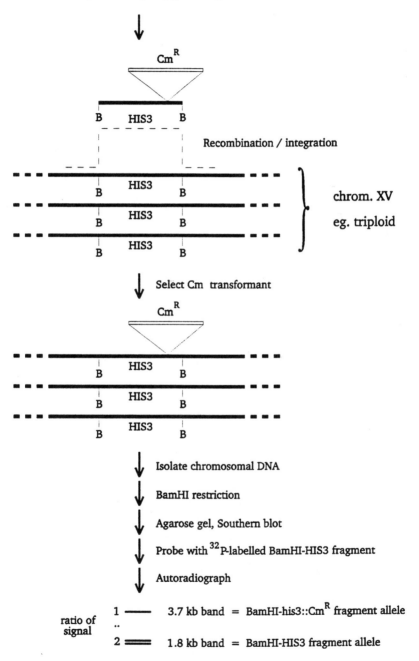

Fig. 2. Method for determining chromosomal locus copy number using a ploidy probe.

is then hybridized with [^{32}P]-labeled wild-type allele fragment—in this case the 1.8-kb BamHI-*HIS3* fragment. This probe will hybridize to both the wild-type and ploidy probe alleles. Furthermore, since the ploidy probe allele contains the same yeast DNA as the wild-type allele, the amount of wild-type fragment probe that hybridizes to each allele should be the same per unit amount of allele DNA blotted onto the filter.

Autoradiography of the filter reveals two bands: a lower wild-type allele fragment band (1.8 kb) and an upper ploidy probe fragment band (3.7 kb). Finally, the amount of probe bound to each fragment allele is measured, either by phosphoimager, scintillation counting, or densitometry, and the ratio between the two alleles determined. In the example shown in Fig. 2, the ploidy probe has integrated into one allele of a triploid chromosome. This leads to a ratio of 1:2 of ploidy probe to wild-type allele. Such a ratio indicates a total of three alleles, showing the chromosome to be triploid at that locus. Similarly, a 1:0 ratio would indicate haploid, 1:1 diploid, 1:3 tetraploid, and so on.

Single ploidy probe integrations occur in the majority of transformants if the amount of transforming fragment DNA is kept low (>0.5 µg). With higher amounts of transforming DNA (>5 µg), multiple integrations occur. However, even in these circumstances it is still possible to make a ploidy determination, provided that a number of transformants are examined; we will discuss this later. First, we will describe the method using conditions that favor single-copy ploidy probe fragment integrants being obtained. This is the most efficient way of employing the method.

2. Materials

1. DNA of a plasmid containing a ploidy probe for the chromosome to be probed. As mentioned earlier, this is simply prepared from a plasmid-cloned chromosomal DNA fragment, into which is inserted a dominant selectable marker, capable of functioning effectively in single copy, in the manner exemplified in Fig. 1. Suitable dominant markers are chloramphenicol-resistance *(4)*, G418-resistance *(5)*, and resistance to the herbicide sulfometuron methyl *(6)*. Ploidy probes for chromosomes XV, V, and IV are described in ref. *(3)*. In this case, we will continue to describe the method with specific reference to the chromosome XV *HIS3* locus probe described in Fig. 1.
2. Materials for dominant marker transformation and selection (*see* Chapter 16, by Mount and colleagues in this volume, and *see also* Note 2).

3. Materials for agarose-gel electrophoresis, Southern blotting, the production of [^{32}P]-labeled DNA fragment probes, hybridization, and autoradiography; see, for example, Chapters 9 and 18 of this volume, refs. *7,8,* and vols. 2 and 4 of these series. It is important to use a hybridization solution that effectively prevents background radioactivity bound to the filter. We recommend: 1% BSA (fraction V), 1 m*M* EDTA, 0.5*M* phosphate buffer, pH 7, 7% SDS (added last in a 20% w/v solution, after the other components have dissolved) *(9).* Prehybridize the filter in this solution for a min of 5 minutes before adding the probe.
4. A means of measuring the amount of radioactivity bound to the Southern blot bands. This requires either a phosphoimager or scintillation counter set to monitor counts per minute for [^{32}P], plus scintillation vials and nonaqueous scintillation fluid (e.g., LKB OptiScint "T"), or a densitometer for analyzing autoradiographs.
5. Radioactive ink for scintillation counter method: Prepare by adding a small amount of [^{32}P]-deoxynucleotide to 500 µL of ink, to give in the region of 5×10^4 Cerenkov counts/min.

3. Method

1. Liberate the ploidy probe fragment from the vector plasmid by cleavage with the appropriate restriction endonuclease. In our example, 2.5 µg of pJK103 was cut with *Bam*HI: 0.5 µg of this was then removed and used to check that digestion was complete, by running it on an agarose-gel, next to a 0.5-µg sample of uncut plasmid DNA; the remainder of the cut plasmid DNA was kept on ice. (If details of the basic technique of gel analysis of DNA are needed, *see* Section 3.2. in Chapter 18, by Bignell and Evans.)
2. Competent yeast cells are then transformed with the cut plasmid DNA containing the liberated ploidy probe fragment (i.e., there is no need to separate the probe fragment from the rest of the vector). The level of competence will vary from strain to strain and between batches, but the object is to use the minimum amount of fragment DNA to obtain transformants. Transform aliquots of 4×10^7 competent cells (by the lithium acetate method; for details, *see* Chapter 16, by Mount et al., and *see also* Note 3) with 0.1, 0.5, and 1.0 µg of cut plasmid DNA (*see* Notes 4 and 5). Plate on selective agar to select transformants (again, *see* Chapter 16, by Mount et al.).
3. Select 6 Cm^R transformant colonies from the plate containing transformants resulting from the *lowest* amount of fragment DNA. Prepare total genomic DNA of the transformants and the untransformed strain (*see* Chapter 11, by Piper, or Chapter 18, by Bignell and Evans).
4. Restrict 2 µg of each of the total genomic DNA samples with *Bam*HI. This restriction has to be complete, yet yeast DNA preparations are notoriously

difficult to restrict (*see* also Chapter 20, by Jordan et al.). Avoid incomplete digestion by adopting three strategies.
   a. Ensure that the genomic DNA sample has been treated to remove polysaccharides, which inhibit restriction (e.g., by treatment with DNase-free α-amylase and glucoamylase, as described in Chapter 20, by Jordan et al.).
   b. Add a second restriction enzyme that cuts outside of the ploidy probe region, to help break down the DNA. This increased breakdown of the DNA acts to enhance the effectiveness of the *Bam*HI restriction.
   c. Incubate the restriction digestion overnight to give what would be a calculated 20- to 40-fold overdigestion if the DNA was pure (which it is not). To prevent evaporation, which would alter the buffer conditions, undertake the restriction digestion either in a sealed glass capillary, or in a microcentrifuge tube, overlayed with paraffin oil.
5. Electrophorese the genomic DNA samples on a 0.8% agarose gel; then Southern blot. Isolate the 1.8-kb *Bam*HI-*HIS3* fragment (produced by digestion of a plasmid clone of the yeast DNA fragment) from the gel and [^{32}P]-label (*see* Chapter 20, by Jordan et al. for protocol; originally described in ref. *10*). Hybridize this probe with the Southern blot (*see* Chapter 20, by Jordan et al., and vol. 2 in this series).
6. After washing the filter, blot it dry between sheets of Whatman (Maidstone, UK) 3MM paper and then place it on a sheet of cling-film. If quantifying with a densitometer or scintillation counter, place four small self-adhesive paper labels at different points around the periphery of the filter. Make a mark on each label with radioactive ink, which can be applied with a toothpick. Allow the ink to dry, then place another sheet of cling film over the top, sandwiching the filter and labels. Expose to X-ray film (use intensifying screen and exposure at –70°C, as necessary). Make a range of autoradiographic exposures—probably from a few hours to a few days—in order to ensure getting one or more in which both bands are within the reactive region of the film.
7. Two bands should be visible in the tracks that contained the transformant DNAs; the upper band represents the ploidy probe allele; the lower band should correspond with the single band in the untransformed control track, and corresponds to the wild-type allele. Examples are shown in Fig. 3; also *see* Note 7.
8. Determine the relative amount of radioactivity bound to the ploidy probe and wild-type allele bands. A phosphoimager will do this directly. Alternatively, use one of the following methods (*see* Note 6).
   a. Scan each transformant gel track on the X-ray film using a densitometer. Two density peaks are obtained (as exemplified in Fig. 4A) and the density units under them recorded. Calculate the ratio of ploidy probe to wild-type allele density units.

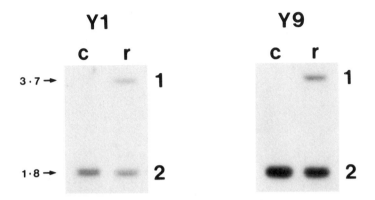

Fig. 3. Autoradiographs showing ploidy probe (3.7 kb) and wild-type allele (1.8 kb) bands for two brewing strains of yeast, Y1 and Y9 (top- and bottom-fermenting ale yeasts, respectively). Track **c** contains *Bam*HI-restricted total genomic DNA of the untransformed control strain; track **r** contains *Bam*HI-restricted total genomic DNA of a chloramphenicol-resistant isolate following transformation with the *Bam*HI-*his3*::Cm^R ploidy probe described in Fig. 1. The Southern blots were probed with [^{32}P]- labeled *his3* fragment. In both cases, for the **r** track, the ratio of ploidy probe to wild-type allele fragment allele is 1:2.

    b. Take an autoradiographic exposure that clearly shows the allele bands and the radioactive ink markings. Use this to mark off the location of the allele bands on the Southern filter, as follows: Cut out each of the bands on the X-ray film, by piercing closely around the periphery with a needle; then place the film over the filter and position by aligning the radioactive ink markings with their images on the film; use the holes in the film as a template to mark off the positions of the bands with a marker pen upon the cling film covering the Southern filter. Cut out each band, remove any attached cling film; place in a scintillation vial and add 5 mL of scintillation fluid. Measure [^{32}P] in the scintillation counter, calculate the ratio of counts per minute for the ploidy probe to wild-type alleles (as exemplified in Fig. 4B).
9. The ploidy probe to wild-type allele ratio enables the calculation of the chromosomal locus ploidy (1:0, haploid; 1:1, diploid; 1:2, triploid; 1:3, tetraploid; and so on). It is advisable to base a ploidy determination on at least three independent transformant isolates, as incorrect ratios can occur owing to multiple ploidy probe fragment integrations. The chance of such occurrences is minimized by utilizing the lowest possible amount of transforming DNA in the first instance to obtain the transformants.

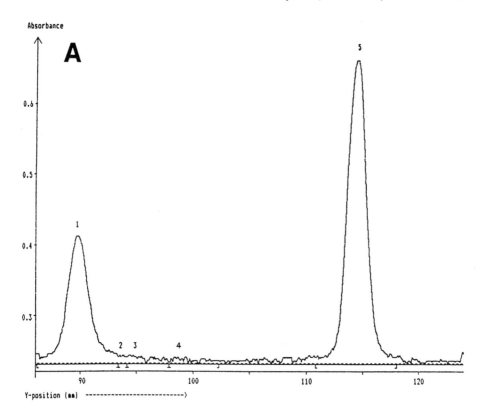

PEAK	POSI-TION	HEIGHT	AREA	REL. AREA	:	PEAK	POSI-TION	HEIGHT	AREA	REL. AREA
£	mm	AU	AU*mm	%	:	£	mm	AU	AU*mm	%
1	89.8	0.18	0.44	32.0	:	4	98.6	0.01	0.02	1.5
2	93.5	0.01	0.01	0.6	:	5	114.4	0.43	0.89	64.1
3	94.7	0.01	0.02	1.8	:					

B Ratio of density of upper band (peak 1) to lower band (peak 5) = 0.44 : 0.89
= 1 : 2.0

b.

Yeast strain	band cpm		ratio
	upper	lower	
Y1	560	1008	1 : 1.8
Y9	820	1919	1 : 2.3

4. Notes

1. With regard to the selection of ploidy probes, it should be noted that some industrial yeast strains are hybrids and may contain non-*S. cerevisiae* chromosomal DNA. This may be in the form of whole chromosomes or translocated insertions. For these regions, new ploidy probes can be created based upon cloned DNA fragments from the hybrid regions.
2. Before constructing or utilizing a ploidy probe based on a particular dominant marker, it is essential to determine that the yeast strain under investigation is sensitive to the selection agent and what concentration is necessary in the medium to prevent growth of untransformed cells. To do this, follow the guidelines set in Chapter 16, by Mount et al. Further information is given in papers by Hadfield and colleagues *(4,5)* and Casey, Xaio, and Rank *(6)*.
3. Lithium acetate transformation is widely applicable. We have used it successfully with over 20 industrial strains of *Saccharomyces*, including brewing, baking, spirit, wine, cider, and sake yeasts. Spheroplast transformation is known to work as an alternative. We have no experience as to how successfully electroporation can be adapted to this technique.
4. The most crucial part of the whole procedure is the initial fragment transformation and choosing of the transformants for analysis. The occurrence of single or multiple fragment integrations is concentration dependent. By using the minimum amount of transforming fragment, single-copy integrant should be obtained almost exclusively.
5. If too high an amount of transforming ploidy probe fragment is used, multiple integrations will occur. However, a ploidy determination can still usually be made provided a sufficient number of isolates are examined (minimum about six). Under such circumstances, a mixed population of transformant types occurs. Among these there is usually a fair percentage of single integrants. Additionally, for a diploid a 2:0 ratio can be obtained, for an integration at both alleles; this is indistinguishable from a 1:0 ratio for a haploid, unless other isolates show 1:1. Similarly, a triploid could give 2:1 or 3:0 (*see* Fig. 5A); a tetraploid, 2:2, 3:1, or even 4:0. A second type of multiple insertion, which happens less frequently than insertions into different alleles, is tandem insertions into a single allelic locus. This gives rise to an incorrect ratio as illustrated in Fig. 5B.

Fig. 4. *(preceding page)* Examples of band densitometry and scintillation counting data. **(A)** Densitometer trace for a gel track similar to the **r** tracks in Fig. 3. Peaks 1 and 5 represent the ploidy probe and wild-type allele fragment bands, respectively. AU = absorbance units; AU*mm = area under peak. **(B)** Scintillation counting data for the excised upper (3.7 kb) and lower (1.8 kb) bands in **r** tracks of Fig. 3.

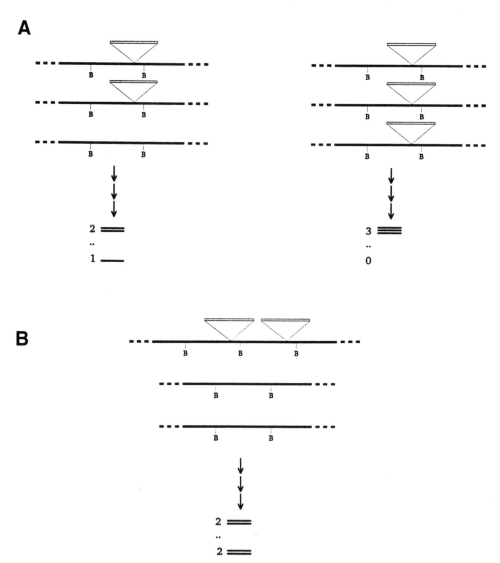

Fig. 5. Various outcomes resulting from increasing the amount of transforming ploidy probe fragment. **(A)** Ploidy probe integrations at more than one allelic locus in a polyploid strain result in altered ratios of ploidy probe to wild-type alleles. **(B)** Tandem insertion of ploidy probes at one allelic locus result in incorrect ratios of ploidy probe to wild-type alleles. **(C)** Interaction at the wrong locus, somewhere else in the genome; this is discernible because of incorrect ploidy probe allele size. (Can also rarely occur even with low amounts of transforming ploidy probe fragment.)

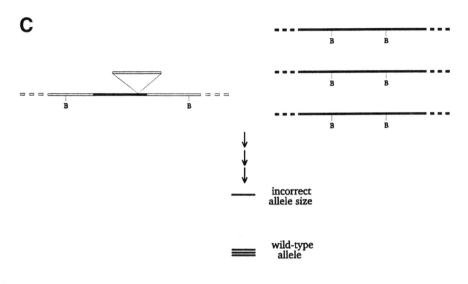

Fig. 5C.

6. Scintillation counting is prone to errors owing to static electricity on the scintillation vials. Use a means of removing static (e.g., an antistatic gun or antistatic chamber); repeat countings three times and look for fluctuations in recorded counts per minute, as this is an indication of static on the vials. For densitometry, always take a range of autoradiographic exposures. Only exposures within the linear reactive region of the film will give an accurate measurement.
7. Reject any samples that do not give the predicted allele sizes. As a rare occurrence, the ploidy probe fragment integrates at the wrong locus, as indicated in Fig. 5C. Such a locus may be a site of partial homology. If all isolates are integrations at the wrong locus, this indicates that the host strain does not contain the allele used to make the ploidy probe. (This should have been established at the start.) Alternatively, a frequent occurrence is that the genomic DNA preparation is not sufficiently pure to be restricted properly, and this gives an aberrant band pattern.
8. Additional dual probe hybridization methodology has been developed to help extend the ploidy analysis to the rest of the genome (*see* ref. *3*).

References

1. Kielland-Brandt, M. C., Nilsson-Tillgren, T., Litske Petersen, J. G., Holmberg, S., and Gjermansen, C. (1983) Approaches to the genetic analysis and breeding of brewer's yeast, in *Yeast Genetics, Fundamental and Applied Aspects.* (Spencer, J. F. T., Spencer, D. M., and Smith, A. R. W., eds.), Springer-Verlag, New York, pp. 421–438.

2. Snow, R. (1983) Genetic improvement of wine yeast, in *Yeast Genetics, Fundamental and Applied Aspects.* (Spencer, J. F. T., Spencer, D. M., and Smith, A. R. W., eds.), Springer-Verlag, New York, pp. 439,440.
3. Hadfield, C., Harikrishna, J. A., and Wilson, J. A., (1995) Determination of chromosome copy numbers in *Saccharomyces cerevisiae* strains via integrative probe and blot hybridization techniques. *Curr. Genet.* **27,** 217–228.
4. Hadfield, C., Cashmore, A. M., and Meacock, P. A. (1986) An efficient chloramphenicol-resistance marker for *Saccharomyces cerevisiae* and *Escherichia coli. Gene* **45,** 149–158.
5. Hadfield, C., Jordan, B. E., Mount, R. C., Pretorius, G. H. J., and Barak, E. (1990) G418-resistance as a dominant marker and reporter for gene expression in *Saccharomyces cerevisiae. Curr. Genet.* **18,** 303–313.
6. Casey, G. P., Xaio, W., and Rank, G. H. (1988) A convenient dominant selection marker for gene transfer in industrial strains of *Saccharomyces* yeast: *SMR1* encoded resistance to the herbicide sulfometuron methyl. *J. Inst. Brew.* **94,** 93–97.
7. Sambrook, J., Fritsch, E. F., and Maniatis, T. (1982) *Molecular Cloning: A Laboratory Manual.* 2nd ed., Cold Spring Harbor Laboratory, Cold Spring Harbor, NY.
8. Boulnois, G. J. (ed.) (1987) *Gene Cloning and Analysis: A Laboratory Guide.* Blackwell, Oxford, UK.
9. Church, G. M. and Gilbert, W. (1984) Genomic sequencing. *Proc. Natl. Acad. Sci. USA* **81,** 1991–1995.
10. Feinberg, A. P. and Vogelstein, B. (1984) A technique for radiolabelling DNA restriction fragments to high specific activity. *Anal. Biochem.* **137,** 266–267.

CHAPTER 22

Chromosome Engineering in Yeast with a Site-Specific Recombination System from a Heterologous Yeast Plasmid

Yasuji Oshima, Hiroyuki Araki, and Hiroaki Matsuzaki

1. Introduction

The targeted deletion or inversion of a large chromosomal segment and targeted recombination between two nonhomologous chromosomes can be created in *Saccharomyces cerevisiae* by using the site-specific recombination system from a heterologous yeast plasmid *(1)*. The method consists of the insertion of specific recombination sites from a yeast plasmid with the aid of a yeast integrative vector (YIp) and the introduction of another plasmid carrying the gene for the corresponding site-specific recombination enzyme. This enzyme is stringently regulated by an inducible promoter. When cells are grown under inducing conditions, the site-specific recombination enzyme is produced, stimulating recombination between the two specific recombination sites, and creating a modified chromosome(s).

Numerous chromosomal aberrations result when the 2-µm DNA molecule is integrated into the *S. cerevisiae* chromosome, owing to recombination between the newly inserted 2-µm DNA and the resident 2-µm DNA molecule *(2)*. For this reason, the 2-µm DNA recombination system cannot be used for the manipulation of the *S. cerevisiae* chromosome by the above procedure. Therefore we use a heterologous

From: *Methods in Molecular Biology, Vol. 53: Yeast Protocols*
Edited by: I. Evans Humana Press Inc., Totowa, NJ

plasmid, pSR1, isolated from *Zygosaccharomyces rouxii* (3). The pSR1 plasmid is a 6251-bp circular DNA molecule organized similarly to that of the 2-μm DNA, but without homology to either the 2-μm DNA or *S. cerevisiae* genomic DNA. The pSR1 molecule has a pair of inverted repeats, each composed of a 959-bp sequence, dividing the plasmid into two unique regions of 2654 and 1679 bp (Fig. 1A). The intramolecular recombination initiates by binding the R protein at specific 12-bp inverted elements separated by a 7-bp spacer sequence (4) in the inverted repeats (5). The pSR1 site-specific recombination system operates efficiently in *S. cerevisiae* and the R protein is strictly specific for 12-bp element with the 7-bp spacer. The FLP protein, which catalyzes the similar recombination of 2-μm DNA, has no specificity for this recombination site (RS) (6). Thus, the resident 2-μm DNA in an *S. cerevisiae* cell does not have to be removed to apply the pSR1 system. This recombination system has been successfully applied to plant cells (7).

In the practical application of the pSR1 system, a 2.1-kb DNA fragment (Fig. 1A) bearing the specific RS sequence on the inverted repeats of pSR1 was inserted using YIp vectors at target sites on a single or two different *S. cerevisiae* chromosomes (1). The cells were then transformed with a YEp (or YRp) plasmid bearing the *R* gene under control of the *GAL1* promoter (Fig. 1B). When the transformants were cultivated in galactose medium, *R* expression induced chromosome modification owing to recombination between two specific RSs inserted on the chromosome(s). This procedure has been used successfully for the physical determination of gene-order of two or more closely linked genes (8) or gene and centromere (9).

2. Materials

Note: Basic molecular genetic techniques, e.g., plasmid DNA manipulation, yeast transformation, and Southern blottings are detailed in Chapters 9, 16, and 18.

1. Yeast strains with several selectable markers, e.g., the *his3, leu2, trp1,* and *ura3* mutant alleles.
2. YIp vectors bearing a *S. cerevisiae* DNA fragment (e.g., fragments A, B, or E; Figs. 2 and 3) to target the integration, the RS sequence from pSR1 (Fig. 1), as well as appropriate selective marker(s) (SM1 and SM2; Figs. 2 and 3).
3. The *R* expression plasmid (RE plasmid; e.g., pHM153; Fig. 1B).

Chromosome Engineering in Yeast

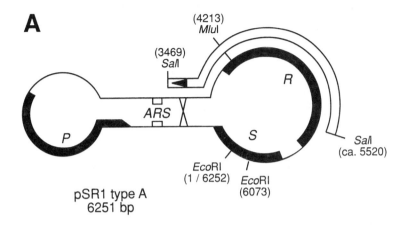

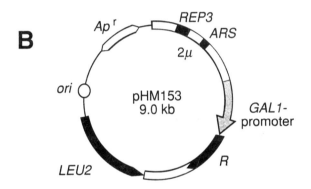

Fig. 1. Structure of pSR1, RS fragment, and RE plasmid (pHM153). **(A)** Structure of pSR1 and of the RS fragment. The linear portions of pSR1 represent the inverted repeats, the thick lines marked *P, R,* and *S,* indicate the open reading frames and each tapered end is 3' end of the frame. The open box with *ARS* represents autonomously replicating sequence. A cross connection between the inverted repeats shows the approximate position of the specific recombination site. Numbers at the restriction site represent their positions with respect to G of the *Eco*RI site as position 1. The 2.1-kb *Sal*I fragment extending from nucleotide position 3469 to ca 5520 (exact position is not determined) was used as the RS fragment and the arrowhead in the fragment indicates the specific RS site. Both *Sal*I sites were created by linker insertion *(5)*. **(B)** Structure of the *R* protein-producing plasmid, pHM153. The thin line with *Ap^r* and *ori* is a DNA fragment from pBR322. Double lines with closed boxes marked *LEU2* and *GAL1*-promoter are from *S. cerevisiae*, *R* is from pSR1, and *REP3* and *ARS* are from 2-μm DNA.

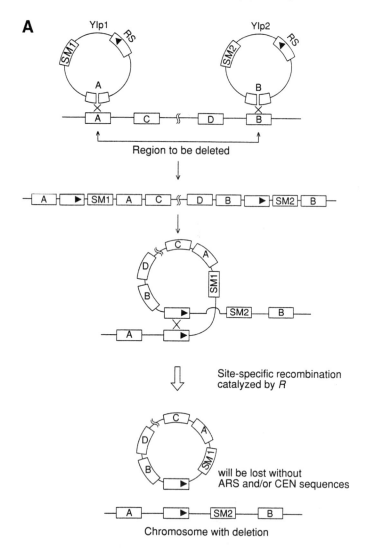

Fig, 2. Schematic illustration of events creating deletion (**A**) and inversion (**B**) of a chromosomal segment. Symbols: *RS*, RS fragment; SM1 and SM2, selection markers 1 and 2; A, B, C, and D, genes *A, B, C,* and *D*.

3. Methods (*see* Materials Note)

3.1. Deletion or Inversion of a Targeted Chromosome Segment

1. Construct two YIp vectors with appropriate selectable genetic markers (SM1 or SM2; Fig. 2) bearing the RS fragment and the DNA fragment

Chromosome Engineering in Yeast

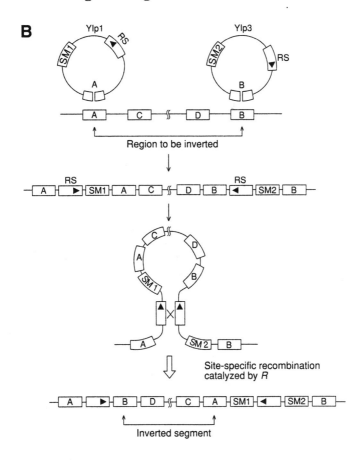

Fig. 2B.

(fragment A or B) for targeted insertion. To delete a chromosomal segment, the two RS fragments should be inserted in the same orientation at two sites (locus A and B on the chromosome) flanking the region to be deleted (Fig. 2A). For inversion, the RS fragments should be inserted in the opposite orientation to each other (Fig. 2B).

2. Transform the yeast cells with these two YIp vectors, one by one sequentially, each linearized at a site within each chromosomal segment (A or B) with an appropriate restriction enzyme. Select for the SM1 and SM2 markers. Confirm the proper plasmid integration in the transformants by Southern analysis.
3. Transform yeast cells with the proper chromosomal integration of the RS sites with the RE plasmid. Select transformants with both the RE plasmid

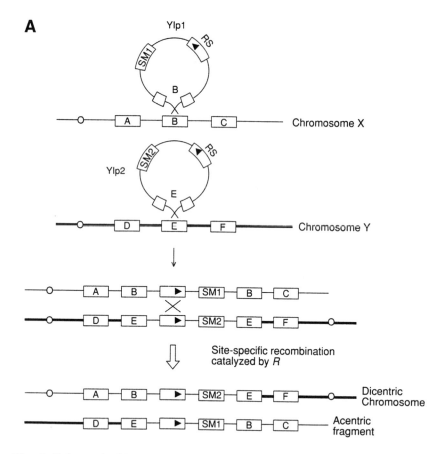

Fig. 3. Schematic illustration of events causing recombination between two nonhomologous chromosomes. (**A**) Creation of two recombinant chromosomes. (**B**) Creation of dicentric and acentric chromosomes. Open circle represents centromere. The other symbols are the same as Fig. 2.

phenotype and those of the integrated YIp vectors. All these manipulations should be done in glucose medium.

4. Inoculate a transformant into a selective medium for the RE plasmid containing galactose as the carbon source (galactose medium) and incubate at 30°C overnight.
5. Spread the galactose-grown cells on a glucose nutrient plate after appropriate dilution for single colonies and incubate the plates at 30°C for 2–4 d. Select several colonies at random and streak them on a glucose nutrient agar.
6. Examine the chromosomal DNAs by pulsed field gel electrophoresis (PFGE, see Chapter 8) or analyze appropriate genetic markers C and D

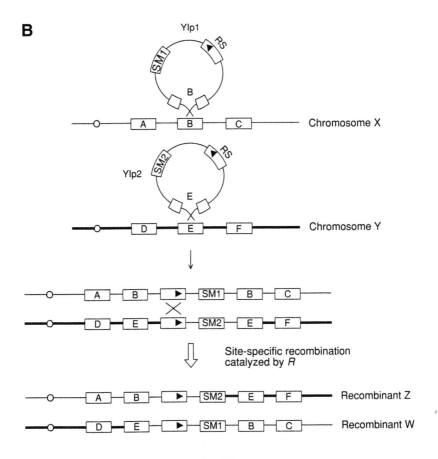

Fig. 3B.

from the region between the two inserted RSs on chromosome in order to detect a deletion (Fig. 2A). To detect the inversion (Fig. 2B), the structure of the chromosomal DNA at the relevant region should be examined by cutting the genomic DNA with appropriate restriction enzymes and Southern blotting. It would be expected that more than 80% of the yeast clones isolated at random have a deletion at the target region, whereas 20–30% of the isolate would have an inversion at the targeted region.

3.2. Recombination Between Two Nonhomologous Chromosomes

1. Insert two RSs on two different chromosomes with the YIP vectors as described above, each one at the site to be recombined. The orientation of the RSs should be the same with respect to each centromere, as illustrated

in Fig. 3A. Otherwise, recombination between the two chromosomes will result in one acentric and one dicentric chromosome (Fig. 3B), a lethal event.
2. Introduce the RE plasmid, induce the expression of R protein by cultivation of the transformant in galactose medium, and isolate single clones by plating the galactose-grown cells on glucose nutrient plates, as described earlier (Section 3.1.).
3. Analyze the chromosomal DNAs by PFGE. On the average, about 20% of the isolated clones have recombinant chromosomes.

4. Notes

1. A diploid strain must be used as the host for deletion of an essential chromosomal segment.
2. The reaction catalyzed by the R protein is reversible and low expression of the R protein in noninducible medium occurs. Therefore, we recommend that the RE plasmid must be cured by growth on nonselective medium after the recombination event is established, to prevent reversion to the original state.
3. By the reason of instability of a foreign DNA in a host cell and for the reasons mentioned in Note 2, insertion of large DNA molecules by the reverse reaction of the deletion of a chromosomal segment is rather difficult.
4. Southern hybridization of a PFGE gel with an appropriate probe DNA is recommended for detection of a short chromosome or a chromosome overlapping with another.
5. Vidal and Gaber *(10)* have recently published on selectable marker replacement in *S. cerevisiae*.

Acknowledgment

The authors thank Christine C. Dykstra, Department of Pathology, University of North Carolina, Chapel Hill, NC, for critical reading of the manuscript.

References

1. Matuzaki, H., Nakajima, R., Nishiyama, J., Araki, H., and Oshima, Y. (1990) Chromosome engineering in *Saccharomyces cerevisiae* by using a site-specific recombination system of a yeast plasmid. *J. Bacteriol.* **172**, 610–618.
2. Falco, S. C., Li, Y., Broach, J. R., and Botstein, D. (1982) Genetic properties of chromosomally integrated 2 μ plasmid DNA in yeast. *Cell* **29**, 573–584.
3. Araki, H., Jearnpipatkul, A., Tatsumi, H., Sakurai, T., Ushio, K., Muta, T., and Oshima, Y. (1985) Molecular and functional organization of yeast plasmid pSR1. *J. Mol. Biol.* **182**, 191–203.
4. Araki, H., Nakanishi, N., Evans, B. R., Matsuzaki, H., Jayaram, M., Oshima, Y. (1992) Site-specific recombinase, R, encoded by yeast plasmid pSR1. *J. Mol. Biol.* **225**, 25–37.

5. Matsuzaki, H., Araki, H., and Oshima, Y. (1988) Gene conversion associated with site-specific recombination in yeast plasmid pSR1. *Mol. Cell. Biol.* **8,** 955–962.
6. Serre, M-C., Evans, B. R., Araki, H., Oshima, Y., and Jayaram, M. (1992) Half-site recombinations mediated by yeast site-specific recombinases Flp and R. *J. Mol. Biol.* **225,** 621–642.
7. Onouchi, H., Yokoi, K., Machida, C., Matsuzaki, H., Oshima, Y., Matsuoka, K., Nakamura, K., and Machida, Y. (1991) Operation of an efficient site-specific recombination system of *Zygosaccharomyces rouxii* in tobacco cells. *Nucleic Acids Res.* **19,** 6373–6378.
8. Bun-ya, M., Harashima, S., and Oshima, Y. (1992) Putative GTP-binding protein, Gtr1, associated with the function of the Pho84 inorganic phosphate transporter in *Saccharomyces cerevisiae. Mol. Cell. Biol.* **12,** 2958–2966.
9. Kawasaki, H., Matsuzaki, H., Nakajima, R., and Oshima, Y. (1991) The *PHO80/TUP7* locus in *Saccharomyces cerevisiae* is on the left arm of chromosome XV: mapping by chromosome engineering. *Yeast* **7,** 859–865.
10. Vidal, M. and Gaber, R. F. (1994) Selectable marker replacement in *Saccharomyces cerevisiae. Yeast* **10,** 141–149.

CHAPTER 23

Insertional Mutagenesis by Ty Elements in *Saccharomyces cerevisiae*

David J. Garfinkel

1. Introduction

The development of genetically marked Ty elements that transpose at high levels has made Ty mutagenesis a useful tool in yeast genetics *(1,2)*. Ty mutagenesis is useful for several reasons:

1. Mutations made by insertion of a marked Ty element into a gene permit the rapid cloning of that gene into *Escherichia coli*.
2. A tagged locus can be rapidly mapped to a chromosome by hybridization to chromosomes separated by electrophoresis.
3. Mutations caused by Ty elements have useful phenotypes.

This chapter will cover the basic protocol used to generate Ty-induced mutations. Further information on the biology of Ty elements and their properties as insertional mutagens can be found in a recent review *(3)*.

2. Materials

Note: Basic molecular genetic techniques not detailed here (e.g., plasmid isolation, yeast transformation, meiotic analysis), are fully covered in other chapters in the book.

1. pGTy plasmid DNA (*see* Note 1).
2. Appropriate *GAL* yeast strains (*see* Note 2).
3. GAL-indicator medium (for identifying *GAL* strains; *see* Note 2): 10 g yeast extract, 20 g peptone, 20 g agar, 20 g galactose, and 80 mg bromthymol blue/L. Galactose and bromthymol blue (prepared as stock

solutions of 20% [w/v] and 4 mg/mL, respectively, and filter sterilized) are added to the medium after autoclaving. Plates are stored in the dark.
4. YEPD: 10 g yeast extract, 20 g peptone, 20 g glucose, and 20 g agar/L.
5. YEPD-G418: Filter-sterilized G418 is added to YEPD just before pouring the plates. Stock solutions of G418 are made at 25–50 mg/mL and stored at –20°C. The plates are stable if stored at 4°C.
6. Synthetic Complete (SC): 5 g ammonium sulfate, 1.5 g Bacto-yeast nitrogen base without amino acids or ammonium sulfate, 20 g agar, 20 g dextrose or galactose, and 2 g "amino acid" (AA) mix/L. The AA mix contains one part each of the L-amino acids alanine, arginine, asparagine, aspartic acid, cysteine, glutamine, glutamic acid, glycine, histidine, isoleucine, lysine, methionine, phenylalanine, proline, serine, threonine, tryptophan, tyrosine, and valine, two parts L-leucine, one part myo-inositol, 0.2 parts para-aminobenzoic acid, 0.5 parts adenine sulfate, and 1 part uracil. The AA mix, Bacto-yeast nitrogen base, and ammonium sulfate are mixed together with the appropriate volume of water (minus the volume of the relevant sugar solution) and autoclaved, then added to an equal volume of molten 4% (w/v) agar. The sugar is added to SC just prior to pouring the plates. Media lacking any defined nutrient can be made by omitting the nutrient from the AA mix; hence, SC-URA is synthetic complete minus uracil.
7. SD: 1.5 g Bacto-yeast nitrogen base, 5 g ammonium sulfate, 20 g agar, and 20 g glucose/L.
8. SC + FOA: Plates are prepared the same way as SC-URA plates with the following modifications. Solid uracil (50 mg/L of medium) and 5-fluoro-orotic acid (FOA), 800 mg, is added to the 2X SC-URA solution, allowed to dissolve, and filter sterilized (*see* Note 3).
9. L-α-aminoadipic acid medium: 0.2% (w/v) L-α-aminoadipic acid, 0.16% Bacto-yeast nitrogen base w/o amino acids and w/o ammonium sulfate, 2% glucose, 30 mg/L L-lysine, 2% Bacto agar. A 6% (w/v) L-α-aminoadipic acid stock solution is prepared by dissolving 6 g of L-α-aminoadipic acid in 100 mL of distilled water and adjusting the pH to 6.0 with 10M KOH; then filter sterilize *(4)*.

3. Methods (Summarized in Fig. 1)

Note: Basic molecular genetic techniques not detailed here (e.g., plasmid isolation, yeast transformation, and meiotic analysis) are fully covered in Chapters 9, 16, and 18.

3.1. Introduction of pGTy Plasmids into Yeast

1. Introduce a pGTy plasmid bearing a *HIS3*-marked Ty element on a *URA3*-based pGTy1 plasmid into a suitable strain using the lithium acetate trans-

Ty Element Mutagenesis

formation method of Ito et al. *(5)*. Cells are plated on SC-URA (glucose) medium and incubated for 2–3 d at 30°C.
2. Clonally purify several transformants by streaking for single colonies on SC-URA (glucose).
3. Single colonies are streaked on SC-URA (glucose), then replica plated to SC-HIS (glucose) to check for the presence of the *HIS3* gene on the pGTy plasmid. Only Ura$^+$, His$^+$ transformants are picked for further study. In addition, the *URA3* and *HIS3* plasmid and Ty markers, respectively, should cosegregate after nonselective growth in YEPD.

3.2. Ty Transposition Assay

1. Streak several transformants from a fresh patch grown on SC-URA (glucose) for single colonies on SC-URA (galactose) solid medium and incubate at 20°C for 5 d (*see* Note 4).
2. The resulting colonies are streaked for single colonies on SC-URA (glucose) to abolish transcription from the *GAL1* promoter and to select cells that retain pGTy1-H3*HIS3* for the entire galactose induction (*see* Note 4).
3. Colonies are picked and then grown on nonselective YEPD plates to allow segregation of the pGTy plasmid.
4. Ura$^-$ segregants are identified by replica plating to SC-URA (glucose) or to medium containing SC + FOA (*see* Note 3).
5. Transpositions are scored after 1 d at 30°C on SC-HIS (glucose) plates if *HIS3* is the Ty marker (*see* Note 5). The transformant that has the highest transposition efficiency is used for Ty mutagenesis.

3.3. Ty Mutagenesis (Isolation of lys2 and lys5 Mutants as Examples, see Note 6)

1. Streak strain DG662 (*MATα his3-Δ200 ura3-167 trpl Δ1 leu2Δ GAL* [pGTy1-H3*HIS3*]) for single colonies on several SC-URA (galactose) plates and incubate at 20°C for 5 d.
2. Replica plate to selective medium containing L-α-aminoadipic acid and incubate at 30°C for 14 d (*see* Note 7).
3. Clonally purify the *lys* mutants and determine if they are lysine auxotrophs and if the lysine requirement is temperature sensitive. Mutants that show any lysine auxotrophy should be saved for further analysis.
4. Determine if the mutation is caused by Ty1*HIS3* insertional inactivation by meiotic segregation analysis (*4; see* Note 8).
5. Identify the mutated *LYS* gene by complementation tests or molecular analyses (*see* Note 9).

4. Notes

1. An effective way to study Ty transposition is to use a series of expression plasmids called pGTy plasmids *(1)*. These consist of the regulated yeast *GAL1* promoter fused to a Ty element at its transcription initiation site and cloned in a multicopy shuttle plasmid. A Ty1 or Ty2 element can be marked by inserting foreign DNA in a nonessential site adjacent to the 3' long terminal repeat (LTR) *(1,6)*. In the presence of galactose, transposition of marked pGTy elements and chromosomal Ty elements increases to high levels *(1,7)*. When cells are propagated on medium containing glucose, the Ty transposition system is repressed. Because the pGTy vector is a plasmid that can be lost by segregation, cells that have experienced a marked Ty transposition can be identified as those that have lost the marker carried on the backbone of the plasmid but retain the marker carried within the Ty element.

   Ty1 and Ty2 elements have been marked with truncated yeast *HIS3* (the *URA3*-based plasmid pGTy1-H3*HIS3* is used in the example presented earlier) and *TRP1* genes, the *E. coli* miniplasmid πN, or the Tn903 *neo* gene *(2,8,9)*. The yeast *HIS3* or *TRP1* genes present in Ty1-H3 allow direct selection in *his3* or *trp1* yeast strains and *hisB* or *trpC E. coli* strains. The *neo* gene confers dominant resistance to the antibiotic G418 in yeast *(10)*, and to neomycin and kanamycin in *E. coli*. The Ty2-917πN element can be used to recover any Ty917-induced mutation directly, since it contains sequences required for selection and replication in *E. coli (2)*. Recently, a derivative of the yeast *HIS3* gene (m*his3*AI) has been developed that is activated by retrotransposition *(11)*. The pGTym*his3*AI vectors should be useful for Ty mutagenesis because plasmid loss is not required to detect marked transpositions phenotypically.

2. The major requirements of any strain used for Ty mutagenesis are that it be Gal$^+$ and contain the relevant mutant alleles for plasmid selection and tracking the Ty marker. Two tests are commonly used for determining the strength of galactose induction. The first involves growing the strains on galactose indicator plates containing the dye bromthymol blue. Optimum results are obtained if relatively few cells are inoculated onto the plates. Strong Gal$^+$ strains turn the agar yellow after incubation for several days at 30°C. The second assay is to compare the level of β-galactosidase produced in various strains containing a p*GAL1-lacZ* expression plasmid pCGS286 (kindly provided by J. Schaum and J. Mao). Standard biochemical methods are used to detect β-galactosidase activity in yeast *(12,13)*. Strong Gal$^+$ strains yield 10,000–20,000 U of activity.

The chromosomal markers used for plasmid selection and tracking Ty transposition should be nonreverting. A dominant marker that has no specific strain requirement is the Tn903 *neo* gene. Presently, resistance to G418 can only be scored on YEPD plates *(14)*. The levels of endogenous resistance to G418 vary among yeast strains. Most strains are sensitive to G418 concentrations in the range of 50–100 µg/mL, but each strain should be tested individually. Occasionally, a strain has a high endogenous resistance to several hundred micrograms of G418/mL. Strains containing Ty*neo* transpositions are routinely scored at 100–400 µg of G418/mL. In genetic crosses, the high and low endogenous resistance phenotype segregates as a multigenic trait. Ty*neo* transpositions can be followed in crosses if colonies are replica plated to YEPD containing different G418 concentrations *(6)*.
3. These plates are very useful for selecting against cells containing the *URA3* gene *(15)*. Although less FOA may be used in the selective plates, this results in more background growth of Ura$^+$ cells. Supplemented SD plates can also be used for FOA selection.
4. The efficiency of Ty1-*HIS3* transposition using the pGTy system is highest at incubation temperatures of 20°C or less. The transposition efficiency drops about fivefold at 30°C, the optimum growth temperature for *S. cerevisiae*, and is practically undetectable at 37°C. Most of the single colonies that form are relatively small, although some larger colonies also appear. Cells taken from large colonies as well as cells taken from confluent areas of growth have fewer marked transpositions.

Transposition also occurs at about the same efficiency in SC-URA (galactose) liquid cultures. Marked transpositions are detected earlier if the cells are pregrown in a nonrepressing carbon source such as glycerol (3% [v/v] final concentration) or raffinose (2% [w/v] final concentration), and then diluted about 20- to 40-fold into SC-URA (galactose) liquid medium. Cells are plated for single colonies on SC-URA (glucose) and processed as described in Section 3.2. Under these conditions, maximum transposition frequencies are observed after incubation with aeration at 20°C for about 16 h.
5. The transposition efficiency is defined as the total number of His$^+$, Ura$^-$ segregants divided by the total number of colonies analyzed. This assay gives an underestimate of the transposition frequency in that some of the cells have multiple insertions of the marked element. In strong Gal$^+$ strains, at least half of the cells that retain the pGTy plasmid for the entire induction period have at least one marked Ty1*HIS3* transposition.
6. There are many variables that can be manipulated to optimize the level of Ty transposition for a particular mutant search. These include temperature, length of induction, type of pGTy plasmid, cell ploidy, and other geno-

typic features. In instances where a selection exists for a particular mutant class, such as in the selection of *lys2* and *lys5* mutants described above, the induction plates can be replica plated directly and the resulting mutant colonies can be analyzed for marked transpositions. Any His$^+$, Ura$^-$ colony has a potential Ty1*HIS3*-induced mutation in the target gene. An alternative is to first enrich for cells containing at least one Ty1*HIS3* insertion and then perform the specific mutant selection or screen. Thousands of independent Ty1*HIS3* insertions can be collected by making use of the FOA selection and the instability of the pGTy plasmid under galactose-inducing conditions. Only those cells containing a Ty1*HIS3* transposition and lacking the plasmid are able to grow on SC + FOA plates that also lack histidine (FOA-HIS; this medium is prepared the same way as SC + FOA, except a nutrient mix is used that does not contain histidine). Cells from the galactose-induction plates can be directly replica plated to FOA-HIS, or cells taken from the galactose plates can be suspended in water, diluted, and plated on FOA-HIS. Transposition libraries can be constructed by making permanent stocks (15% glycerol [v/v], stored at −70°C) from the FOA-HIS cultures. To obtain more marked transpositions per cell, induction plates can be replica plated to fresh SC-URA (galactose) plates. However, multiple genetic crosses are then required to isolate the relevant marked insertion (*see* Notes 8 and 9). Figure 1 presents a general scheme for Ty mutagenesis.

Mutations can be caused by unmarked Ty elements and by other spontaneous events as well as by Ty1*HIS3*. The background of unwanted mutations depends on several factors, including the length of galactose induction, the particular screen or selection, and whether cells without a marked transposition are still present in the population. It would be advantageous to reduce the background created by unmarked Ty elements. In the present study, 12% of the *lys2* (6/52) and *lys5* (1/8) mutants are caused by Ty1*HIS3*, and 21% (11/52) of the *lys2* mutants are caused by unmarked chromosomal Ty elements. It should be possible to virtually eliminate the chromosomal Ty transpositions by inducing transposition in an *spt3* mutant background. The *SPT3* gene was originally isolated as an extragenic suppressor of Ty-induced mutations *(16)*. It is required for transposition of chromosomal elements, but transposition of the *GAL1*-promoted Ty elements is relatively unaffected (Table 1) *(17)*. However, *spt3* mutants also affect diploid formation and sporulation *(18)*. These pleiotropic defects in normal yeast physiology may have unforeseen consequences in various mutant searches.

7. This selection yields mostly *lys2* mutants (≥90%) and a few *lys5* mutants *(19)* and has been used successfully to recover Ty insertions at *LYS2* in

normal cells (1–5% are Ty-induced mutations) *(20,21)*, and in cells undergoing high levels of Ty transposition (30–40% are Ty-induced) *(1)*.

8. In the example presented earlier, this cross indicates if the *lys* mutant contains multiple marked transpositions, and if the mutant phenotype is linked to a Ty1*HIS3* insertion. Each mutation caused by the marked Ty1*HIS3* element should carry a functional *HIS3* gene genetically linked to the new mutation. Two types of *lys* mutants are presented as examples: One in which the marked Ty is the only copy of Ty1*HIS3* in the genome, and another in which there are two copies of Ty1*HIS3* in the genome. In three different Ty1*HIS3*-induced mutants at *LYS2* (*lys2-941*, *lys2-956*, and *lys2-923*), in which the marked insertion is the only copy in the genome, *HIS3* segregates as a gene tightly linked to *lys2* (Table 1). Multiple unlinked Ty1*HIS3* transpositions should assort independently during meiosis. As a result, the ratio of His$^+$:His$^-$ segregants should increase as the number of unlinked Ty1*HIS3* transpositions increases. For example, if two unlinked copies of Ty1*HIS3* are present in the genome, the ratio of His$^+$:His$^-$ progeny should be 3:1, if three copies are present the ratio should be 7:1, and so on. To test these predictions, I show the segregation pattern of a *lys2* mutant that contains two Ty1*HIS3* transpositions (Table 2). In 13 tetrads, the ratio of His$^+$:His$^-$ segregants approaches 3:1 (38:14), but there are no His$^-$, Lys$^-$ segregants present (Table 2). These results suggest that there are two unlinked Ty1*HIS3* insertions in the genome, and one of these has mutated *LYS2*.

In the absence of a *LYS5* hybridization probe, segregation of the Ty1-*HIS3* transpositions and the *lys5* mutation has been analyzed by tetrad analysis. When crossed with a suitable strain, seven of the eight *lys5* mutants are not marked by Ty1*HIS3* because His$^-$, Lys$^-$ segregants appear among the progeny. One cross shows a different segregation pattern (Table 2). Even though two *HIS3* genes segregate, there is an association between the *lys5* mutation and one copy of the *HIS3* gene. No Lys$^-$, His$^-$ progeny appear in the cross. Subsequent analysis of a segregant that contains only the Ty1*HIS3*-induced *lys5* mutation *(lys5-973)* confirms the linkage of the His$^+$ and Lys$^-$ phenotypes and their 2:2 segregation (Table 1).

9. Standard complementation tests to establish allelism and dominance relationships can be done with Ty-induced mutants. In the example presented here, *lys2* and *lys5* tester strains are readily available for performing complementation tests. However, the influence of cell type on Ty-induced mutations, called *ROAM* mutations, that activate adjacent gene expression should be kept in mind *(22)*. These mutations result in constitutive gene expression in *MAT*a, α, a/a, or α/α cells, but are repressed in a/α cells. Therefore, mutations appear recessive in a/α diploids and dominant in a/a

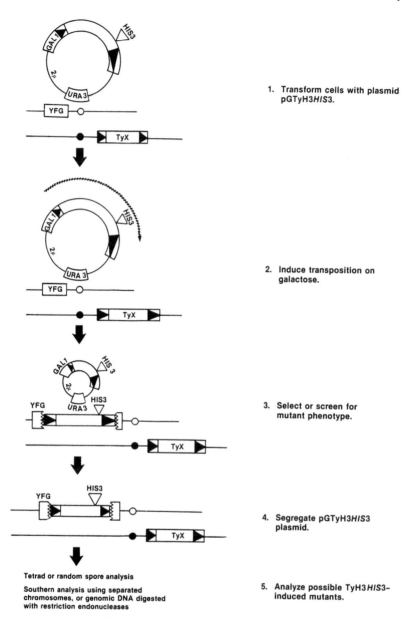

Fig. 1. Flowsheet for isolating marked Ty-induced mutations. The pGTy1-H3*HIS3* plasmid is shown at the top. Yeast chromosomes are represented below the pGTy plasmid. The thin circular line represents pBR322 sequences. Boxed segments represent yeast genes. The arrows represent Ty LTRs and point in the direction of transcription. The *HIS3* marker gene is in the

Table 1
Tetrad Analysis of Single Ty1*HIS3* Transpositions

Target gene	Gene pair	Ascus type[a]		
		PD	NPD	T
LYS2	lys2-941/HIS3	20	0	0
LYS2	lys2-956/HIS3	18	0	0
LYS2	lys2-923/HIS3	14	0	0
LYS5[b]	lys5-973/HIS3	18	0	0

[a]PD, Parental ditype; NPD, nonparental ditype; T, tetratype. Only tetrads with four viable spores were included. These asci showed 2:2 segregation for both markers.
[b]Represents the tetrads from two different His⁺, Lys⁻ segregants.

Table 2
Tetrad Analysis of Ty1*HIS3*-Induced *lys2* or *lys5* Mutants That Contain an Additional Marked Transposition

Target gene	Tetrads analyzed[a]	Spore phenotype				His⁺:His⁻[b]
		His⁺,Lys⁺	His⁺,Lys⁻	His⁻,Lys⁺	His⁻,Lys⁻	
LYS2	13	12	26	14	0	38:14 (2.7:1)
LYS5	32	33	64	31	0	97:31 (3.1:1)

[a]Only tetrads with four viable spores are included.
[b]The total number of His⁺:His⁻ spores present.

or α/α diploids. Because the mutant phenotype may be suppressed or altered in a/α diploids, the mutations should be characterized in diploids with an a or α phenotype. There are several genetic tricks used to create a/a or α/α that can be incorporated easily into the mutant analysis. For example, if mutagenesis is done in an α strain, *MAT*a and *mat*a testers can be used in the complementation analysis. In the resulting diploids, *ROAM* mutants are recessive in the first case (*MAT*a/*MAT*α), and dominant in the second case (*mat*a/*MAT*α) because a/α repression does not occur.

---

same transcriptional orientation as the TyH3. The wavy line represents *GAL1*-promoted Ty transcription. On the chromosomes are a hypothetical target gene *YFG*, and a chromosomal element, TyX. Mutations in *YFG* can be caused by TyH3*HIS3* or TyX or by other mutagenic events.

There are several ways to clone a Ty-induced mutation and/or the wild-type gene. Any recessive Ty-induced mutation can be cloned by complementation using standard techniques *(23)*. Since *ROAM* mutations are dominant, these Ty-induced mutations can be cloned by constructing a library from the mutant strain. Marked Ty element insertions can be isolated from a clone bank made from a mutant strain by using the Ty marker as a probe in colony or plaque hybridizations. Such flanking sequences can then be used to isolate the wild-type gene. Ty2-917πN vectors have also been used to directly recover several random transpositions in *E. coli (2)*.

Acknowledgments

Research sponsored by the National Cancer Institute, DHHS, under contract no. N01-C0-46000 with ABL. The contents of this publication do not necessarily reflect the views or policies of the Department of Health and Human Services, nor does the mention of tradenames, commercial products, or organizations imply endorsement by the US Government.

The author is grateful to J. N. Strathern for his enthusiasm and enlightened conversations, to N. Sanders for expert technical assistance, and to A. Arthur, P. Hall, and J. Hopkins for preparing the manuscript.

References

1. Boeke, J. D., Garfinkel, D. J., Styles, C. A., and Fink, G. R. (1985) Ty elements transpose through an RNA intermediate. *Cell* **40**, 491–500.
2. Garfinkel, D. J., Mastrangelo, M. F., Sanders, N. J., Shafer, B. K., and Strathern, J. N. (1988) Transposon tagging using Ty elements in yeast. *Genetics* **120**, 95–108.
3. Boeke, J. D. and Sandmeyer, S. B. (1991) Yeast transposable elements, in *The Molecular and Cellular Biology of the Yeast Saccharomyces: Genome Dynamics, Protein Synthesis, and Energetics*, vol. 1 (Broach, J. R., Pringle, J., and Jones, E., eds.), Cold Spring Harbor Laboratory, Cold Spring Harbor, NY, pp. 193–261.
4. Sherman, F., Fink, G. R., and Hicks, J. B. (1986) *Laboratory Course Manual for Methods in Yeast Genetics*, Cold Spring Harbor Laboratory, Cold Spring Harbor, NY.
5. Ito, H., Fukuda, Y., Murata, K., and Kimura, A. (1983) Transformation of intact yeast cells treated with alkali cations. *J. Bacteriol.* **153**, 163–168.
6. Curcio, M. J., Sanders, N. J., and Garfinkel, D. J. (1988) Transcriptional competence and transcription of endogenous Ty elements in *Saccharomyces cerevisiae*: Implications for regulation of transposition. *Mol. Cell. Biol.* **8**, 3571–3581.
7. Garfinkel, D. J., Boeke, J. D., and Fink, G. R. (1985) Ty element transposition: Reverse transcriptase and virus-like particles. *Cell* **42**, 507–517.
8. Boeke, J. D., Xu, H., and Fink, G. R. (1988) A general method for the chromosomal amplification of genes in yeast. *Science* **239**, 280–282.
9. Eichinger, D. J. and Boeke, J. D. (1988) The DNA intermediate in yeast Ty1 element transposition copurifies with virus-like particles: Cell-free Ty1 transposition. *Cell* **54**, 955–966.

10. Jimenez, A. and Davies, J. (1980) Expression of a transposable antibiotic resistance element in *Saccharomyces cerevisiae:* a potential selection for eukaryotic cloning vectors. *Nature* **287,** 869–871.
11. Curcio, M. J. and Garfinkel, D. J. (1991) Single-step selection for Ty1 element retrotransposition. *Proc. Natl. Acad. Sci. USA* **88,** 936–940.
12. Rose, M., Casadaban, M. J., and Botstein, D. (1981) Yeast genes fused to β-galactosidase in *Escherichia coli* can be expressed normally in yeast. *Proc. Natl. Acad. Sci. USA* **78,** 2460–2464.
13. Guarente, L., and Ptashne, M. (1981) Fusion of *Escherichia coli lacZ* to the cytochrome C gene of *Saccharomyces cerevisiae. Proc. Natl. Acad. Sci. USA* **78,** 2199–2203.
14. Webster, T. D. and Dickson, R. C. (1983) Direct selection of *Saccharomyces cerevisiae* resistant to the antibiotic G418 following transformation with a DNA vector carrying the kanamycin-resistance gene of Tn903. *Gene* **26,** 243–252.
15. Boeke, J. D., Lacroute, F., and Fink, G. R. (1984) A positive selection for mutants lacking orotidine-5'-phosphate decarboxylase activity in yeast: 5-fluoro-orotic acid resistance. *Mol. Gen. Genet.* **197,** 345,346.
16. Winston, F., Chaleff, D. T., Valent, B., and Fink, G. R. (1984) Mutations affecting Ty-mediated expression of the *HIS4* gene of *Saccharomyces cerevisiae. Genetics* **107,** 179–197.
17. Boeke, J. D., Styles, C. A., and Fink, G. R. (1986) *Saccharomyces cerevisiae SPT3* gene is required for transposition and transpositional recombination of chromosomal Ty elements. *Mol. Cell. Biol.* **6,** 3575–3581.
18. Winston, F., Durbin, K. J., and Fink, G. R. (1984) The *SPT3* gene is required for normal transcription of Ty elements in *S. cerevisiae. Cell* **39,** 675–682.
19. Chattoo, B. B., Sherman, F., Azubalis, D. A., Fjellstedt, T. A., Mehvert, D., and Ogur, M. (1979) Selection of *lys2* mutants in the yeast *Saccharomyces cerevisiae* by the utilization of α-aminoadipate. *Genetics* **93,** 51–65.
20. Eibel, H. and Philippsen, P. (1984) Preferential integration of yeast transposable element Ty into a promoter region. *Nature* **307,** 386–388.
21. Simchen, G., Winston, F., Styles, C. A., and Fink, G. R. (1984) Ty-mediated expression of the *LYS2* and *HIS4* genes of *Saccharomyces cerevisiae* is controlled by the same *SPT* genes. *Proc. Natl. Acad. Sci. USA* **81,** 2431–2434.
22. Errede, B., Cardillo, T. S., Sherman, F., Dubois, E., Deschamps, J., and Wiame, J. M. (1980) Mating signals control expression of mutations resulting from insertion of a transposable repetitive element adjacent to diverse yeast genes. *Cell* **22,** 427–436.
23. Rose, M. D., Novick, P., Thomas, J. H., Botstein, D., and Fink, G. R. (1987) A *Saccharomyces cerevisiae* genomic plasmid bank based on a centromere-containing shuttle vector. *Gene* **60,** 237–243.

CHAPTER 24

Reporter Gene Systems for Assaying Gene Expression in Yeast

Robert C. Mount, Bernadette E. Jordan, and Christopher Hadfield

1. Introduction

Reporter gene constructs are used to monitor the expression of genes whose natural product is not readily assayed. The most convenient reporters to use encode enzymes that catalyze a reaction that is easily monitored. Three useful reporter systems for intracellular expression will be described here: β-galactosidase *(1)*, chloramphenicol acetyl transferase (CAT) *(2,3)*, and aminoglycoside phosphotransferase (APT) *(4)*. Each reporter has advantages and disadvantages according to the simplicity or the accuracy of the assay.

Reporter sequences can be introduced into the subject gene as either a transcriptional or a translational fusion (Fig. 1). Transcriptional fusions are constructed by replacing the entire coding sequence of the subject gene with that of the reporter gene; the reporter provides both the translational start (ATG) and stop (TAA) sequences. Translational fusions require the reporter coding sequence (lacking the ATG) to be fused in-frame to the ATG of the subject gene. The construction of such fusions is now facilitated by the use of PCR techniques.

The β-galactosidase-encoding sequence is often used as a translational fusion, although versions are also available for transcriptional fusion. The enzyme encoded hydrolyses β-D-galactosides, two compounds of which provide suitable chromogenic substrates: *o*-nitrophenyl-β-D-galactoside (ONPG) and 5-bromo-4-chloro-3-indoyl-β-D-galactosidase

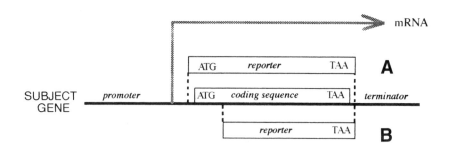

Fig. 1. Reporter gene fusion constructions. (**A**) Transcription fusion; e.g., CAT (both ATG and TAA provided by the reporter). (**B**) Translational fusion; e.g., β-galactosidase, fused in frame to the subject gene.

(X-gal). ONPG hydrolysis releases *o*-nitrophenol, which is yellow and easily measured spectrophotometerically. When ONPG is present in excess, the amount of *o*-nitrophenol produced is proportional to the amount of β-galactosidase present. Hydrolysis of X-gal releases 5-bromo-4-chloro-indigo, which is blue and can be detected easily on agar plates when screening bacteria; however, with yeast the blue color is not so readily obtained. *Saccharomyces cerevisiae* is poorly permeable to X-gal, so coloration is slow. However, colored colonies may be detected after several days' prolonged storage at 4°C, following growth. There is no problem with assaying for the activity in cell protein extracts, however, as the enzyme is released from the cells into assay buffer for reaction with ONPG, and the product is detected spectrophotometrically.

Chloramphenicol resistance can be conferred on *Escherichia coli* and yeast by the expression of the *CAT* gene, making it ideal for the use in shuttle vectors. CAT function is reportedly poor as a fusion protein, and it is therefore better used as a transcriptional fusion. CAT catalyzes the acetylation of chloroamphenicol using acetyl-CoA, which liberates a free CoA sulfhydryl group. The reaction of this reduced CoA with 5,5'-dithiobis-2-nitrobenzoic acid (DTNB) produces a disulfide of CoA and thionitrobenzoic acid and a free 5-thio-2-nitrobenzoate, which can be measured spectrophotometrically.

Like *CAT*, *APT* encodes a resistant phenotype that enables in vivo detection. The in vitro assay for APT described here is very powerful and sensitive. Furthermore, *APT* has the advantage that it can be used as either a transcriptional or a translational fusion. APT catalyzes the specific transphosphorylation of aminoglycoside antibiotics, such as kanamycin, neomycin, and G418. The phosphate is donated by cellular adenosine triphosphate (ATP) and is transferred to the 2-deoxystreptamine moiety at the 3'-hydroxyl group. This reaction can be assayed in vitro using [γ-^{32}P]ATP as the phosphate donor, which enables the amount of phosphorylated aminoglycoside to be quantified. There are, however, other similar cellular reactions that must be minimized for the assay to be both reproducible and highly sensitive. Methods have been described that partially purify the APT *(5)*, or separate ATP from the cellular proteins by gel electrophoresis, followed by blotting to detect the APT activity *(6)*. Described herein is an alternative, easier method that removes background phospholabeled proteins, arising because of protein kinase activity in cell protein extracts, by protease digestion, and washing; thereby a suitably low background is provided against which APT transphosphorylation labeling of G418 can be determined accurately.

2. Materials

1. Stock YEP broth: 1% w/v yeast extract and 2% w/v bactopeptone, dissolved in distilled, deionized (dd) water and autoclaved at 15 psi for 15 min. This can be stored at room temperature for up to 2–3 mo. For YEPD broth, add glucose to 2% w/v from a sterile stock solution (40% w/v, sterilized by autoclaving at 10 psi for 20 min).
2. Liquid minimal medium (SD): Prepare by mixing together appropriate separate sterile stocks. Starting with sterile dd water (e.g., 400 mL), the following are added: Difco (Detroit, MI) yeast nitrogen base to 1X concentration from 10X stock (prepared according to manufacturer's instructions), glucose (or other carbon source) to 2%, plus required supplements (amino acids to 40 µg/mL [from 4 mg/mL stock solutions], bases to 25 µg/mL [from 1 mg/mL stock solutions]).
3. Chloramphenicol selective medium (YEPGE-Cm): Prepare from YEP broth, to which is added glycerol to 1.5% (v/v) (from 60% sterile stock) and ethanol to 1% (v/v), from an absolute ethanol stock, in which the chloramphenicol has been dissolved (*see* step 11). For haploid laboratory strains, use a final concentration of 2 mg/mL chloramphenicol in this medium.
4. G418 selective medium (YEPD-G418): Prepare from YEPD broth, to which solid G418 has been directly added (owing to low solubility, a concentrated stock solution cannot be made). For haploid laboratory strains, use a final concentration of 0.5 mg/mL G418 in this medium.

5. β-galactosidase assay and extraction buffer (Z buffer): 60 mM Na$_2$HPO$_4$, 40 mM NaH$_2$PO$_4$, 10 mM KCl, 1 mM Mg$_2$SO$_4$, pH 7.0. Autoclave at 15 psi for 15 min store at room temperature. Just prior to use, add β-mercaptoethanol to 50 mM (e.g., from a 1M stock solution, which can be stored frozen at –20°C).
6. CAT extraction buffer: 50 mM Tris-HCl, pH 7.5, 100 mM NaCl, 0.1 mM EDTA, 0.1 mM chloramphenicol, 1 mM phenylmethylsulfonylfluoride (PMSF).
7. APT extraction buffer: 10 mM Tris-HCl, pH 7.0, 10 mM MgCl$_2$, 1 mM dithiothreitol, 400 mM NaCl, 1 mM PMSF.
8. Acid-washed glass beads: Glass beads (0.5-mm diameter) are prepared by soaking in concentrated nitric acid for a minimum of 1 h. They are then extensively washed with sterile dd water to a neutral pH and thoroughly dried in a baking oven. To maintain sterility, avoid the use of nonsterile spatulas in the stock.
9. CAT assay buffer (CAB): 50 mM Tris-HCl, pH 7.5, 100 mM EDTA, 1 mM 5,5'-dithiobis-2-benzoic acid, 0.1 mM chloramphenicol. Store at 4°C for up to 4 wk. Add acetyl-CoA to 0.4 mM prior to use.
10. APT assay buffer (AAB): 10 mM Tris-HCl, pH 7.5, 10 mM MgCl$_2$, 1 mM dithiothreitol, 400 mM NaCl, 30 mM NH$_4$Cl, 0.3 mg/mL G418, 1 mM mixed hot/cold ATP, pH 7.0 (molecular ratio 1:1976 of [γ^{-32}P]ATP:coldATP). The [γ^{-32}P]ATP used must be of known specific activity (*see* Note 1).
11. *o*-nitrophenol-β-D-galacloside (ONPG): Dissolve ONPG at 4 mg/mL in 100 mM potassium phosphate buffer, pH 7.0 (*see* Note 2); filter sterilize and store frozen.
12. Sodium carbonate: 1M solution in dd water; autoclave at 15 psi for 15 min and store at room temperature.
13. Chloramphenicol: *For CAT assay buffer,* prepare as a 100 mM solution in 50% ethanol; store frozen at –20°C. *For culture medium,* prepare as a 100 mg/mL solution in absolute ethanol; store at –20°C.
14. Sodium dedocyl sulfate (SDS): Make as a 10% w/v solution in dd water.
15. Protease K: 10 mg/mL solution in dd water; filter sterilize and store at –20°C.
16. Scintillation fluid: Nonaqueous type; e.g., OptiScint HiSafe from EG & G Wallac (Gaithersburg, MD).
17. Protein assay: Protein Assay Kit I or II from Bio-Rad (Hercules, CA), composed of protein reagent dye, plus a protein standard (bovine gamma globulin or bovine serum albumin (BSA), respectively).

3. Methods

The first step is to make soluble cell protein extracts; then determine the concentration of protein in the extracts; and finally, assay the enzyme activity present.

3.1. Soluble Protein Extracts

The preparation of cell extracts is essentially the same for each different reporter gene described earlier, the main difference being the extraction buffer employed.

1. Pellet ~10^8 reporter-gene-containing cells from a liquid culture. The mid-logarithmic phase of culture growth is often used as the standard reference point for reporter assays. The following culture volumes can be harvested for assay at this point: 10 mL of cells at 1×10^7/mL in YEPD, YEPD-G418 or YEPGE-chloramphenicol, grown at 30°C with shaking; or 40 mL of cells at 2.5×10^6/mL in minimal medium, grown at 30°C with shaking (see Note 3).
2. Pellet the culture in universal bottles (e.g., using an Heraeus Christ [South Plainfield, NJ] microbiological swing-out centrifuge) at 4000 rpm ($2300g$) for 5 min at 4°C.
3. Pour off the supernatant and resuspend the pellet in 5 mL ice cold extraction buffer (use the buffer specific for the reporter gene—i.e., Z, CAT, or APT). Pool into one universal bottle, if more than one was used.
4. Pellet the culture as before.
5. Resuspend the pellet in 0.25 mL of the appropriate ice-cold extraction buffer; transfer to a 5-mL disposable, capped test tube and keep on ice.
6. Add 0.5 g acid-washed glass beads.
7. Use a benchtop vortex mixer to vortex the mixture for three periods of 30 s, interspaced with 30 s on ice.
8. Confirm cell breakage by microscopic examination. If intact cells remain, repeat the vortex treatment.
9. Add a further 0.75 mL of the appropriate extraction buffer and vortex for 10 s to mix thoroughly. (This step can be omitted, or a smaller volume added, if a more concentrated sample is required; e.g., if the expression level is expected to be very low.)
10. Centrifuge at 5000 rpm ($3600g$) in a swing-out centrifuge, at 4°C, for 5 min.
11. Remove the supernatant with a Pasteur pipet and transfer to a 1.5-mL microcentrifuge tube (kept on ice). Spin at 13,000 rpm ($11,500g$) for 5 min, at 4°C, in a microfuge.
12. Collect the supernatant, which contains the soluble cell proteins. Store on ice or freeze at –20°C (see Note 4).

3.2. Total Protein Estimation by Bio-Rad Microassay (see Note 5)

1. Make dilutions of the protein standard in dd water to 2.5, 5.0, 7.5, and 10.0 µg/mL. (For the relationship between the BSA or gamma globulin standards and the soluble cell protein extract, see Note 6.)

2. Place 0.80 mL of the standards into 1-mL plastic disposable cuvets. Include two blanks (dd water only).
3. Add 0.20 mL Bio-Rad dye reagent to each cuvet.
4. Seal the top of each cuvet with parafilm and mix the contents by inverting several times.
5. Incubate at room temperature for a minimum of 5 min or a maximum of 60 min.
6. Measure the absorbance at 595 nm against a blank. To do this, initially use the other blank to set the initial absorbance reading to zero, before taking readings for the protein standard solutions.
7. Plot the absorbance against the amount of standard protein (*see* Fig. 2).
8. The amount of cellular protein can now be determined by measuring the absorbance at 595 nm of a suitable dilution of the extract as described for the standards. The standard curve is used to convert the A_{595} to µg protein. Multiply by the dilution factor to determine the exact protein concentration.

3.3. Reporter Assay

3.3.1. β-Galactosidase Enzyme Assay

1. For each cell protein extract sample, set up two reactions in 1.5-mL microfuge tubes: 100 µL extract + 900 µL Z buffer, and 50 µL extract + 950 µL Z buffer. Also set up a zero control, containing 1000 µL Z buffer only.
2. Add 200 µL of 4 mg/mL ONPG, vortex for 5 s and place in a 30°C water bath. Commence timing.
3. Visually monitor the reaction to a medium yellow color (optimally, this should take between 30 min and 4 h).
4. Stop the reaction by the addition of 500 µL $1M$ Na_2CO_3 and note the time.
5. Measure the absorbance at 420 nm. Ideally, this should be between 0.3 and 0.7.
6. Calculate the activity according to the following equation:

$$U = [A_{420}/(t \times p \times 0.0045)] \text{ nmol of ONPG converted/mg/min}$$

where A_{420} = Absorbance of *o*-nitrophenol at 420 nm; t = reaction time in minutes; p = total protein used in mg; 0.0045 = molar extinction coefficient.

Typical levels of β-galactosidase activity expected depend on the type of construct used and culture conditions. Levels as high as 5000 U may be achieved with a highly inducible *GAL1* promoter fusion and as low as 1 to 3 U for noninducible gene fusions.

3.3.2. CAT Enzyme Assay (see Note 7)

1. In a 0.5-mL quartz cuvet (1 cm path length) aliquot 490 µL CAB and 10 µL of cell extract (0.2–5.0 µg protein). Immediately place a piece of parafilm over the top of the cuvet and quickly invert the cuvet three times to thoroughly mix the contents (without causing foaming).

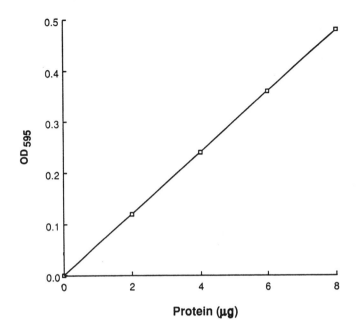

Fig. 2. Protein standard curve. BSA (standard II) assayed by the Bio-Rad microassay technique. *Note:* Micrograms of protein standard = concentration (μg/mL) × volume (0.8 mL).

2. The *initial* rate of increase in absorption at 412 nm is measured (ΔA_{412}/min). See Note 8.
3. The activity is calculated as follows:

$$\Delta A_{412}/\text{min} \times 2 = \Delta A_{412}/\text{min/mL}$$
$$(\Delta A_{412}/\text{min/mL})/(\mu g \text{ protein} \times 10^3) = \Delta A_{412}/\text{min/mL/mg protein}$$
$$(\Delta A_{412}/\text{min/mg})/0.136 = \text{CAT U/mg protein}$$

where 0.136 = M extinction coefficient of DTNB at 412 nm × 10^{-3} (for 1 mL vol).

One enzyme unit is defined as the amount of enzyme that catalyzes the production of 1 μmol DNTB/min. The units can be expressed relative to the cellular protein. Typical levels of expression from high copy number plasmids are 10–20 CAT U/mg protein.

3.3.3. APT Enzyme Assay (see Note 9)

1. Each reaction typically contains: 10 μg extract in a final volume of 33 μL AAB, in a 1.5-mL microfuge tube. Background controls are provided by using extracts of untransformed cells or vector-only transformed cells.

2. Incubate the reaction for a specific time (5–60 min depending on the APT concentration, sufficient to produce a linear response).
3. Add 5 µL 10% SDS and 5 µL protease K (10 mg/mL). (This stops the reaction and degrades kinase-labeled proteins.) Incubate for a further 30 min at 35°C.
4. Spot the reaction onto 1 cm^2 squares of Whatman (Maidstone, UK) P81 paper. For subsequent identification, the paper squares should be marked in advance, using a pencil or ballpoint pen.
5. Leave for 30 s to allow the G418 to adsorb to the P81 paper; then, using forceps, transfer the paper squares to a beaker containing at least 1 L of dd water at 80°C. Wash for 2 min.
6. Then follow with four further similar washes with water at 65°C.
7. Dry the papers under an infrared lamp for 20 min.
8. Transfer the papers to separate scintillation vials containing 4 mL of nonaqueous scintillation fluid. Measure the incorporated [^{32}P] by scintillation counting (*see* Note 10).
9. Subtract the background counts obtained with the negative control from the counts obtained for each assay sample.
10. Convert the cpm/10 µg protein to APT U/mg protein using cpm per pmol [γ-^{32}P]ATP as the conversion factor (*see* Note 1). This is determined by spotting an aliquot of [γ-^{32}P]ATP (of known concentration) onto a P81 paper square, and drying and counting as described earlier.

Typically, a plasmid-encoded APT gene under a strong promoter will produce 2000–3000 APT U/mg, whereas an integrated copy will produce a few hundred U/mg.

4. Notes

1. The specific activity of the [γ-^{32}P]ATP used needs to be known, as this value is used to calculate the pmol amount of [γ-^{32}P]ATP transferred to the antibiotic. Therefore, use [γ-^{32}P]ATP as close as possible to the manufacturer's stated activity date. Also, for reliable standardization, always use [γ-^{32}P]ATP-labeled to the same specific activity by the same manufacturer; e.g., [γ-^{32}P]ATP at 110 TBq/mmol (at reference date), 370 Mbq/mL in aqueous solution from Amersham (Arlington Heights, IL).
2. 100 mM potassium phosphate buffer, pH 7.0 is prepared by mixing together 30.5 mL 100 mM K$_2$HPO$_4$ and 19.5 mL 100 mM KH$_2$PO$_4$ (total vol 50 mL).
3. In principle, reporter gene expression assays can be undertaken after cell growth in any medium. However, in practice, a number of variables influence the choice of cell culture medium. A common one is the stability of the vector. An integrating vector will be highly stably inherited and so growth can be undertaken in any selected medium without loss. On the other hand, episomal plasmids are not so stable, and a selective medium

may be chosen to ensure retention of the plasmid during growth (e.g., minimal, G418-containing or chloramphenicol-containing media). However, different media will yield different levels of expression; and a selective medium may not be appropriate for other reasons. In such cases with an unstable vector, a starter culture can initially be prepared in selective medium and used to inoculate the nonselective medium. Thus, episomal plasmids derived from the 2-µm plasmid are sufficiently stable for 10–15 cell doublings in nonselective medium for instability to be a very minor factor for such a stratagem.

4. Best results are obtained with fresh extracts. Freezing can affect reproducibility in assay. Fresh extracts can be stored satisfactorily on ice for at least 24 h prior to use without any apparent loss of activity or reproducibility.
5. The protein concentration of the extract is required as a parameter to which the activity of the reporter gene can be referred. Any standard protein quantification assay may be employed. However, that of Bradford *(7)* is recommended because of its simplicity and reproducibility, and the Bio-Rad version of the assay is given here.
6. For each batch of protein extracts, a new standard curve should be made for accurate estimation of cellular protein concentration. The best standard for any particular protein is a purified form of the protein itself, but for total cellular protein it is sufficient to use a relative standard, such as BSA (Bio-Rad standard II) or bovine gamma globulin (Bio-Rad standard I). For protein estimation, bovine albumin standard (II) absorbs approximately twice that of average soluble cell proteins. Gamma globulin absorbs the same as soluble proteins and therefore represents a more convenient standard to use.
7. This method is essentially as described by Hadfield et al. *(2,3)*. The assay must be executed at a given controllable temperature, 25°C for example, to enable comparison of different extracts.
8. The slope should be linear. If it curves off rapidly, too much CAT activity is present, indicating that a more dilute protein extract sample should be used. If the rate of increase is very slow, more protein extract can be used, provided that it is concentrated enough. Alternatively, the reaction can simply be left incubating for a longer period of time.
9. This method is as described by Hadfield el al. *(4)*. The reaction utilizes [^{32}P]-radiolabeled ATP and consequently should be completed in a designated radioactive area abiding by local radioactive safety regulations.
10. Scintillation vials often carry electrostatic charges, which provide variable background counts. This is a significant source of possible error, particularly when low APT activities are being measured. Therefore, use a scintillation counter with a built-in antistatic device and pass the samples through three times, and look for reproducible readings.

11. As an alternative to the cytoplasmic reporter proteins described here, β-lactamase has recently been used as a secreted reporter of yeast promoter functions *(8)*.

References

1. Guarente, L. (1983) Yeast promoters and *lacZ* fusions designed to study expression of cloned genes in yeast. *Methods Enzymol.* **101,** 181–191.
2. Hadfield, C., Cashmore, A. M., and Meacock, P. A. (1986) An efficient chloramphenicol-resistance marker for *Saccharomyces cerevisiae* and *Escherichia coli*. *Gene* **45,** 149–158.
3. Hadfield, C., Cashmore, A. M., and Meacock, P. A. (1987) Sequence and expression characteristics of a shuttle chloramphenicol-resistance marker for *Saccharomyces cerevisiae* and *Escherichia coli*. *Gene* **52,** 59–70.
4. Hadfield, C., Jordan, B. E., Mount, R. C., Pretorius, G. H. J., and Burak, E. (1990) G418-resistance as a dominant marker and reporter for gene expression in *Saccharomyces cerevisiae*. *Curr. Genet.* **18,** 303–318.
5. Jiminez, A. and Davies, J. (1980) Expression of a transposable antibiotic resistance element in *Saccharomyces*. *Nature* **287,** 869–871.
6. Reiss, B., Sprengel, R., Will, H., and Schaller, H. (1984) A new sensitive method for qualitative and quantitative assay of neomycin phosphotransferase in crude cell extracts. *Gene* **30,** 211–218.
7. Bradford, M. M. (1976) A rapid and sensitive method for the quantitation of microgram quantities of protein utilizing the principle of protein-dye binding. *Anal. Biochem.* **72,** 248–254.
8. Cartwright, C. P., Li, Y., Zhu, Y.-S., Kang, Y.-S., and Tipper, D. J. (1994) Use of β-lactamase as a secreted reporter of promoter function in yeast. *Yeast* **10,** 497–508.

CHAPTER 25

Preparation and Use of Yeast Cell-Free Translation Lysate

Alan D. Hartley, Manuel A. S. Santos, David R. Colthurst, and Michael F. Tuite

1. Introduction

The yeast *Saccharomyces cerevisiae* is now a standard model organism for the dissection of eukaryotic transcriptional and translational mechanisms (e.g., *1*). The study of the mechanism and control of eukaryotic translation requires an in vitro system that will reflect in vivo events and yet be amenable to manipulation. To this end several protocols for preparing cell-free translation systems from the yeast *S. cerevisiae* have been published *(2–4)*.

Here the authors describe an optimized method for the study of translation initiation, elongation, and termination events in yeast cell-free lysates. The method is based on that summarized by Tuite and Plesset *(4)* in which spheroplasts are lysed and a postpolysomal supernatant, containing all components necessary for protein synthesis, is obtained by centrifugal fractionation. The system is made dependent on exogenous K^+, Mg^{++}, amino acids, and nucleotides by gel filtration through Sephadex G-25. A mild treatment of the S100 obtained with micrococcal nuclease removes endogenous messenger RNA *(5)*, allowing the system to be programmed with exogenous RNA templates.

2. Materials

2.1. Cell Disruption and Cell-Free Lysate Preparation

1. 6X 500 mL YEPD in 2-L conical flasks. One liter of YEPD contains 10 g of yeast extract (Oxoid), 10 g of bactopeptone (Difco [Detroit, MI]), and 20 g of glucose.

2. *Saccharomyces cerevisiae:* We use the diploid strains SKQ2N or ABYS1, the latter a multiple protease-deficient strain *(6)* obtained from D. H. Wolf, Biochemisches Institut (Freiburg, Germany).
3. One liter sterile deionized water.
4. 0.5M ethylenediaminetetraacetic acid, disodium salt (EDTA), adjusted to pH 7.6 with 2M sodium hydroxide.
5. β-mercaptoethanol.
6. 500 mL 1M sterile sorbitol.
7. 500 mL 1.2M sterile sorbitol.
8. 500 mL of YM5/0.4M magnesium sulfate. YM5 is made as a 2X stock containing 4% (w/v) glucose, 0.4% (w/v) bactopeptone, 0.2% (w/v) succinic acid, 1.2% (w/v) sodium hydroxide and 1.34% (w/v) yeast nitrogen base with amino acids (Difco). YM5 is sterilized by autoclaving at 10 lb/in^2 for 30 min and then diluted 1:1 with sterile 0.8M magnesium sulfate.
9. 250 mL lysis buffer: 20 mM Hepes-KOH, pH 7.4, 100 mM potassium acetate, 2 mM magnesium acetate. Sterilized by autoclaving at 15 lb/in^2 for 15 min, then cooled to 4°C before adding dithiothreitol (DTT) and phenylmethylsulfonyl fluoride (PMSF) to 2 mM and 0.5 mM, respectively.
10. Lyticase (partially purified grade, Sigma, St. Louis, MO).
11. Column buffer: As for lysis buffer except made 20% (v/v) glycerol.
12. 40-mL Dounce glass/glass homogenizer (Jencons Scientific Ltd., Leighton Buzzard, UK). Bake at 230°C for at least 4 h before use.
13. Pasteur pipets and two 30-mL Corex tubes baked as described.
14. 10% (w/v) sodium dodecylsulfate (SDS).
15. Sephadex G-25 (medium) equilibrated with column buffer (*see* step 11) in a column (h = 30 cm, d = 1.5 cm, bed vol = 53 mL). Store at 4°C.
16. 100 mL of glass beads (BDH [Dorset, UK], 40 mesh, 0.4 mm diameter). The glass beads should be soaked in concentrated HCl for 4 h, thoroughly washed with water (check the pH after each wash to ensure all traces of acid are removed), and bake at 230°C for at least 4 h (*see* Note 1).
17. Two sterile 40-mL screwcap centrifuge tubes and two 10-mL polycarbonate ultracentrifuge tubes.

2.2. In Vitro Translation Assays

(All components that are required for cell-free translation should be stored at –20°C in small, single-use aliquots.)

1. Hepes/dithiothreitol buffer: 400 mM Hepes-KOH, pH 7.4, 40 mM DTT.
2. 100 mM magnesium acetate (MgOAc).
3. 2M potassium acetate (KOAc).
4. "Energy-mix" solution: This is 10X stock containing 250 mM creatine phosphate, 1 mM GTP, and 5 mM ATP.

5. Complete amino acid stock solution minus phenylalanine or methionine depending on which radiolabeled amino acid is to be used (*see* step 11). The final concentration of each of the 20 amino acids in the stock solution should be 2.5 mM.
6. Creatine phosphokinase (CPK, 350 U/mg), 4 mg/mL.
7. Sterile, diethyl pyrocarbonate([DEPC], *Caution:* Hazardous chemical)-treated H_2O. To treat the H_2O with DEPC add 100 µL DEPC/100 mL H_2O and stir for 3–4 h (or overnight) in a loosely capped bottle before autoclaving for 15 min at 15 lb/in^2.
8. 25 mM calcium chloride.
9. Micrococcal nuclease 1 mg/mL. The source of enzyme is important. We routinely use the enzyme supplied by Worthington Biochemicals (Freehold, NJ).
10. 62.5 mM ethylene glycol *bis*-(β-aminoethyl ether) N,N,N',N'-tetraacetic acid (EGTA), stock solution, pH 7.3, adjusted with 1M HCl.
11. [^{35}S]-methionine (specific activity >1000 Ci/mmol) or [^{3}H]-phenylalanine (specific activity > 50 Ci/mmol).
12. Whatman (Maidstone, UK) 3MM, 2.5 cm cellulose filters.
13. Trichloroacetic acid (TCA) 5% w/v (aqueous). *Caution:* Caustic solution!
14. Scintillant: Scintillation vials and vial inserts.
15. 5X SDS sample buffer. Make up as follows: 10 mL 1.25M Tris-HCl, pH 6.8, 20 mL glycerol, 10 mL β-mercaptoethanol, 40 mL 10% (w/v) SDS, and 15 mg bromophenol blue; make up to a final volume of 100 mL with dH2O, aliquot and store at –20°C.

3. Methods
3.1. Cell Disruption and Cell-Free Lysate Preparation

1. Grow a starter yeast culture to stationary phase (ca 2×10^8 cells/mL) in 50 mL YEPD. Use a 200-mL conical flask and incubate the culture at 30°C for 48 h. This stock culture can be stored at 4°C for up to 1 mo without significant loss of viability.
2. Inoculate 6X 500 mL YEPD in 2-L flasks using 100 µL of the starter culture for each flask. Incubate overnight at 30°C with vigorous agitation (150 rpm) until a cell density of $1–2 \times 10^7$ cells/mL (OD_{600} = 1) is reached. This depends on the strain and age of the stock culture. Normally, it should take between 14 and 16 h for the strains listed earlier.
3. Harvest cells by centrifugation at 4400g for 5 min at 4°C. Combine cell pellets in 200 mL cold sterile distilled water and transfer to a preweighed 500-mL centrifuge tube. Centrifuge as before.
4. Discard the supernatant and measure the wet weight of the cell pellet (usually 5–7g from a 3-L culture).

5. Resuspend the cell pellet in 100 mL freshly prepared 10 mM β-mercaptoethanol, 2 mM EDTA, pH 8.0. Incubate for 30 min at 20–25°C with gentle agitation (50 rpm). This ensures more efficient spheroplast formation.
6. Harvest the cells (see step 3) and wash once with 100 mL cold, sterile 1M sorbitol.
7. Resuspend the cells in 100 mL sterile 1.2M sorbitol and add 6 mg of solid lyticase (7). Incubate at 25°C for 60–90 min with gentle agitation (50 rpm). Monitor spheroplast formation every 15 min by diluting 20 µL of the cell suspension in 1 mL 0.1% (w/v) SDS and comparing the OD_{600} against another 20-µL aliquot of the suspension diluted in 1 mL 1M sorbitol. Use H_2O as the blank. Spheroplast formation should be stopped when the difference in OD_{600} reaches 70–80%. If the formation of spheroplasts is poor, add a further 4 mg of lyticase and incubate for a further 30 min. If sufficient spheroplasts do not form after this period of time, an alternative method of cell disruption should be used (see Note 1).
8. Pellet spheroplasts at 3000g for 10 min at 4°C. Wash once with 200 mL 1.2M sorbitol at 25°C, then *gently* resuspend in 250 mL YM5/0.4M magnesium sulfate by stirring with a sterile 10-mL pipet. Incubate at 25°C with slight agitation for 60–90 min.
9. Harvest spheroplasts by centrifugation as described in step 8. At this stage, the spheroplast pellet can be rapidly frozen in a dry ice/ethanol bath and stored at −70°C for 1–2 d.
10. Resuspend spheroplast pellet on ice in 0.5 mL/g original wet wt, of ice cold lysis buffer, and pour into a prechilled, baked, Dounce glass/glass homogenizer, which should also be kept on ice. Full lysis of the spheroplasts usually takes between 20 and 150 strokes of the homogenizer depending on how tight-fitting the homogenizer is. Lysis should be monitored every 20–25 strokes by microscopic examination. At least 80% lysis should be achievable by this method. Poor lysis will result in a preparation of a cell-tree lysate with a low translational activity.
11. Transfer the lysate to a baked 30-mL Corex tube previously kept on ice and centrifuge, using a prechilled rotor, at 30,000g for 10 min at 4°C.
12. With a cold, sterile glass Pasteur pipet, aspirate most of the opaque-yellow fraction into a cold, sterile, ultracentrifugation tube. Be sure to avoid the top lipid layer and the flocculent material above the cell pellet.
13. Centrifuge the opaque-yellow fraction at high speed using a prechilled fixed angle rotor, at 4°C for 30 min at 100,000g.
14. Pre-equilibrate a Sephadex G-25 column by running cold column buffer through it for 30 min prior to sample loading. The flowrate should be adjusted to around 0.5 mL/min. Pre-equilibration and fractionation of the lysate should be conducted in a cold room or cabinet at 4°C.

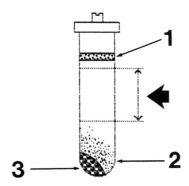

Fig. 1. The fractionation of a yeast cell-free lysate by centrifugation. The top fatty layer (1), the fluffy pellet (2), and clear polysomal pellet (3) are to be avoided. Only the middle layer shown by the large arrow should be used.

15. Carefully remove the tube from the rotor and aspirate the middle layer (*see* Fig. 1), into a baked, cold 15-mL Corex tube and put on ice. The polysomal pellet that should remain, can be frozen at −70°C in the tube and used if necessary to prepare polysomal-associated mRNA *(8)*.
16. Remove the buffer layer from the Sephadex G25 column and quickly load the lysate (maximum loading volume 3 mL). Allow the sample to enter into the Sephadex G25 matrix before carefully replenishing the buffer layer. Reconnect the reservoir and collect 1-mL fractions.
17. Split each fraction into 200-µL aliquots in cold sterile 1.5-mL microfuge tubes. Leave a few microliters of each 1-mL fraction to determine the OD_{260} and thereby locate the 1-mL fractions with highest OD_{260}. In general, we find that only the three 1-mL fractions with the highest OD_{260} are useful for in vivo translation assays (*see* Note 2).
18. Rapidly freeze the aliquots in a dry ice/ethanol bath and store at −70°C. Aliquots stored in this manner should retain more than 80% of their activity over a period of 1 yr (*see* Note 3).
19. Fractions with an OD_{260} of more than 20 should be assayed for translational activity (*see* Section 3.2.).

3.2. In Vitro Translation Assays

This method describes the preparation of 12 × 25-µL reactions for the translation of a natural mRNA using [^{35}S]-methionine to label the in vitro synthesized protein (*see* Fig. 2). For elongation assays with synthetic templates, e.g., poly (U), *see* Note 4.

Fig. 2. Autoradiograph showing the [^{35}S]-labeled translation products of homologous and heterologous mRNA translated in the *Saccharomyces cerevisiae* cell-free translation system. Lane 1 Brome Mosaic Virus RNA4 (coat protein); lane 2, *Saccharomyces cerevisiae* polysome-associated mRNA. Lane 1 exhibits a single protein band (BMV coat protein) and so demonstrates the high activity and low background of this type of cell-free translation system.

1. Prepare the following translation cocktail immediately before performing the translation reactions: Hepes/DTT buffer, 22.5 µL; 100 mM MgOAc, 12.5 µL; 2M KOAc, 31 µL; energy mix, 42 µL; amino acid mix minus methionine, 22.5 µL; creatine phosphokinase, 8.5 µL; DEPC-treated H$_2$O, 7 µL.
2. Pipet 100 µL of the translation cocktail into a cold microfuge tube containing 100 µL of an undiluted cell-free lysate. If using a frozen lysate, this should be thawed first; the remaining 100 µL of the lysate aliquot can be refrozen once and stored at –70°C without significant loss of activity. Add 8 µL of 25 mM CaCl$_2$, 10 µL DEPC-treated H$_2$O and 6 µL of micrococcal nuclease. Gently mix and incubate in a 20°C water bath for 10 min.
3. Stop the micrococcal nuclease activity by adding 8 µL 62.5 mM EGTA, then add 50 µCi of [^{35}S]-methionine diluted in H$_2$O to a total volume of 24 µL.
4. Split the reaction mixture into 12 × 19 µL aliquots. To each aliquot add the template mRNA, adjust the volume of each reaction to 25 µL with DEPC-treated H$_2$O and mix gently. Prepare a negative control for each experiment by replacing the template mRNA with the appropriate volume of water. If the optimal amount of template mRNA to add is not known, assay a range of concentrations (1–20 µg/25 µL reaction). We usually add 3–8 µg of each RNA to be translated.

5. Incubate the reactions in a 20°C water bath for 1 h. *Note:* The temperature is critical.
6. Remove 5 µL from each reaction and spot onto 2.5-cm, 3MM Whatman cellulose filters, prelabeled in pencil. Include two blank filters as background controls. Let the filters dry at room temperature. Meanwhile, treat the remainder of the reactions as described in step 11.
7. Once the filters have dried, drop them into a beaker containing 200 mL of ice-cold 5 % (w/v) TCA. Place the beaker on ice for 10 min, periodically stirring with a glass rod to keep the filters separated.
8. Carefully pour off the cold TCA and wash the filters with 200 mL of room temperature 5% (w/v) TCA. Pour off the TCA and place the filters into 200 mL of boiling 5% (w/v) TCA. Leave for 5 min then pour off the hot TCA and wash the filters with 200 mL TCA at room temperature. (*Remember:* The TCA washes will remove unincorporated radioactivity, so handle and dispose of waste solutions with care.)
9. Swirl the filters in 100 mL of absolute ethanol at room temperature, then in 100 mL of ethanol:acetone (50:50 v/v), and finally in 100 mL of absolute acetone. After pouring away the acetone, place the filters individually onto aluminum foil and dry them at 70°C for 30 min.
10. Place each filter in a scintillation vial insert, add 4 mL of scintillant, cap the tubes, and count the incorporated radioactivity.
11. Stop the in vitro translation reactions in the remaining lysate by adding 10 µL of SDS sample buffer to each tube. Vortex and place on ice.
12. Pierce a small hole in the cap of the microfuge tube with a needle and denature the samples in boiling water for 5 min. These samples can either be run on SDS-PAGE immediately or they can be stored at −80°C for several weeks.
13. Load samples onto a suitable SDS-polyacrylamide gel, electrophorese, and autoradiograph the gel (*see* Fig. 2 and Chapter 26, by Grant, Fitch, and Tuite). The incorporation of radioactivity estimated by scintillation counting will give an approximate idea as to the volume of sample to be loaded. If more than 20,000 cpm/5 µL then the whole reaction mixture should be loaded and the signal enhanced by fluorographic methods *(9)*, e.g., by soaking the gel for 15–30 min in Amplify (Amersham) prior to gel drying.

4. Notes

1. Some yeast strains do not spheroplast efficiently and require mechanical disruption with glass beads. This is achieved as follows: Starting at step 4 of Section 3.1., the cell pellet should be resuspended in half volume to weight of cold lysis buffer and transferred to a sterile, 40-mL, screwcap, Oakridge tube kept on ice. Do not exceed more than 12 mL per tube. Add

cold, baked, acid-washed glass beads to just below the meniscus. Put the tubes on ice for 15 min. Agitate vigorously by hand for several periods of 30 s each, placing the tubes back on ice for 1 min between each period. Estimate microscopically the degree of disruption after each period of agitation. Aim for a 70–80% lysis. Insert a sterile plastic pipet to the bottom of the tube and withdraw as much liquid as possible into a baked 30-mL Corex tube. Wash the glass beads with 1 mL cold lysis buffer and pool this wash with the lysate. Continue the lysate preparation as described in step 11 of Section 3.1.

2. The peak of translational activity generally coincides with the peak OD_{260}. In our experience, the translational activity of the fractions occurring after the peak of OD_{260} falls off rapidly because of the existence of a small, heat-stable, translation initiation inhibitor in these fractions (A. Hartley and M. F. Tuite, unpublished).

3. The yeast cell-free translation system is remarkably stable at –70°C when compared with other cell-free systems. If your system appears to have perished after only a few months storage, we strongly recommend that you remake all other labile components of the translation system *before* discarding the lysate or you check the temperature of your freezer.

4. Elongation assays with the synthetic template poly(U) are performed essentially as described in Section 3.2., with the following exceptions: The amino acid stock should lack phenylalanine and 25 µCi (in 25 µL) of [^{3}H]-phenylalanine used in place of [^{35}S]-methionine. The concentration of the magnesium stock solution should be 400 mM (to give a final concentration of 12 mM in the assay). Poly(U) should be added to a final concentration of 10 µg/assay. [^{3}H] is a low energy β-emitter, so translational activity is best quantified by spotting the entire volume of each reaction (25 µL) onto Whatman filters and incorporation of radioactivity determined by scintillation counting (*see* steps 6–8, Section 3.2.).

References

1. Stansfield, I. and Tuite, M. F. (1994) Polypeptide chain termination in *Saccharomyces cerevisiae. Curr. Genet.* **25,** 385–395.
2. Gasior, E., Herrera, F., Sadnick, I., McLaughlin, C. S., and Moldave, K. (1979) The preparation and characterization of a cell-free system from *Saccharomyces cerevisiae* that translates natural mRNA. *J. Biol. Chem.* **254,** 3965–3969.
3. Moldave, K. and Gasior, E. (1983) Preparation of a cell-free system from *Saccharomyces cerevisiae* that translates exogenous messenger ribonucleic acids. *Methods Enzymol.* **101,** 644–655.
4. Tuite, M. F. and Plesset, J. (1986) mRNA-dependent yeast cell-free translation systems: theory and practice. *Yeast* **2,** 35–52.

5. Pelham, H. R. B. and Jackson, R. J. (1976) An efficient mRNA-dependent translation system from reticulocyte lysates. *Eur. J. Biochem.* **67,** 247–256.
6. Achstetter, T., Emter, O., Ehmann, C., and Wolf, D. H. (1984) Proteolysis in eukaryotic cells. Identification of multiple proteolytic enzymes in yeast. *J. Biol. Chem.* **259,** 13334–13343.
7. Scott, J. H. and Schekman, R. (1980) Lyticase: endoglucanse and protease activities that act together in yeast cell lysis. *J. Bacteriol.* **142,** 414–423.
8. Gallis, B. M., McDonnell, J. P., Hopper, J. E., and Young, E. T. (1975) Translation of poly (ribonucleic acid)-enriched messenger RNAs for the yeast, *Saccharomyces cerevisiae,* in heterologous cell-free systems. *Biochemistry* **14,** 1038–1046.
9. Laskey, R. A. (1980) The use of intensifying screens or organic scintillators for visualizing radioactive molecules resolved by gel electrophoresis. *Methods Enzymol.* **65,** 363–371.

CHAPTER 26

Electrophoretic Analysis of Yeast Proteins

Christopher M. Grant, Ian T. Fitch, and Michael F. Tuite

1. Introduction

Polyacrylamide gel electrophoresis (PAGE) of proteins provides a powerful qualitative and quantitative tool for studying yeast protein synthesis. Complex mixtures of proteins can be separated purely on the basis of molecular weight by means of sodium dodecylsulfate-polyacrylamide gel electrophoresis (SDS-PAGE) *(1)*, or alternatively on the basis of a combination of both molecular weight and electric charge using two-dimensional electrophoresis (2D SDS-PAGE). Two methods of 2D SDS-PAGE have been described, namely isoelectric focusing (IEF) gels *(2)* and nonequilibrium pH gradient (NEPHGE) gels *(3)*. NEPHGE gels give better resolution of basic proteins compared to IEF gels (Fig. 1), and this is particularly useful for analyzing yeast proteins that contain many basic amino acids, e.g., ribosomal proteins and histones.

The use of sensitive detection methods allows for the rapid characterization and quantification of yeast proteins synthesized in vitro. In addition, pulse-labeling experiments, using a suitable radiolabeled amino acid *(see later)* can be used to quantify the rate of protein synthesis during time course experiments. The choice of separation technique and detection method used will thus depend on the physical properties of the protein(s) of interest.

Radiolabeling of yeast proteins allows for particularly sensitive detection of both major and minor abundance proteins. To label yeast proteins to a high specific activity requires that the radioactive amino acid used is

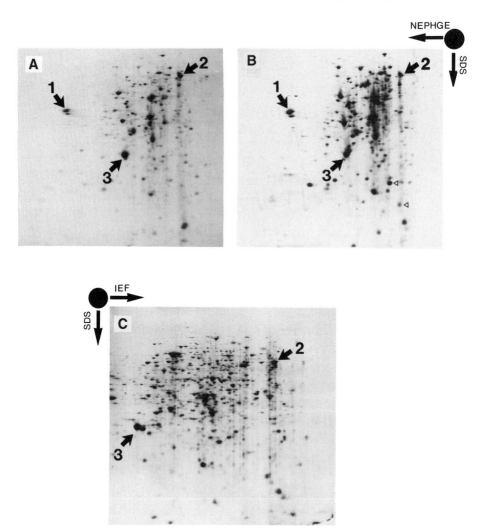

Fig. 1. Two-dimensional SDS-PAGE analysis of proteins synthesized in vivo in the yeast *Saccharomyces cerevisiae*. (A) Exponentially growing cells pulse-labeled for 10 min with [^{35}S]-methionine, then separated in the first dimension by a nonequilibrium pH gradient gel (NEPHGE), and in the second dimension by 15% SDS-PAGE (SDS). (B) As for (A), except labeled with a [^{14}C]-amino acid mixture. Open arrow heads (◁) indicate proteins not labeled with [^{35}S]-methionine. (C) As for (A), except protein samples were run in a first dimension isoelectric focusing gel (IEF). The proteins indicated are: 1, elongation factor EF-1α; 2, the heat shock protein Hsp70 family of proteins; 3, glyceraldehyde-3-phosphate dehydrogenase.

incorporated efficiently into the protein(s) of interest. Thus, the choice of amino acid as the radioactive precursor is critical, and will depend on the intracellular pools of that amino acid and on its frequency of occurrence in proteins. The intracellular pools of amino acids can differ by as much as 250-fold; for example, in nitrogen-rich medium, the intracellular concentration of serine is 5000 µmol/10 g cells, whereas cysteine is only present at around 20 µmol/10 g cells *(4)*. Clearly, the larger the intracellular pool of an amino acid the longer it will take for the radiolabeled form of that amino acid to reach isotopic equilibrium. In the case of the relatively abundant amino acid lysine it takes 70–80 min to reach isotopic equilibrium, whereas methionine, present at relatively low intracellular concentrations, takes less than 30 s *(5)*.

Radiolabeled methionine (L-[^{35}S]-methionine) is used extensively for labeling cellular proteins, since it is commercially available at very high specific activity (>1000 Ci/mmol), it is used to initiate protein synthesis (although the N-terminal methionine is usually cleaved from yeast proteins), and also β-emission is a convenient energy form for radioactive detection and autoradiography. L-[^{35}S]-methionine is also available formulated with [^{35}S]-cysteine (e.g., Trans ^{35}S label, ICN, Biomedicals, High Wycombe, UK) allowing for increased radioactive incorporation, particularly in proteins that do not contain many methionine residues. After methionine, radiolabeled leucine is the most common choice of labeled amino acid for protein synthesis studies, since it is present at low amounts intracellularly, and it is also one of the most prevalent amino acids found in proteins. Leucine, as for most other amino acids, is available in both the [^{3}H]- and [^{14}C]-labeled forms, and the choice of label used will depend on the application. Since the specific activity of [^{3}H]- and [^{14}C]-labeled amino acids (usually 10–100 Ci/mmol) is much lower than that of [^{35}S]-methionine, they are not suitable for the detection of very small amounts of proteins. The use of [^{3}H]- or [^{14}C]-labeled amino acid mixtures allows for the maximum possible incorporation of labeled material into proteins, as shown in Fig. 1.

The electrophoretic analysis of radiolabeled proteins involves three basic steps; incorporating the radioisotope into proteins, preparing protein samples for electrophoresis and the electrophoretic analysis itself. Each of these stages is covered separately in the sections that follow. A recent example of the use of this analytic approach occurs in our study of the mistranslation of human phosphoglycerate kinase in yeast in the presence of paromomycin *(6)*.

2. Materials
2.1. Radiolabeling and Extraction of Proteins

1. Yeast strains should be grown in either YEPD medium (2% w/v glucose, 1% w/v bactopeptone, 1% w/v yeast extract) or, for radiolabeling, in CM medium (0.67% w/v yeast nitrogen base without amino acids, 2% w/v glucose-supplemented with amino acids and bases as required if using autotrophic strains).
2. Appropriate radiolabeled amino acids: For example, we use L-[^{35}S]-methionine (specific activity >1000 Ci/mmol), L-[4,5-^{3}H]-leucine (specific activity 120 Ci/mmol) or [U-^{14}C]-amino acid mixture (activity 0.1 Ci/mL).
3. 20 mM phosphate buffer, pH 7.0. The protease inhibitor phenylmethylsulfonyl fluoride (PMSF) should be added to this buffer to a final concentration of 0.5 mM. PMSF can be stored as a 50 mM stock solution in ethanol at 4°C.
4. Glass beads (0.4–0.45 mm diameter, BDH, Dorset, UK). Glass beads must be cleaned by acid washing in hydrochloric acid, thoroughly rinsed in distilled water, and baked at 200°C for at least 3 h before use.

2.2. SDS-Polyacrylamide Gel Electrophoresis (SDS-PAGE)

1. Acrylamide stock solution: 40% (w/v) acrylamide plus 1% (w/v) bisacrylamide, filtered (0.45 µm filter) and kept at 4°C in the dark for up to 1 mo.
2. 1M Tris-HCl, pH 8.7.
3. 1M Tris-HCl, pH 6.8.
4. 20% w/v sodium dodecyl sulfate (SDS).
5. 10% w/v ammonium persulfate (APS): Made fresh on day of use.
6. N,N,N',N',-tetramethylethylenediamine (TEMED).
7. Running buffer (make as a 10X stock solution): 10 g/L SDS, 144 g/L glycine, 30.3 g/L Tris.
8. Staining solution: 0.1% w/v Coomassie blue in 10% v/v methanol/7.5% v/v acetic acid.
9. Destain: 5% v/v methanol/7.5% v/v acetic acid.
10. Sample buffer (made as a 5X stock solution): 125 mM Tris-HCl, pH 6.8, 20% (w/v) glycerol, 10% (v/v) β-mercaptoethanol, 4% (w/v) SDS, 0.015% (w/v) bromophenol blue. Sample buffer is aliquoted and stored at −20°C.

2.3. Iso-Electric Focusing Gels (IEF) Gels

1. Acrylamide stock solution: 30% (w/v) acrylamide plus 0.8% (w/v) bisacrylamide filtered (0.45 µm filter) and keep at 4°C in the dark for up to 1 mo.
2. Urea (ultrapure grade).
3. 10% v/v Nonidet P-40.

4. Ampholines, pH 3–10, pH 5–7, pH 5–8.
5. 10% w/v ammonium persulfate (APS).
6. N,N,N',N',-tetramethylethylenediamine (TEMED).
7. Cathode solution: 2 mM NaOH.
8. Anode solution: 10 mM H_3PO_4.
9. 2X IEF sample buffer: 9.5M urea, 0.8% (v/v) Nonidet P-40, 10% (v/v) β-mercaptoethanol, 2% (v/v) ampholines, pH 3–10. Aliquot and store at −20°C.
10. SDS-sample buffer: 0.4% (w/v) sodium dodecylsulfate, 0.5M Tris-HCl, pH 6.8, 5% (v/v) β-mercaptoethanol, 10% (w/v) glycerol. Store at 4°C.
11. Agarose solution: SDS-sample buffer plus 1% (w/v) agarose, 0.002% (w/v) bromophenol blue.
12. Electrophoresis system: Standard glass tubing for the first dimension should typically be 140 mm × 3 mm internal diameter and hold approximately 0.5 mL vol.
13. Power supply capable of up to 1000 V and 60 mA.
14. 20-mL syringe and a long small-bore needle at least as long as the glass tubes (i.e., >140 mm).
15. Parafilm.
16. Tygon tubing and a 10-mL syringe.

3. Methods
3.1. Radiolabeling and Extraction of Proteins

1. Pulse-labeling of yeast cellular proteins is achieved by adding a radiolabeled amino acid to an actively growing yeast culture for a short period of time using either 1 µCi L-[^{35}S]-methionine, 0.1 µCi [U-^{14}C]-amino acid mixture or 3 µCi L-[4,5,-^{3}H]-leucine per 10^6 cells in 10-min pulse-labeling experiments. Terminate radioactive incorporation by adding the protein synthesis inhibitor cycloheximide (to a final concentration of 50 µg/mL) together with crushed ice. Harvest the cells (10,000g at 4°C for 10 min), wash them once with 20 mM phosphate buffer, pH 7.0, and store as cell pellets at −70°C.
2. Cell breakage is achieved by rounds of freeze-thawing the cells and subsequent vigorous vortexing with glass beads. Rapidly thaw frozen cell pellets by the addition of 100 µL 20 mM phosphate buffer, pH 7.0, and mix by vortexing. Add glass beads to a level just below the meniscus so that the cells are still free to swirl, and break open cells by vortexing for 1.5 min. Freeze cells at −70°C for 15 min, then thaw slowly on ice for approx 2 h. Remove cell debris and glass beads by a 30-s spin in a microfuge (13000 rpm) or by spinning for 10 min to prepare a soluble cell extract. Remove the supernatant using a micropipet and store at −70°C. Protein yields should typically be in the order of 5 mg/mL.

3.2. SDS Polyacrylamide Gel Electrophoresis (PAGE)

1. Thoroughly clean glass plates and Teflon spacers with ethanol to remove any grease and assemble the apparatus ready to pour the gel as per manufacturer's instructions.
2. Table 1 lists the components necessary for making 5, 10, and 15% polyacrylamide resolving gels. Prepare the gel mixture containing everything except the ammonium persulfate and TEMED, which should be added immediately prior to pouring the gel. Pour the gel using a 50-mL syringe, leaving space for the sample well comb. Take care not to introduce any air bubbles. Overlay the gel with water and leave to polymerize for approximately 30 min at room temperature.
3. The upper "stacking gel" mixture is a 2.5-mL acrylamide stock solution, 2.5 mL $1M$ Tris-HCl, pH 6.8, 0.1 mL 20% w/v SDS, 14.8 mL distilled water, 0.1 mL APS (10% w/v) and 16 µL TEMED. Briefly mix the solutions and pour onto the top of the resolving gel. Use a sample well comb to form wells for sample loading. Polymerization should be complete in 15–30 min at room temperature.
4. Remove the sample well comb and wash the wells extensively with 1X running buffer to remove any unpolymerized acrylamide. Load the gel into the electrophoresis tank and fill the tank with 1X running buffer.
5. Dilute protein samples in a 20-mM phosphate buffer, pH 7.0, and add sample buffer to a concentration of 1X. Denature the samples by boiling for 2 min in a boiling water bath and either load immediately using a Hamilton syringe or store at −70°C until required for future analysis.
6. Electrophorese the gels until the bromophenol blue is approx 1 cm from the bottom of the gel. Separate the gel plates, discard the stacking gel, and place the resolving gel gently into a staining solution. Stain the gels for 1 h with constant agitation and then transfer to destain and leave for 2–24 h.
7. Radioactive gels should be dried onto 3MM filter paper (Whatman, Maidstone, UK) under vacuum on a suitable gel drier (e.g., Bio-Rad [Hercules, CA] Model 543 slab gel drier), at 60°C for 60–90 min. Expose labeled gels to X-ray film (e.g., Fuji NIFRX, Amersham [Arlington Heights, IL] Hyperfilm).

3.3. Iso-Electric Focusing Gel Electrophoresis

1. A gel mixture sufficient for 12 gels should be prepared containing 1.34 mL acrylamide stock solution, 5.5 g urea, 2 mL Nonidet P-40 (10% v/v), 0.1 mL ampholines, pH 3–10, 0.2 mL ampholines, pH 5–7, 0.2 mL ampholines, pH 5–8 and 2 mL H_2O. Gently warm the gel solution to dissolve the urea, taking care not to allow the temperature to rise above 37°C.

Table 1
Composition of Polyacrylamide Resolving Gels

Components	Final Percent (w/v) acrylamide		
	5	10	15
Acrylamide stock, 40% w/v, + bisacrylamide, 1% w/v	5 mL	10 mL	15 mL
H_2O	19.6 mL	14.6 mL	9.6 mL
$1M$ Tris-HCl, pH 8.7	15 mL	14 mL	15 mL
SDS, 20% w/v	0.2 mL	0.2 mL	0.2 mL
APS, 10% w/v	0.2 mL	0.2 mL	0.2 mL
TEMED	30 μL	30 μL	30 μL

2. Seal one end of the clean, dry glass tubes with two layers of Parafilm.
3. Add 20 μL ammonium persulfate (10% w/v) and 14 μL TEMED to the gel mixture, mix, and quickly introduce into the glass tubes to a level within 10 mm of the top using the 20-mL syringe and long small bore needle. Take care not to trap any bubbles at the side or bottom of the tube. Overlay gels with $8M$ urea, hold in a vertical position, and leave to polymerize for approx 30 min at room temperature.
4. Once the gel has polymerized, remove the Parafilm, discard the urea overlay, and set up the tubes in the gel apparatus with the anode solution.
5. Dilute the protein samples to a volume of 50 μL with 20 mM phosphate buffer (pH 7.0) in 1.5 mL microfuge tubes and add solid urea until the solution is saturated. Added urea should be dissolved by gently tapping the Eppendorf tube. When the mixture is saturated, grains of solid urea will be seen at the bottom of the tube. Add 50 μL of 2X IEF sample buffer to the samples and either use immediately or store at −20°C.
6. Microfuge the samples (30 s at 13000 rpm) prior to loading, then load onto the gels with a micropipet and overlay with 50 μL of 3/4X IEF sample buffer. The cathode solution should then be carefully added so as not to disturb the samples.
7. Electrophorese at 325 V for 15 h and then at 1000 V for 1 h.
8. After electrophoresis, remove the gels from the glass tubes using a 10-mL syringe filled with water, and connect to the top of the IEF tube with Tygon TM tubing. Apply slow even pressure to force the gels out of the tube. Equilibrate the gels with 5 mL of SDS-sample buffer for one hour by gently shaking at room temperature and then either load immediately onto a 2D SDS-polyacrylamide gel or store in the SDS-sample buffer at −70°C for 2–4 wk.

9. Run rod gels in the second dimension on flattop SDS-PAGE gels as follows: prepare SDS-PAGE gels as described, but instead of using a comb to form sample wells, overlay the stacking gel with 3 mL of water so as to form a flat surface. Fix the stick gels to the stacking gel with hot (85°C) 1% (w/v) agarose solution using a Pasteur pipet. Take care not to introduce bubbles between the stacking gel and the stick gel. Electrophorese as described in Section 3.2.

4. Notes

1. The 10-min pulse-labeling procedure described here will typically lead to the incorporation of >20000 cpm/µL lysate for [^{35}S]-methionine or >4000 cpm/µL lysate for [^{14}C]-amino acids in early log phase cells (1–3 × 10^6 cells/mL). The amount of radiolabeled amino acid used and length of pulse-label will depend on the choice of label and of amino acid (*see* Introduction).
2. Protein concentration can be readily determined by the method of Bradford *(7)*. To determine radioactive incorporation, spot small aliquots (1–2 µL) of protein extract onto 2.5-cm diameter Whatman 3MM filter discs, oven dry, and sequentially wash in ice-cold 5% (v/v) TCA for 10 min, at room temperature in 5% TCA for 1 min, in boiling 5 % TCA for 5 min, at room temperature 5 % TCA for 1 min, and, finally, rinse in absolute ethanol. Filters should be dried before counting in a scintillation counter.
3. The amount of protein or number of counts to load on an SDS-PAGE gel depends on the size of the gel and on the abundance of the protein of interest. We use gels that measure 15 × 17 cm and routinely load 20–100 µg of protein and 10^6 cpm total counts.
4. For greater sensitivity in protein staining, silver staining procedures *(8)* can be employed.
5. Polyacrylamide gradient gels (e.g., 5–20% acrylamide gels) can be used for greater resolution of proteins of both high and low molecular weight.
6. NEPHGE gels *(3)* are particularly useful for the analysis of yeast proteins, since they allow for the resolution of both acidic and basic proteins. They can be run essentially as described for IEF gels, but with the following alterations: Cathode and anode solutions are reversed and electrophoresis should be performed at 400 V for 4 h.
7. Glass tubes must be very clean to allow for easy removal of the gels. They are best cleaned by soaking in 2*M* chromic acid, rinsing extensively in H$_2$O, soaking in absolute ethanol, rinsing in H$_2$O and, finally, oven dried at moderate temperature (e.g., 55°C).
8. Triton X-100 can be used as a substitute for Nonidet P-40.
9. Typical loadings are 50–100 µL protein or 10^5–10^6 cpm per stick gel.

10. Ampholine concentrations usually represent 2% (v/v) of the total gel volume. When working at pH ranges outside neutrality, it is advisable to add a pH 6–8 or pH 5–8 or pH 5–7 ampholine. The effect of adding a certain range of ampholines is to decrease the slope of the pH gradient for that region.

References

1. Laemmli, U. K. (1970) Cleavage of structural proteins during the assembly of the head of bacteriophage T4. *Nature* **227,** 680–685.
2. O'Farrell, P. H. (1975) High resolution two-dimensional electrophoresis of proteins. *J. Biol. Chem.* **250,** 4007–4021.
3. O'Farrell, P. Z., Goodman, H. M., and O'Farrell, P. H. (1977) High resolution two-dimensional electrophoresis of basic as well as acidic proteins. *Cell* **12,** 1133–1142.
4. Chan, P. Y. and Cossins, F. A. (1976) General properties and regulation of arginine transporting systems in *Saccharomyces cerevisiae*. *Plant Cell Physiol.* **17,** 341–349.
5. Gross, K. J. and Pogo, O. (1974) Control mechanism of ribonucleic acid synthesis in eukaryotes. *J. Biol. Chem.* **249,** 568–576.
6. Grant, C. M. and Tuite, M. F. (1994) Mistranslation of human phosphoglycerate kinase in yeast, in the presence of paromomycin. *Curr. Genet.* **26,** 95–99.
7. Bradford, M. (1976) A rapid and sensitive method for the quantitation of microgram quantities of protein utilizing the principle of dye-binding. *Anal. Biochem.* **72,** 248–254.
8. Merril, C. R., Goldman, D., Sedman, S. A., and Ebert, M. H. (1981) Ultrasensitive stain for proteins in polyacrylamide gels show regional variations in cerebrospinal fluid. *Science* **211,** 1437–1438.

Chapter 27

Preparation of Total RNA

Alistair J. P. Brown

1. Introduction

RNA analysis is central to a detailed understanding of gene expression. The expression of many nuclear genes from a wide range of organisms has been studied at the protein level using reporter genes (*lacZ*, for example), and changes in expression levels in response to specific stimuli are frequently presumed to be mediated at the transcriptional level. However, there are an increasing number of examples of fungal and vertebrate genes that are post-transcriptionally regulated. It is therefore prudent to investigate expression at the RNA and protein levels. This chapter will discuss the isolation of total RNA from yeast. The reader should refer to ref. *(1)* for a detailed description of the methods for preparation of ribosomal and small RNAs from yeast.

RNA preparation is reputed to be difficult, but the method itself is very straightforward. It involves the shearing of yeast cells with glass beads and the extraction of nucleic acid by phenol:chloroform extraction. The integrity of the RNA is then checked usually by agarose gel electrophoresis. Difficulties often arise through failure to prevent contamination with exogenous ribonucleases (RNases). Therefore, the key to success lies in the preparation of RNase-free materials for RNA preparation and in the prevention of subsequent contamination by RNase. Unfortunately, RNases are widespread and frequently extremely stable, and hence extra precautions must be taken compared with DNA work.

Two related methods are described; one for "large"-scale, and the other for "small"-scale RNA preparation. The former, which is particularly useful for the accurate quantitation of multiple mRNAs under identical conditions, is adapted from the method of Lindquist *(2)*. The small scale procedure, which facilitates the rapid analysis of a large number of samples, is adapted from those described by Dobson and coworkers *(3)*.

Having isolated total RNA from yeast, individual RNAs may be analyzed by Northern blotting. Generally, the RNA is subjected to agarose gel electrophoresis following denaturation with formaldehyde *(4)*, glyoxal *(5)*, or methylmercuric hydroxide *(6)*, blotted onto a nitrocellulose or nylon membrane *(5,7)*, and then hybridized with a specific radiolabeled probe *(7)*. Under the appropriate conditions, Northern blotting can be used to quantitate mRNA levels, but dot-blotting is frequently used for the rapid analysis of multiple samples and hence to gain more accurate measurements. These methods are discussed further in Chapter 28.

The procedures for the purification of polyadenylated (poly[A]+) RNA from total RNA preparations are not specific to yeast, and have been described previously. Therefore, these methods, which are based on affinity chromatography using either oligo(dT)-cellulose *(8)* or poly(U)-Sepharose *(9)*, are not discussed here.

2. Materials
2.1. Large-Scale RNA Preparation

All materials must be rendered RNase-free (*see* Section 4.).

1. Glass beads (size approx 0.4-mm diameter) are soaked in concentrated nitric acid for about 1 h, rinsed thoroughly in distilled water, dried overnight in a glassware oven, and then baked at 200°C overnight. Beads are dispensed into RNase-free bottles after heat baking to prevent these bottles from cracking.
2. $1M$ Tris.
3. $1M$ Tris-HCl, pH 7.5.
4. $1M$ LiCl.
5. 10% (w/v) SDS.
6. $3M$ Sodium acetate.
7. RNase-free, distilled water.
8. Phenol: Solid phenol (500 g) is melted at 60°C for about 30 min, and brought to pH 8 by adding RNase-free un-pHed $1M$ Tris. Solid hydroxyquinoline is then added to a final concentration of about 0.02% (w/v). To

saturate the phenol phase, RNase-free distilled water (about 200 mL) is added and mixed by shaking. Upon standing, the phases should separate, with the phenol on the bottom. Store at 4°C (*see* Note 11).
9. Extraction buffer: $0.1M$ LiCl, $0.01M$ dithiothreitol (add fresh), $0.1M$ Tris-HCl, pH 7.5.
10. MOPS buffer (10X): $0.2M$ morpholinopropanesulfonic acid, $0.05M$ sodium acetate, $0.01M$ Na_2EDTA, pH 7.0. Store in a dark bottle.
11. T/E: 1 mM Na_2EDTA, 10 mM Tris-HCl, pH 8.
12. MMF: 500 µL formamide, 162 µL formaldehyde (37%), 100 µL 10X MOPS (solution 10), 238 µL H_2O (*see* Note 11).
13. Ethidium bromide: 0.5 mg/mL (*see* Note 11).
14. Gel loading dye: 0.02 (w/v) bromophenol blue in 50% (v/v) glycerol.

2.2. Small-Scale RNA Preparation

As for the large-scale preparation, except:

1. LET buffer: $0.1M$ Tris-HCl, $0.1M$ LiCl, 0.1 mM Na_2EDTA, pH 7.4.
2. TNES buffer (10X): $0.5M$ Tris-HCl, 1.4 mM NaCl, 50 mM Na_2EDTA, 1.0% (w/v) SDS, pH 7.4.
3. $3M$ potassium acetate.

These materials must be made RNase-free (*see* Notes).

3. Methods
3.1. Cell Storage

It is possible to store cells before RNA preparation. This is particularly useful when performing numerous large-scale RNA preparations over short time courses (for example, in mRNA half-life determinations; *see* Chapter 28 by Brown and Sagliocco). Note that for small-scale RNA preparations, cell pellets can be stored at –70°C before use (*see* Section 2.2., step 3).

1. Add 2 vol of absolute ethanol to cultures, and store at –20°C for a maximum of 24 h. (Storing the cells under ethanol at –20°C for longer periods reduces the yield of RNA.)
2. Harvest the cells from this ethanolic mix by centrifugation at 5000g at 0°C for 5 min and discard the supernatant.
3. Resupend the cell pellet in 10 mL extraction buffer at 0°C, centrifuge at 5000g at 0°C for 5 min, and discard the supernatant.
4. At this point the cell pellet can be stored for long periods at –70°C. Alternatively, prepare the RNA by resuspending the cell pellet in 5 mL extraction buffer and continuing at step 3 of Section 3.2.

3.2. Large-Scale RNA Preparation

1. Prepare tubes containing 14 g glass beads, 1.0 mL 10% SDS, 5 mL phenol, and 5 mL chloroform before harvesting the yeast cells (*see* Note 11).
2. Harvest 100 mL of yeast culture in midexponential growth phase (A_{600} = 0.4–0.6) by centrifugation at 5000g for 5 min.
3. Resuspend the cell pellet rapidly in 5 mL of extraction buffer by vortexing, add the suspension to the tube prepared in step 1, and immediately vortex continuously for 5 min. Ensure that vigorous mixing is achieved to ensure efficient cell breakage.
4. Centrifuge the sample at 5000g for 5 min in a swing-out rotor.
5. Carefully remove the upper aqueous phase with a Pasteur pipet (Fig. 1), transferring it to a fresh tube containing 5 mL phenol and 5 mL chloroform. Avoid taking any of the interphase that contains protein (and hence RNase). Vortex for 1 min, and then separate the phases by centrifugation at 5000g for 5 min.
6. Repeat step 5.
7. Avoiding the interphase, transfer the aqueous phase to a fresh tube containing 5 mL chloroform. Vortex for 30 s, and then separate the phases by centrifugation at 5000g for 2 min.
8. Repeat step 7.
9. Transfer the aqueous phase to a fresh tube, and add sodium acetate to a final concentration of 0.1M, then 2 vol of absolute ethanol, and mix thoroughly. Store overnight at –20°C. The RNA is stable indefinitely under ethanol. A yield of 200–500 µg RNA should be obtained.
10. Vortex the sample to make it relatively homogeneous, and then transfer 0.5 mL (approximately one-thirtieth of the sample) to an Eppendorf tube. Centrifuge at 10,000g at 4°C for 5 min. Pour off the supernatant and drain the tube. Vacuum desiccate for 15 min to remove residual ethanol.
11. Resuspend the precipitate in 5 µL T/E.
12. Add 35 µL MMF and 2 µL ethidium bromide (0.5 mg/mL), mix, and incubate at 60°C for 15 min.
13. Add 10 µL gel loading dye and mix. If necessary, perform a clearing spin to remove particulate matter (5 min at 10,000g in an Eppendorf centrifuge). Leave on ice until ready to load agarose gel.
14. Prepare the formaldehyde/agarose gel in a fume hood by melting 1.5 g agarose (Sigma [St. Louis, MO], Type 1, low EEO; cat. number A-6013) in 73 mL of H_2O. Cool to 60°C, add 10 mL of 10X MOPS and 16.2 mL 37% formaldehyde, mix, and pour gel immediately. Allow the gel to set for at least 30 min (*see* Note 11).
15. Load about one-half of the sample onto the gel and run at 100–150 V for 1–2 h, as necessary, while circulating the electrophoresis buffer (1X MOPS). Visualize the RNA on a UV transilluminator. **Caution:** Screen

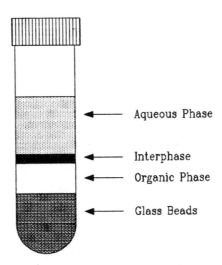

Fig. 1. Separation of phases during phenol:chloroform extraction.

yourself from the UV. Crisp, strong bands of 25S and 18S ribosomal RNA (in about a 2:1 ratio), and 5S and tRNA are visible in intact RNA preparations, with little evidence of smearing (Fig. 2).

3.3. Small-Scale RNA Preparation

1. Harvest yeast cells (about 10–20 mL) in midexponential growth phase (A_{600} = 0.4–0.6) by centrifugation at 5000g for 5 min at 4°C.
2. Resuspend the cell pellet in 1 mL LET Buffer at 4°C, transfer to an Eppendorf tube, and centrifuge for 20 s in an Eppendorf centrifuge at 4°C.
3. Repeat step 2, keeping the samples at 4°C. The cell pellet should not exceed approximately 100 µL. At this stage, the cell pellet can be stored at –70°C.
4. Resuspend the cell pellet vigorously in 100 µL of phenol and 200 µL of LET buffer at 4°C using a vortex mixer.
5. Add glass beads to just below the level of the meniscus, and vortex the sample continuously for 1 min.
6. Place the sample at room temperature, and add 100 µL chloroform, 100 µL of H_2O, and 30 µL of 10X TNES buffer (prewarmed to 55°C). Vortex the samples vigorously for 2 min. Centrifuge the sample at 10,000g for 2 min.
7. Transfer the upper aqueous phase to a fresh Eppendorf tube containing 150 µL phenol and 150 µL chloroform, taking care to avoid the interphase. Vortex the sample for 30 s, and then centrifuge at 10,000g for 2 min.
8. Repeat the phenol:chloroform extraction (step 7) at least two times until the interphase disappears.

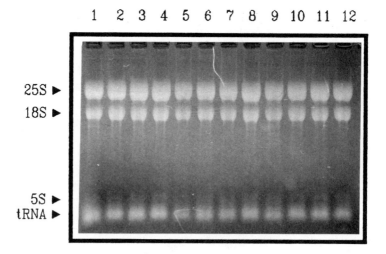

Fig. 2. Agarose gel electrophoresis of yeast RNA. Total yeast RNA prepared according to the "large-scale method" was electrophoresed on a 1.5% (w/v) agarose gel containing formaldehyde as described in the text (provided by Zaf Zaman). Lanes 1–12 contain RNA from several strains of *S. cerevisiae* grown on glucose or raffinose. Ethidium bromide-stained bands corresponding to 25S rRNA, 18S rRNA, 5S RNA, and tRNA are highlighted.

9. Transfer the aqueous phase to a fresh Eppendorf tube containing 800 µL ether, vortex for 20 s, and centrifuge at 10,000*g* for 5 s.
10. Discard the upper ether phase, and add 0.1 vol of 3*M* potassium acetate and 1 mL of absolute ethanol to the sample and mix. Store the sample at –20°C overnight.
11. Assess the integrity of the RNA preparation by gel electrophoresis as described in steps 10–15 of the large-scale RNA preparation, except that about one-tenth of the sample should be used.

4. Notes

All materials must be made RNase-free. It is advisable to prepare an excess of materials in case of a mishap, because the preparation takes time and yet the method should be done quickly.

1. Wrap glassware (Pasteur pipets and pipets) in aluminum foil and bake at 200°C overnight.
2. Only take disposable plasticware (tips, Eppendorf tubes, and so on) from new bags that have been set aside for RNA work, and autoclave.

3. Soak nondisposable plasticware (for example, 50-mL Oakridge centrifuge tubes) in 0.1% (v/v) diethylpyrocarbonate for at least 1 h and then rinse in RNase-free water or incubate at 80°C.
4. Fresh, previously unopened chemicals can be considered to be RNase-free. Set aside a set of chemicals specifically for RNA work to prevent careless contamination by other workers. Do not use spatulas to weigh out these chemicals, because they are a probable source of RNase contamination. Instead, carefully tap these chemicals out of the bottles.
5. To render RNase-free solutions that do not contain amines, add diethylpyrocarbonate to a final concentration of 0.1% (v/v), stand for at least 1 h, and then autoclave to inactivate the remaining diethylpyrocarbonate.
6. Solutions containing amines (for example, Tris or EDTA) are not treated with diethylpyrocarbonate. Instead, extra precautions are taken to prevent contamination with RNase. Add RNase-free chemicals to diethylpyrocarbonate-treated distilled water and pH using concentrated acid or $10M$ NaOH. Monitor the pH by dropping small samples of the solution onto pH papers to prevent RNase contamination of the main solution from a pH electrode. Adjust to the desired final volumes using volumetric markings on the RNase-free solution bottle.
7. Wear gloves throughout.
8. Use only high quality phenol that is colorless on melting. Pink or brown phenol should not be used.
9. Diethylpyrocarbonate is not miscible with water. Shake solutions well to disperse the diethylpyrocarbonate. Diethylpyrocarbonate covalently modifies amine groups, and, therefore, it cannot be used to inactivate RNase in solutions containing Tris or EDTA, for example (*see* Note 6). Diethylpyrocarbonate decomposes to CO_2 and H_2O with a short chemical half-life at room temperature (about 30 min). Solutions can be rendered free of diethylpyrocarbonate before use in RNA preparation by autoclaving or by incubating overnight at room temperature.
10. Increasing the ratio of RNA solution to MMF (step 12 of the large-scale RNA preparation) causes diffuse bands on Northern blotting.
11. **Caution:** *Phenol* is highly corrosive and toxic. Wear gloves and safety glasses. *Ethidium bromide* is a potential carcinogen. Wear gloves and dispose of appropriately. *Diethylpyrocarbonate* should be handled in a fume cupboard. *Formaldehyde* should be handled in a fume cupboard.

Acknowledgments

The author is grateful to Zaf Zaman for providing one of his RNA gels for this chapter, and for not being too ecstatic over the defeat of Scotland by Egypt at soccer during his visit to the author's laboratory.

References

1. Rubin, G. R. (1975) Preparation of RNA and Ribosomes from Yeast. *Meth. Cell Biol.* **12,** 45–64.
2. Lindquist, S. (1981) Regulation of protein synthesis during heat shock. *Nature* **293,** 311–314.
3. Dobson, M. J., Mellor, J., Fulton, A. M., Roberts, N. A., Bowen, B. A., Kingsman, S. M., and Kingsman, A. J. (1986) The identification and high level expression of a protein encoded by the yeast Ty element. *EMBO J.* **3,** 1115–1119.
4. Lehrach, R. H., Diamond, D., Wozney, J. M., and Boedtker, H. (1977) RNA molecular weight determinations by gel electrophoresis under denaturing conditions, a critical re-examination. *Biochemistry* **16,** 4743–4751.
5. Thomas, P. S. (1980) Hybridization of denatured RNA and small DNA fragments transferred to nitrocellulose. *Proc. Natl. Acad. Sci. USA* **77,** 5201–5205.
6. Bailey, J. M. and Davidson, N. (1976) Methylmercury as a reversible denaturing agent for agarose gel electrophoresis. *Anal. Biochem.* **70,** 75–85.
7. Sambrook, J., Fritsch, E. F., and Maniatis, T. (1982) *Molecular Cloning: A Laboratory Manual.* Cold Spring Harbor Laboratory, Cold Spring Harbor, NY.
8. Aviv, H. and Leder, P. (1972) Purification of biologically active globin messenger RNA by chromatography on oligothymidylic acid-cellulose. *Proc. Natl. Acad. Sci. USA* **69,** 1408–1412.
9. Brown, A. J. P. and Hardman, N. (1980) Utilisation of poly(A)$^+$ mRNA during growth and starvation in *Physarum polycephalum. Eur. J. Biochem.* **110,** 413–420.

CHAPTER 28

mRNA Abundance and Half-Life Measurements

Alistair J. P. Brown and Francis A. Sagliocco

1. Introduction

The expression of many genes is regulated at multiple levels. For most genes, the predominant control is at transcription, and this is usually inferred from changes in mRNA levels that respond to specific stimuli. However, changes in mRNA stability may contribute to observed variation in mRNA abundance, and many such changes in mRNA stability probably remain undetected. The relative contributions of posttranscriptional regulatory circuits can only be detected by careful RNA analysis.

There are a great number of examples in the literature where significant changes in the abundance of a specific mRNA are detected by Northern analysis, and where the signals obtained on autoradiography are used to estimate levels of induction or repression. However, such estimates can be extremely misleading because of the nonlinearity in the response of X-ray film under most conditions and the difficulty in obtaining truly equal loadings of RNA in individual lanes on RNA gels.

Possible problems with unequal RNA loading can be overcome by analyzing control RNAs, but what are good controls? Obviously, RNAs whose levels remain constant throughout the experiment represent ideal controls. Therefore, 18S rRNA is used frequently as an internal control in mRNA stability measurements *(1)*. Similarly, the actin mRNA is frequently used as an internal loading control for mRNA abundance measurements *(2,3)*. The problem arises in defining suitable RNA controls for the analysis of mRNA abundances in cells grown under different conditions. The amount of rRNA per cell is not constant; it varies in different growth media, for example, with fermentative or nonfermentative carbon sources. No ideal control RNAs are available for many experiments. The investigator can only optimize the design of an experiment. For example, it may be more appropriate to measure changes in mRNA abundance over short time scales during which the levels of rRNA are unlikely to alter significantly, rather than studying steady-state mRNA levels under different growth conditions.

This chapter describes methods for the accurate quantitation of mRNA abundances by dot-blotting or Northern analysis. An extremely important consideration in these methods is the use of probes in molar excess. mRNA abundance measurements are an essential part of procedures for the measurement of mRNA stability, and two procedures for measuring mRNA stability are described in this chapter. The first, which involves the use of a transcriptional inhibitor, is based on procedures described by Santiago and coworkers *(1)*. The second method uses an *rpb1-1* yeast strain in which RNA synthesis is blocked at the nonpermissive temperature; this method is adapted from the procedures of Herrick and coworkers *(4,5)*.

2. Materials
2.1. Dot Blotting *(see Section 4.)*

1. T/E: 1 mM Na$_2$EDTA, 10 mM Tris-HCl, pH 8.
2. MOPS buffer (10X): 0.2M morpholinopropanesulfonic acid, 0.05M sodium acetate, 0.01M Na$_2$EDTA, pH 7.0. Store in a dark bottle. Do not autoclave.
3. MMF: 500 µL formamide, 162 µL formaldehyde (37%), 100 µL 10X MOPS, 238 µL H$_2$O.
4. Ethidium bromide: 0.5 mg/mL.
5. Gel loading dye: 0.02% (w/v) bromophenol blue in 50% (v/v) glycerol.
6. 20X SSC: 3M NaCl, 0.3M trisodium citrate.
   This solution is diluted by the appropriate amount to make other SSC solutions. For example, 2X SSC is a tenfold dilution of 20X SSC.
7. 20X SSCP: 20X SSC containing 38 g Na$_2$HPO$_4$ and 19.2 g of NaH$_2$PO$_4$/L, pH 7.0.
   This solution is diluted by the appropriate amount to make other SSCP solutions. For example, 0.5X SSCP is a 40-fold dilution of 20X SSCP.
8. 50% (w/v) dextran sulfate.

9. 100X Denhardt's solution: 2g BSA, 2 g Ficoll (mol wt 400,000), 2 g polyvinyl pyrrolidone (mol wt 360,000) in 100 mL H_2O.
10. 10% (w/v) SDS.
11. 0.25M EDTA, pH 8.0.

2.2. Northern Analysis

The following solution is required in addition to those described earlier for dot-blotting (*see* Section 2.1.). A lower concentration of ethidium bromide is needed for Northern blotting because high concentrations of ethidium bromide can reduce the signals obtained upon Northern blotting.

1. Ethidium bromide: 0.1 mg/mL (*see* Section 4.).

2.3. mRNA Stability Measurements

1. YPD: 2% (w/v) glucose, 2% (w/v) bactopeptone, 1% (w/v) yeast extract, (2% [w/v] agar).
2. GYNB: 2% (w/v) glucose, 0.65% (w/v) yeast nitrogen base (without amino acids), 50 µg/mL each auxotrophic supplement (2% [w/v] agar).
3. 1,10-phenanthroline (1000X): 100 mg/mL in ethanol (Sigma [St. Louis, MO] P 9375).
4. RNA extraction buffer: 0.1M LiCl, 0.01M dithiothreitol (add fresh), 0.1M Tris-HCl, pH 7.5.
5. Other materials for RNA preparation are described in Chapter 27.
6. Materials for dot-blotting and gel analysis of RNA preparations are described in Section 2.1.
7. Materials for Northern analysis are described in Section 2.2.

3. Methods

3.1. mRNA Abundance Measurements: Dot-Blotting

RNA samples are prepared according to the methods described in Chapter 27. The yield and integrity of each RNA sample is then estimated by gel electrophoresis, add equivalent amounts of RNA from each sample are used for dot-blotting. The latter procedures are described in this chapter.

3.1.1. Gel Analysis of RNA Samples

1. Prepare RNA samples using the large-scale procedure described in Chapter 27, and store these as ethanol precipitates at −20°C to prevent possible degradation by exogenous RNases.
2. Vortex each ethanolic sample to make them relatively homogeneous, and then transfer 0.5 mL (approximately one-thirtieth) of each sample to

separate Eppendorf tubes. Centrifuge at 10,000g at 4°C for 5 min. Pour off the supernatants and drain the tubes. Vacuum desiccate for 15 min to remove residual ethanol.
3. Resuspend each precipitate in 5 µL T/E.
4. Add 35 µL MMF and 2 µL ethidium bromide (0.5 mg/mL), mix, and incubate at 60°C for 15 min.
5. Add 5 µL gel loading dye and mix. If necessary, perform a clearing spin to remove particulate matter (5 min at 10,000g in an Eppendorf centrifuge). Leave on ice until ready to load the agarose gel.
6. Prepare the formaldehyde/agarose gel in a fume hood by melting 1.5 g agarose (see Note 11) in 73 mL of H_2O. Cool to 60°C, add 10 mL of 10X MOPS and 16.2 mL of 37% formaldehyde, mix, and pour gel immediately. Allow the gel to set for at least 30 min.
7. Load about one-half of the sample onto the gel and run at 100–150 V for 1–2 h, as necessary, while circulating the electrophoresis buffer (1X MOPS). Visualize the RNA on a UV transilluminator and photograph it. (**Caution:** Screen ourself from the UV light.) Bands of 25S and 18S ribosomal RNA, and 5S and tRNA are visible in intact RNA preparations, with little evidence of smearing (Chapter 27).
8. Estimate roughly the relative yield of RNA in each preparation from the gel photograph.

3.1.2. Dot-Blotting of RNA Samples

The following procedure gives the volumes of RNA required for the analysis of a single mRNA. To correct for differences in yields between each RNA preparation, at least one control mRNA should be analyzed. The abundance of the experimental mRNA is measured relative to the level of the control mRNA in each preparation. Therefore, the following quantities of materials should be doubled to allow for the analysis of one experimental mRNA and one control mRNA, and increased further if more mRNAs are to be analyzed. If rRNA is to be used as the control, the RNA dilution series (step 4) should be prepared at 1000-fold lower concentrations because of the high intracellular abundance of these RNAs relative to most mRNAs.

1. Vortex each sample to make it relatively homogeneous, and then transfer the volume of ethanolic solution equivalent to approximately 25 µg of RNA to an Eppendorf tube. Centrifuge at 10,000g at 4°C for 5 min. Pour off the supernatant and drain the tube. Vacuum desiccate for 15 min to remove residual ethanol.
2. Resuspend the precipitate in 250 µL RNase-free water.
3. Add 150 µL of 20X SSC and 100 µL of 37% (w/v) formaldehyde, mix thoroughly, and incubate at 60°C for 15 min. Place sample on ice.

4. Prepare a dilution series for each sample in the following way, keeping the solutions at 0°C, and using a fresh pipet tip at each stage. Add 200 µL of the sample to 200 µL of 15X SSC (two-fold dilution). Mix thoroughly and add 200 µL of the two-fold dilution to 200 µL of 15X SSC (fourfold dilution). Mix thoroughly and add 200 µL of the fourfold dilution to 200 µL of 15X SSC (eightfold dilution). Keep samples on ice. (Samples can be stored at −70°C at this stage.)
5. Using gloves, cut 9 × 12 cm rectangles of nitrocellulose and write labels on each filter with a biro pen (*see* Note 9). Soak each filter briefly in water and then 20X SSC. Air dry the filters by laying them on Whatman (Maidstone, UK) 3MM paper.
6. Place a filter in a Hybridot or Slotblot Apparatus (BRL) and gently apply a vacuum (approx 10 psi). For each RNA sample, load duplicate 100-µL aliquots of each dilution into a defined array of slots, and wash each slot with 100 µL of 15X SSC.
7. Remove the filter from the apparatus and either bake at 80°C for 2–3 h, or expose the RNA on the filter to short wave UV light for 5 min. (**Caution:** Shield yourself from the UV light.)
8. The filter, which is now ready for prehybridization and hybridization, can be stored at room temperature.

3.1.3. Filter Hybridization

1. Prepare the prehybridization solution as follows, making about 0.1 mL/cm² for each filter (*see* Note 10):

2.5 mL	20X SSCP
4.7 mL	Formamide
0.5 mL	50% (w/v) Dextran sulfate
0.5 mL	100X Denhardt's solution
0.5 mL	10% (w/v) SDS
0.4 mL	0.25M EDTA, pH 8.0
0.9 mL	H_2O
10.0 mL	

2. Wet nitrocellulose filter in 5X SSC, and seal in a plastic bag. Cut one corner of the bag, pipet in 10 mL of prehybridization solution, remove bubbles from the bag, and reseal the bag. Incubate the sample in a shaking water bath at 37°C for at least 1 h.
3. Cut off a second corner of the bag and pipet in a maximum of about 2×10^7 cpm of denatured (e.g., by heating, *see* Section 3.3.11., Chapter 9) [^{32}P]-labeled probe and mix immediately into the bulk of the solution. Remove bubbles from the bag and reseal the corner. Seal the hybridization bag in a

second plastic bag as a precaution to prevent radioactive contamination of the water bath if the hybridization bag leaks. Incubate the sample in a shaking water bath at 37°C for 16–20 h.
4. Remove the bag from the water bath to a sink suitable for disposal of radioactivity. Cut open the bag, dispose of the radioactive hybridization solution using the appropriate safety precautions, and transfer the filter to a plastic box containing 250 mL of 2X SSC containing 0.1% (w/v) SDS at 37°C.
5. Wash the filter with gentle agitation for about 20 min as follows: twice in 2X SSCP containing 0.1% (w/v) SDS at 37°C; twice in 0.5X SSCP containing 0.1% (w/v) SDS at 50°C; twice in 0.25X SSCP containing 0.1% (w/v) SDS at 50°C.
6. Seal the damp filter in a plastic bag and autoradiograph with an intensifying screen at −70°C for an appropriate period (frequently overnight).
7. Following autoradiography to check for background contamination on the filter (Fig. 1), cut out each dot on the blot and subject to scintillation counting (5), or 2D-Radioimaging if available (see Section 3.2.3.).
8. Calculate the average bound radioactivity in duplicate dots, and for each dilution series plot these data against the amount of RNA dotted on the filter (see Section 3.2.3.). If the points lie on a straight line, this indicates that the hybridization has been performed under conditions of probe excess (Fig. 2A). Extrapolate the line back to zero RNA loaded to calculate background radioactivity. This value should be deducted from the data points before calculating the relative abundances of the specific RNA in each RNA preparation. If the points do not lie on a straight line, the data should not be used because the amount of radioactivity bound to the filter is not proportional to the concentration of the target sequence. This is frequently caused by the hybridization not being performed under conditions of probe excess (Fig. 2B).

3.2. mRNA Abundance Measurements: Northern Analysis

RNA is prepared according to the methods described in Chapter 27, and the integrity and yield of the samples are checked as described earlier for the dot-blotting procedure. This section deals with the quantitation of these samples by Northern analysis. Nylon membranes are recommended over nitrocellulose filters because this allows filters to be reprobed efficiently, and hence facilitates the analysis of multiple mRNAs using the same gel.

The following method recommends the staining of RNA on formaldehyde gels with ethidium bromide. This allows the RNA to be checked at each stage up to and following Northern transfer. However, excess ethidium bromide can reduce the signals obtained by Northern analysis. Hence, ethidium bromide staining can be omitted once the procedures have become routine.

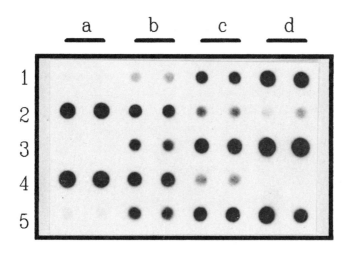

Fig. 1. Quantitation of the pyruvate kinase mRNA by dot-blotting. Four dilutions (a–d) of five separate RNA preparations (1–5) were dotted in duplicate onto nitrocellulose, and the filter baked, prehybridized, hybridized with a probe for the pyruvate kinase mRNA, and washed as described in the text. Dilutions were dotted right-to-left for one RNA sample, then left-to-right for the next to control for even hybridization across the filter. Following autoradiography, each dot was cut out and subjected to scintillation counting. The data for sample 1 are presented in Fig. 2A.

3.2.1. Northern Blotting

1. Vortex each RNA sample to make it relatively homogeneous, and then transfer the volume of ethanolic solution equivalent to approx 20 µg of RNA to an Eppendorf tube. Centrifuge at 10,000g at 4°C for 5 min. Pour off the supernatant and drain the tube. Vacuum desiccate for 15 min to remove residual ethanol.
2. Resuspend the precipitate in 5 µL T/E.
3. Add 35 µL MMF and 1 µL ethidium bromide, mix, and incubate at 60°C for 15 min.
4. Add 5 µL gel loading dye and mix. If necessary, perform a clearing spin to remove particulate matter (5 min at 10,000g in an Eppendorf centrifuge). Leave on ice until ready to load the agarose gel.
5. Prepare the formaldehyde/agarose gel in a fume hood by melting 1.5 g agarose (see Note 11) in 73 mL of H_2O. Cool to 60°C, add 10 mL of 10X MOPS and 16.2 mL of 37% formaldehyde, mix, and pour gel immediately. Allow the gel to set for at least 30 min before removing the comb and transferring to the electrophoresis apparatus.

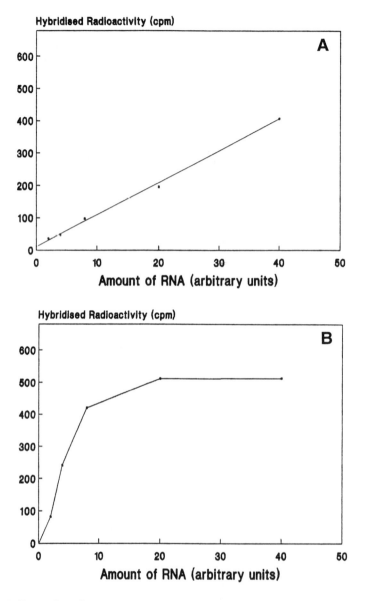

Fig. 2. Effect of probe concentration on mRNA quantitation. Data obtained from dot-blotting (Fig. 1) are plotted as shown. **(A)** The linear response obtained when the probe is in excess in dot-blot hybridization. Note that the line does not intersect the Y-axis at zero; the cpm at RNA = O should be subtracted from the hybridized radioactivity as this represents background radioactivity. **(B)** The nonlinear response obtained when probe is not in excess. These data should not be used for the calculation of mRNA abundance.

6. Load 25 µL of the sample onto the gel and run at 50–100 V for 5–7 h (or at a lower voltage overnight) while circulating the electrophoresis buffer (1X MOPS). Visualize the RNA on a UV transilluminator and photograph it. (**Caution:** Screen yourself from the UV light.) Proceed only if the banding pattern demonstrates that the RNA is intact and that approximately equal amounts of RNA are present in each lane of the gel.
7. Cut out a piece of nylon membrane (for example, Hybond-N; Amersham, Arlington Heights, IL) to about 1 cm larger than the gel in width and length. Write on the filter with a Bic pen to label the filter, to show its orientation, and if necessary to mark the lines along which the filter is to be divided following blotting to allow separate probings. No pretreatment of Hybond-N is necessary before blotting.
8. Blotting is performed using 20X SSC as the transfer buffer either by the old fashioned capillary methods overnight *(5)* or under vacuum for at least 3 h (for example, using the Pharmacia [Uppsala, Sweden] LKB Vacugene apparatus). Ensure that:
    a. The gel and membrane filter are aligned properly with respect to the markings on the filter;
    b. There are no air bubbles trapped between the filter and the gel; and
    c. The capillary action or vacuum cannot bypass the gel and filter.
9. Restain the gel in buffer containing ethidium bromide at 1 mg/mL to check that transfer is complete. Also, visualize the RNA on the filter by shining UV light onto the "RNA side" of the filter. Distinct bands of 25S and 18S rRNA should be apparent. (**Caution:** Shield yourself from the UV light.)
10. If the filter is to be divided and individual sections probed separately, cut the filter along the lines marked in step 7 with a sharp scalpel blade.
11. Bake the filter between two sheets of Whatman 3MM paper at 80°C for 2 h, or wrap the filter in cling-film and fix the RNA to the membrane by treating with long wave UV light following the instructions of the manufacturers of the membrane. The transilluminator should be calibrated since the timing of the UV fixation is important. At this point, membranes can be stored for months in the dark.

3.2.2. Filter Hybridization

Follow the procedure described above for filter hybridization of dot blots (Section 3.1.3.), or follow the instructions of the manufacturers of the membrane. Estimate the probable total amount of target RNA on the filter (in ng) from the amount of total RNA loaded in all lanes, and ensure that the probe is in at least fivefold molar excess. For this calculation, assume that mRNA comprises approximately 10% of total cellular RNA, and that mRNAs of high, medium, and low abundance are approx 1, 0.1, and 0.01% of total mRNA, respectively.

Fig. 3. The AMBIS 2D-radioanalytical system.

3.2.3. Quantitation

Qualitative analysis of the relative intensity of bands on the X-ray film following autoradiography can be very misleading. These signals can be quantitated in several ways:

1. Individual bands on the filter can be cut out and subjected to scintillation counting, but this suffers at least two disadvantages. First, the filter cannot be reused, and hence separate filters must be prepared for experimental and control mRNAs. This reduces the accuracy of the measurements. Second, it can be difficult to align the filter with the X-ray film, and hence to cut out the bands accurately.
2. Densitometer scanning can be used to measure the intensity of the signal on the X-ray film. This allows the filter to be reprobed for additional mRNAs, but the accuracy is entirely dependent on a linear relationship between the amount of radioactivity bound to the filter and the autoradiographic signal. Unfortunately, this is only the case over a relatively narrow range, depending on the type of film and on the autoradiographic conditions.
3. The ideal method involves direct 2D radioimaging from the filter, for example, using an AMBIS 2D-Radioanalytical System (Fig. 3). This apparatus measures the radioactivity being emitted from each part of the filter.

These data can be imaged on the computer and then used to quantitate accurately the amount of radioactivity in specific bands on the Northern blot (Fig. 4). Having filed this information on the computer, the filter can be repeatedly stripped, reprobed, and requantitated. Hence, the relative abundance of many mRNAs can be measured accurately from the same filter, thus reducing the errors involved in the preparation of multiple filters.

3.2.4. Reprobing Filters

Having completed the analysis of one mRNA, the first radioactive probe is stripped from the nylon membrane, and the membrane reprobed for a second mRNA. This cycle can be repeated many times if the membrane is kept damp. Although the following method will strip a hybridized probe effectively, it will not remove background radioactivity efficiently.

1. Place the nylon membrane in a plastic tray and pour 500 mL of boiling 0.1% (w/v) SDS into the tray. Gently rock the tray until the solution has reached about 37°C.
2. Pour off the solution and repeat step 1.
3. Seal the damp membrane in a plastic bag and autoradiograph at −70°C with an intensifying screen for an appropriate period to ensure that the radioactivity has been removed from the membrane.
4. Prehybridize and hybridize according to the procedure described earlier (Section 2.3., steps 17–22).

3.3. mRNA Stability Measurements: Method 1

This method measures the rate at which specific mRNAs decay following the inhibition of transcription using the zinc ion chelator, 1,10-phenanthroline (Fig. 5). The method was developed by Santiago and coworkers for the yeast strain DBY746 (*MATα, his3, leu2, trp1, ura3*; Berkeley, CA) when grown to mid-exponential growth phase in YPD *(1)*. Careful controls were performed to determine the optimum concentration of 1,10-phenanthroline for mRNA half-life measurements *(1)*, but this concentration is strain-dependent. For example, a concentration of 80–100 µg/mL 1,10-phenanthroline in the medium is best for DBY746 *(1,6)*, but a concentration of 250 µg/mL has been used for mRNA stability measurements in another yeast strain *(7)*. Therefore, controls should be performed to determine the optimal concentration of 1,10-phenanthroline when new strains or growth conditions are to be employed.

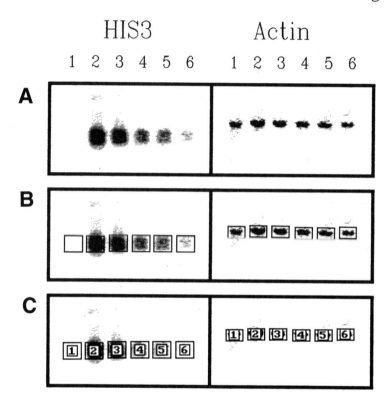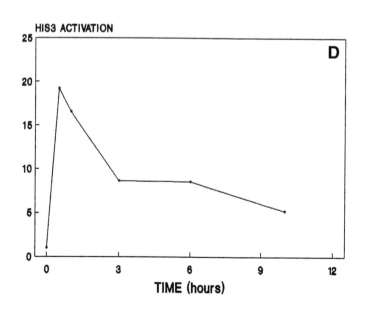

The method is divisible into four independent stages:
1. Cell growth and RNA preparation (steps 1–7);
2. Gel electrophoresis to check the yield and integrity of each RNA preparation (Section 3.1.1.);
3. Quantitation of specific mRNAs in each RNA preparation by dot-blotting or Northern analysis as described in Sections 3.1. and 3.2.; and
4. The calculation of mRNA half-lives (step 10).

The method is as follows:

1. Prepare the materials for at least six large-scale RNA preparations (*see* Chapter 27).
2. Inoculate 300 mL of YPD with the appropriate strain and incubate overnight at 30°C with shaking at 200 rpm until the culture reaches mid-exponential growth phase (A_{600}:0.4–0.6).
3. Immediately before starting the time course, remove 50 mL of the culture and immediately add this sample to 100 mL of cold ethanol in a 250-mL centrifuge pot. Mix and store at –20°C. (Using the procedures described above, this sample can be used to measure the abundance of specific mRNAs at time zero.)
4. Add 250 µL of 1000X phenanthroline (time = 0 min) and continue the incubation at 30°C with shaking at 200 rpm.
5. Using an appropriate range of time points (Fig. 5), quickly remove 50 mL of the culture, and immediately add each of these samples to separate 250-mL centrifuge pots each containing 100 mL of cold ethanol. Mix and store these samples at –20°C.
6. Later the same day, and preferably within 24 h of taking these samples, RNA is isolated from each of the samples using the large-scale procedure

Fig. 4. *(preceding page)* Quantitation of mRNA Levels from Northern blots using the AMBIS 2D-radioanalytical system: Induction of the *HIS3* mRNA in response to amino acid starvation. RNA was isolated from yeast at various times following the transfer of cells from amino acid-rich medium to amino acid-poor medium. Approximately equal amounts of RNA from each sample were subjected to Northern analysis and probed for the actin mRNA control. The filter was scanned using the AMBIS 2D-Radioanalytical system **(A)**, the area corresponding to the actin mRNA in each lane highlighted **(B)**, each box numbered **(C)**, the radiation detected in each box quantitated, and this data stored in the computer. The filter was then stripped, reprobed for the *HIS3* mRNA, and the process repeated. Having subtracted background, the signals obtained for *HIS3* were corrected for loading errors using the actin data, and the corrected *HIS3* mRNA levels calculated relative to the uninduced level observed in sample 1 **(D)** *(9)*.

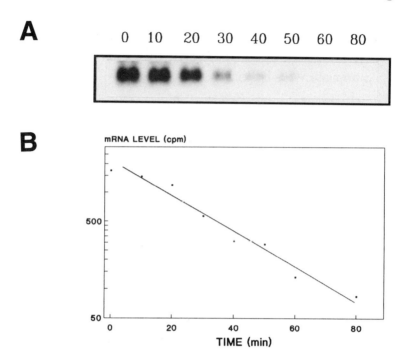

Fig. 5. mRNA half-life measurements using 1,10-phenanthroline. (A) RNA was isolated from a yeast culture at 0, 10, 20, 30, 40, 50, 60, and 80 min following the addition of 1,10-phenanthroline to a final concentration of 100 µg/mL. Approximately equal amounts of each sample were subjected to Northern analysis and probed for the *CYC1* mRNA. (B) The Northern blot in (A) was quantitated by 2D-Radioimaging to determine the abundance of this *CYC1* mRNA at various times following the inhibition of transcription using 1,10-phenanthroline as described in the text. These data were plotted semilogarithmically as shown, and the half-life of the mRNA (12.8 min) calculated statistically.

described in Chapter 27. If the RNA preparation must be postponed, cells are harvested from the ethanolic suspension by centrifugation (5000 rpm for 5 min at 4°C) and the supernatant discarded. The cell pellet is then resuspended in 10 mL RNA extraction buffer at 0°C, centrifuged at 5000 rpm for 5 min at 4°C, and the cell pellet stored at –20°C. Protracted storage of cells under ethanol reduces the yield of RNA.

7. RNA samples are stored as ethanol precipitates at –20°C to prevent possible degradation by exogenous RNases.
8. Check the integrity of the RNA, and estimate roughly the relative yield of RNA in each preparation by gel electrophoresis (Section 2.3., steps 1–8).

9. Perform mRNA quantitation either by dot-blotting or Northern analysis, following the appropriate procedure described earlier (Sections 2.3. or 2.4.). Use equivalent amounts of RNA from each sample to increase the accuracy of the method.
10. Measure the concentration of 18S rRNA in each sample by dot-blotting using 1000-fold dilutions of each RNA preparation to account for the high intracellular abundance of the sequence. Alternatively, measure the A_{260} of the samples to estimate the RNA yield (1 A_{260} = 42 µg RNA/mL). However, note that absorbance measurements can be misleading if the RNA is not further purified by repeated ethanol precipitations or by ultracentrifugation through CsCl shelves *(5)*.
11. For each time point, calculate the abundance of the mRNA of interest relative to 18S rRNA. When these values are plotted on a logarithmic scale against time, they should lie on a straight line that represents the exponential decay of the specific mRNA with time (relative to the 18S rRNA control). The half-life of the mRNA is calculated from this straight line (Fig. 5). Do *not* extrapolate to zero time, since there is a short lag period of about 5 min before the 1,10-phenanthroline inhibits transcription.

3.4. mRNA Stability Measurements: Method 2

This procedure measures the rate at which specific mRNAs decay following the inhibition of transcription by a conditional mutation in the *RPB1* gene that encodes the largest subunit of RNA polymerase II *(8)*. The temperature-inducible mutation *rpb1-1* was characterized by Young and coworkers *(8)*, from whom the following strains are available: RY136 *MATalpha, ura3*; RY137 *MATa, ura3, his4, lys2*; RY260 *MATa, ura3, rpb1-1;* RY262 *MATalpha, ura3, his4, rpb1-1*.

The induction of the *rpb1-1* mutant phenotype, and hence the inhibition of transcription, is achieved by shifting yeast cultures from 25°C–37°C *(8)*. Under these conditions of mild temperature shock the stability of some mRNAs are affected *(7)*. Therefore, the decay of an mRNA following temperature upshift in an *rpb1-1* strain should be compared with that in a control *RPB1* strain under similar conditions.

The method is similar to Method 1 in that it is divisible into four independent stages:

1. Cell growth and RNA preparation;
2. Gel electrophoresis to check the yield and integrity of each RNA preparation;
3. Quantitation of specific mRNAs in each RNA preparation by dot-blotting or Northern analysis as described earlier; and

4. The calculation of mRNA half-lives. The methods are adapted from those of Herrick and coworkers *(4)*.

The method is as follows:

1. For each culture, prepare the materials for at least six large-scale RNA preparations (Chapter 27).
2. Inoculate 130 mL of YPD or GYNB (containing the appropriate supplements) with a late exponential starter culture of the appropriate strain and incubate overnight at 25°C with shaking at 250 rpm until the culture reaches midexponential growth phase (A_{600} = 0.4–0.6). Use a 2-L flask.
3. Prewarm 130 mL of the same medium to 49°C.
4. Add the 130 mL of 49°C medium to the flask containing 130 mL of 25°C culture and immediately place in a shaking water bath at 37°C (time = 0).
5. Immediately following the start of the time course, remove 40 mL of the culture, and immediately add this sample to 80 mL of cold ethanol in a 250-mL centrifuge pot. Mix and store at –20°C.
6. At time points of 5, 10, 20, 40, and 60 min quickly remove 40 mL of the culture and immediately add each of these samples to separate 250-mL centrifuge pots, each containing 80 mL of ethanol. Mix and store these samples at –20°C.
7. Within 24 h of taking these samples, RNA is isolated from each sample using the large-scale procedure described in Chapter 27. If the RNA preparation must be postponed, cells are harvested from the ethanolic suspension, by centrifugation at 5000*g* for 5 min at 0°C, the cells washed in 10 mL extraction buffer at 0°C, recentrifuged, and the cell pellet stored at –20°C.
8. RNA samples are stored as ethanol precipitates at –20°C.
9. The integrity and yield of each RNA preparation is checked by gel electrophoresis (*see* Section 3.1.1.).
10. mRNA quantitation is performed either by dot-blotting or Northern analysis using 18S rRNA as the internal loading control as described in Sections 3.1. and 3.2. (Fig. 6).

4. Notes

Materials can be categorized broadly into those that must be made RNase-free, and those for which this is not necessary. Generally, those materials that are used in filter hybridizations need not be RNase-free, whereas those that are used to handle RNA must be treated as described in the following. It is advisable to prepare an excess of RNase-free mate-

mRNA Abundance and Half-Life Measurements

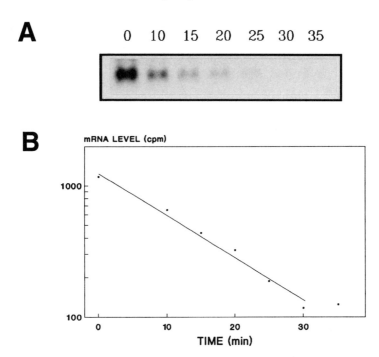

Fig. 6. mRNA half-life measurements using the *rpb1-1* mutant. **(A)** RNA was isolated from a culture of the *rpb1-1* strain at 0, 10, 15, 20, 25, 30, and 35 min following the increase in temperature from 25–37°C. Approximately equal amounts of each sample were subjected to Northern analysis and probed for the *CYC1* mRNA. **(B)** The Northern blot in (A) was quantitated by 2D-Radioimaging to determine the abundance of the *CYC1* mRNA at each time point as described in the text. These data were plotted semilogarithmically as shown, and the half-life of the mRNA (9.0 min) calculated statistically.

rials in case of a mishap, because the preparation takes time and yet the methods should be done quickly.

1. Glassware (Pasteur pipets and pipets) is wrapped in aluminum foil and baked at 200°C overnight.
2. Disposable plasticware (tips, Eppendorf tubes, and so on) is taken only from new bags that have been set aside for RNA work, and autoclaved.
3. Nondisposable plasticware (for example, 50-mL Oakridge centrifuge tubes) is soaked in 0.1% (v/v) diethylpyrocarbonate for at least 1 h and then rinsed in RNase-free water or incubated at 80°C.

4. Fresh, previously unopened chemicals can be considered to be RNase-free. A set of chemicals are set aside specifically for RNA work to prevent careless contamination by other workers. Spatulas are not used to weigh out these chemicals, because they are a probable source of RNase contamination. Instead, these chemicals are carefully tapped out of the bottles.
5. Solutions that do not contain amines are rendered RNase-free by adding diethylpyrocarbonate to a final concentration of 0.1% (v/v), standing for at least 1 h, and then autoclaving to inactivate the remaining diethylpyrocarbonate.
6. Solutions containing amines (for example, Tris or EDTA) are not treated with diethylpyrocarbonate. Instead, extra precautions are taken to prevent contamination with RNase. RNase-free chemicals are added to diethylpyrocarbonate-treated distilled water and pHed using concentrated acid or $10M$ NaOH. The pH is monitored by dropping small samples of the solution onto pH papers to prevent RNase contamination of the main solution from a pH electrode. Final volumes are then adjusted using volumetric markings on the RNase-free solution bottle, and the solution autoclaved.
7. Gloves should be worn throughout.
8. **Caution:** The *phenol* used in RNA preparations is highly corrosive and toxic. Wear gloves and safety spectacles. *Ethidium bromide* is a potential carcinogen. Wear gloves and dispose of appropriately. The *diethyl pyrocarbonate* used in preparing RNase-free materials should be handled in a fume cupboard. *Formaldehyde* should be handled in a fume cupboard. Protect yourself from *UV light* sources.
9. For dot-blotting, nitrocellulose membranes are recommended over nylon filters because nitrocellulose membranes stretch slightly under vacuum in the Hybridot apparatus to form dimples at each dot that are maintained through the hybridization and washing procedures. This makes it very easy to align filters and to cut out individual dots accurately when quantifying the bound radioactivity by scintillation counting. (Filters are not reprobed in this method.) Rectangular nitrocellulose filters of 9 × 12 cm are recommended in the method (*see* Secton 3.1.2.) because this is the size of filter required for the Hybridot Apparatus (Gibco BRL, Gaithersburg, MD).
10. The hybridization and washing conditions described in Section 3.1.3. are suitable for random-primed or nick-translated DNA probes and for RNA probes of greater than 250 bases in length that are homologous to the target sequence. Conditions should be adjusted for end-labeled oligonucleotide probes or for heterologous probes.
11. Agarose used for Northern gels: Sigma [St. Louis, MO], Type 1, low EE0 (cat. no. A-6013). Agarose from other sources, such as IBI and BRL, may also be used.

Acknowledgments

The authors are grateful to Andy Bettany and Zaf Zaman for kindly providing data for this chapter, and to Russell Sykes (LabLogic, Sheffield, UK) for providing the photograph for Fig. 3.

References

1. Santiago, T. C., Bettany, A. J. E., Purvis, I. J., and Brown, A. J. P. (1986) The relationship between mRNA stability and length in *Saccharomyces cerevisiae*. *Nucleic Acids Res.* **14,** 8347–8360.
2. Bettany, A. J. E, Moore, P. A., Cafferkey, R., Bell, L. D., Goodey, A. R., Carter, B. L. A., and Brown, A. J. P. (1989) 5'-secondary structure formation, in contrast to a short string of non-preferred codons, inhibits the translation of the pyruvate kinase mRNA in yeast. *Yeast* **5,** 187–198.
3. Moore, P. A., Bettany, A. J. E., and Brown, A. J. (1990) Expression of a yeast glycolytic gene is subject to dosage limitation. *Gene* **89,** 85–92.
4. Herrick, D., Parker, R., and Jacobson, A. (1990) Identification and comparison of stable and unstable mRNAs in *Saccharomyces cerevisiae*. *Mol. Cell. Biol.* **10,** 2269–2284.
5. Parker, R. Herrick, D., Peltz, S. W., and Jacobson, A. (1991) *Methods Enzymol.* **194,** 415–423.
6. Sambrook, J., Fritsch, E. F., and Maniatis, T. (1982) *Molecular Cloning: A Laboratory Manual.* Cold Spring Harbor Laboratory, Cold Spring Harbor, NY.
7. Lithgow, G. J. (1989) Transcription in the yeast *Saccharomyces cerevisiae*. PhD Thesis, University of Glasgow.
8. Herruer, M. H., Mager, W. H., Raue, H. A., Vreken, P., Wilms, E., and Planta, R. J. (1988) Mild temperature shock affects transcription of yeast ribosomal protein genes as well as the stability of their mRNAs. *Nucleic Acids Res.* **16,** 7917–7929.
9. Nonet, M., Scafe C., Sexton, J., and Young, R. (1987) Eukaryotic RNA polymerase conditional mutant that rapidly ceases mRNA synthesis. *Mol. Cell. Biol.* **7,** 1602–1611.
10. Zaman, Z., Brown, A. J. P., and Dawes, I. W. (1992) A 3' transcriptional enhancer within the coding sequence of a yeast gene encoding the common subunit of two multi-enzyme complexes. *Mol. Microbiol.* **6,** 239–246.

CHAPTER 29

Polysome Analysis

Francis A. Sagliocco, Paul A. Moore, and Alistair J. P. Brown

1. Introduction

During the cyclic process of translation, a small (40S) and large (60S) ribosomal subunit associate with mRNA to form an 80S complex (monosome). This ribosome moves along the mRNA during translational elongation, and then dissociates into the 40S and 60S subunits on termination. During elongation by one ribosome, further ribosomes can initiate translation on the same mRNA to form polysomes. Each polysomal complex can contain from two to over twenty ribosomes, and the mass of each complex is determined primarily by the number of ribosomes it contains. Hence, the population of polysomes within the cell can be size-fractionated by sucrose density gradient centrifugation on the basis of the loading of ribosomes on the mRNA. Also, mRNA that is being actively translated can be fractionated from untranslated mRNA by separating polysomal and monosomal material by centrifugation through sucrose shelves.

The ribosome loading on a population of mRNAs depends primarily on the rates of translational initiation and elongation. These are strongly influenced by a large number of parameters, which include the availability of ribosomal subunits, active translation factors, charged tRNAs, ATP, and GTP. As a result, the translational process is extremely sensitive to the physiological status of the cell, and hence polysome profiles are a sensitive indicator of cell growth. For example, the ratio of polysomes to monosomes decreases markedly as cells enter late exponential growth phase. Also, polysome analysis has been used to study the effects of particular mutations upon translation in yeast *(1)*.

From: *Methods in Molecular Biology, Vol. 53: Yeast Protocols*
Edited by: I. Evans   Humana Press Inc., Totowa, NJ

The ribosome loading upon a particular mRNA depends primarily on the relative rates of translational initiation and elongation on that mRNA. Therefore, the distribution of specific mRNAs across polysome gradients has been measured by Northern and dot-blotting analyses to determine directly the relative translatability of these mRNAs in vivo. For example, this has been applied to investigations of dosage compensation of yeast ribosomal protein genes *(2)*, the general control of amino acid biosynthesis by the GCN and GCD genes *(3)*, and the relationship between mRNA structure and function *(4–7)*.

Two alternative methods are available for cell breakage that is the first step in the preparation of polysomes. The enzymic generation and then lysis of spheroplasts gives consistently high yields of polysomes. However, before lysis, spheroplasts must undergo a period of recovery in osmotically stabilized rich medium *(5)* during which the physiological status of the "cell" returns to normal *(7)*. Therefore, polysome analysis using this procedure is limited to studies in which cells are grown in rich medium *(5)*, but this does not preclude the analysis of plasmid-encoded mRNAs *(6)*. Alternatively, the shearing of cells using glass beads has the advantage that polysome analysis can be performed using different growth media *(2,3)*, but care must be taken to ensure that cell breakage and hence polysome yields are consistent. Both methods have been used to great effect *(2–7)*. Both procedures are described in this chapter along with a method for the preparation of polysomes using sucrose shelves.

2. Materials

2.1. Polysome Gradients: Method 1

1. YPD: 2% (w/v) glucose, 2% (w/v) bactopeptone, 1% (w/v) yeast extract, (2% [w/v] agar).
2. GYNB: 2% (w/v) glucose, 0.65% (w/v) yeast nitrogen base (without amino acids), 50 µg/mL each auxotrophic supplement (2% [w/v] agar).
3. Solution 1: $1M$ Sorbitol, 25 mM Na$_2$EDTA, 50 mM DTT (add fresh on day of experiment).
4. Solution 2: 1.2M Sorbitol, 10 mM Na$_2$EDTA, 100 mM trisodium citrate, pH 5.8 2% (v/v) β-glucuronidase (add fresh on day of experiment). Filter sterilize.
5. 10% (w/v) SDS.
6. YPD/sorbitol: YPD containing 1.2M sorbitol. Filter sterilize.
7. Lysis buffer: 100 mM NaCl, 3 mM MgCl$_2$, 10 mM Tris-HCl, pH 7.4.
8. 5% (w/v) Sodium deoxycholate (Sigma, St. Louis, MO).
9. 5% (w/v) Brij 58 (Aldrich, Milwaukee, WI).

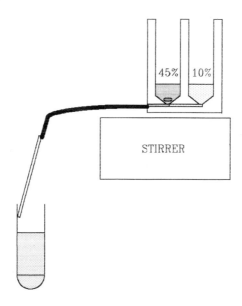

Fig. 1. Preparation of sucrose density gradients for polysome analysis. 18 mL of 45% (w/w) and 10% (w/w) sucrose are placed into the front and back compartments of an RNase-free gradient pourer, respectively. A small magnetic flea is used to mix the solution in the front compartment, which is drained slowly using a peristaltic pump into the RNase-free centrifuge tube (approx 10 min per gradient). Care should be taken to minimize the mixing in the centrifuge tube.

10. Solution 3: 100 mM NaCl, 30 mM MgCl$_2$, 10 mM Tris-HCl, pH 7.5.
11. Sucrose Gradients: The 36-mL 10–45% (w/w) sucrose gradients are prepared in diethylpyrocarbonate-treated polyallomer SW28 tubes (Fig. 1) using 18 mL each of:
    a. 10 g sucrose in 90 mL of solution 3; and
    b. 45 g sucrose in 55 mL of solution 3.
    Cover the gradients with aluminum foil and store at −70°C. Following removal from the −70°C freezer, allow 3 h for the gradients to thaw to 4°C. Before use, balance the gradients with their SW28 centrifuge buckets using 10% sucrose and keep at 0°C.

2.2. Polysome Gradients: Method 2

Media and sucrose gradients are prepared as described in Section 2.1. In addition, the following are required (*see* Section 4.):

1. Glass beads (size approx 0.4 mm diameter) are soaked in concentrated nitric acid for about 1 h, rinsed thoroughly in distilled water, dried overnight

in a glassware oven, and then baked at 200°C overnight. Beads are dispensed into RNase-free bottles after heat baking to prevent these bottles from cracking.
2. Lysis buffer: 10 mM NaCl, 30 mM MgCl$_2$, 10 mM Tris-HCl, pH 7.5. Before use add cycloheximide (prepared fresh as a 5 mg/mL concentrated solution in RNase-free H$_2$O) to a final concentration of 100 µg/mL.

2.3. mRNA Analysis Across Polysome Gradients (see Section 4.)

1. 20X SSC: 3M NaCl, 0.3M trisodium citrate.
2. 1M Tris-HCl, pH 7.5.
3. 10% (w/v) SDS.
4. Phenol: Solid phenol (500 g) is melted at 60°C for about 30 min, and brought to pH 8 by adding RNase-free un-pHed 1M Tris. Solid hydroxyquinoline is then added to a final concentration of about 0.02% (w/v). To saturate the phenol phase, RNase-free distilled water (approx 200 mL) is added and mixed by shaking. Upon standing, the phases should separate, with the phenol on the bottom. Store at 4°C (*see* Note 11).
5. Phenol:CHCl$_3$: Mix 1 vol of phenol (solution 4) with 1 vol of CHCl$_3$.
6. 3M sodium acetate.
7. RNase-free, distilled water.
8. T/E: 1 mM Na$_2$EDTA, 10 mM Tris-HCl, pH 8.

2.4. Preparation of Polysomes

YPD, lysis buffer and glass beads are prepared as described in Section 2.3.

30% (w/v) Sucrose in lysis buffer: 30 g sucrose in 100 mL lysis buffer.

3. Methods

3.1. Polysome Gradients: Method 1

In this procedure, cell breakage is achieved by spheroplasting. Following cell breakage, lysates are centrifuged to generate postmitochondrial supernatants, and these are subjected to sucrose density gradient fractionation to produce polysome gradients *(5)*. The distribution of polysomal material across such gradients is usually monitored spectrophotometrically at 260 nm. If necessary, the gradients are then fractionated for mRNA analysis (*see* Section 3.3.). The procedure for spheroplast preparation described in this section is adapted from that of Beggs *(8)*, and the methods for the generation of polysome gradients are adapted *(5)* from those described by Hutchison and Hartwell *(7)*.

Polysome Analysis

1. Inoculate 5 mL starter cultures with the yeast strains of interest, and grow to stationary phase by incubating at 30°C for 48–72 h with shaking (approx 250 rpm). Grow in rich medium (YPD), or in minimal medium (GYNB) to select for the maintenance of a particular plasmid.
2. Inoculate approx 200 µL of the stationary phase starter culture into 400 mL of YPD and incubate at 30°C with shaking at 250 rpm for 14–16 h until the cells reach midexponential growth phase (A_{600} = 0.4–0.6). Use a 2-L flask.
3. Before starting the experiment, prepare solutions 3, 4, and 6 (Section 2.1.), and remove the sucrose gradients from the −70°C freezer to thaw to 4°C (Section 2.1., item 11). Also, precool the SW28 ultracentrifuge rotor to 0°C.
4. Immediately before harvesting the culture, transfer 1 mL of the culture to a sterile Eppendorf tube and store on ice. Use this sample to measure the proportion of plasmid-containing cells in the culture at the time of harvesting. Dilutions of the sample (usually, 10^{-4}) are plated onto the appropriate media (e.g., GYNB minus leucine, and GYNB containing leucine for a *LEU2*-containing plasmid) and grown for 2–3 d at 30°C.
5. Harvest the cells from the bulk of the 400 mL culture by centrifugation at 3000*g* for 5 min at room temperature.
6. Resuspended the cell pellet in 50 mL of solution 1, and incubate with continuous gentle shaking in a water bath at 30°C for 20 min.
7. To spheroplast the cells, centrifuge at 3000*g* for 5 min at room temperature, resuspend in 50 mL of solution 2 that contains β glucuronidase, and incubate with continuous gentle shaking in a water bath at 30°C for 25 min.
8. At this stage, assess the efficiency of spheroplasting crudely by transferring approx 100 µL of the turbid cell suspension to an Eppendorf tube and estimating the degree of clarification on the addition of 5 µL of 10% SDS. SDS will lyse spheroplasts, but not whole cells, and therefore the formation of a relatively clear, viscous solution when the detergent is added is indicative of efficient spheroplast formation. Alternatively, a more accurate measure of spheroplast formation can be obtained by estimating the extent of cell lysis by the detergent using light microscopy.
9. Harvest the spheroplasts by centrifugation at 3000*g* for 5 min at room temperature.
10. Gently resuspended the pellet in 50 mL of YPD containing 1.2*M* sorbitol, and incubate the spheroplasts for 120 min at 30°C with continuous gentle shaking (approx 75 rpm).
11. Harvest the "recovered" spheroplasts by centrifugation at 3000*g* for 5 min at room temperature. From this stage on, perform steps 12–14 as rapidly as possible, and at 0°C.

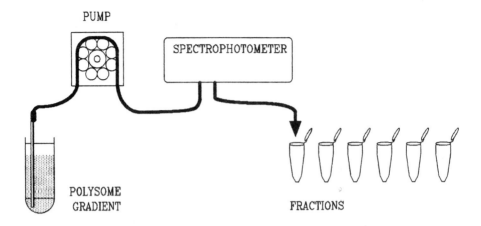

Fig. 2. Fractionation of polysome gradients. Using a peristaltic pump, each gradient is drawn evenly from the bottom of the tube through a spectrophotometer that monitors the absorbance at 260 nm, and then fractionated into Eppendorf or Falcon tubes (*see text* for details). The tubing and flow cell should be made RNase-free before use by rinsing with 0.1% (v/v) diethylpyrocarbonate.

12. Gently resuspend the spheroplast pellet in 500 µL lysis buffer at 0°C and transfer 700 µL to an Eppendorf tube. Add 70 µL of 5% (w/v) sodium deoxycholate, and vortex for 5 s. Then add 70 µL of 5% (w/v) Brij 58 and vortex the sample for 5 s.
13. Centrifuge the lysate in an Eppendorf centrifuge at 14,000g for 1 min at 0°C, and carefully layer exactly 600 µL of the supernatant onto the top of a prethawed and prebalanced 10–45% sucrose density gradient. (The layering of an exact quantity of material onto prebalanced gradients eliminates the time-consuming step of balancing samples immediately prior to ultracentrifugation.)
14. Centrifuge the gradients using a Beckman (Fullerton, CA) SW28 rotor at 25,500 rpm (140,000g) rpm for 170 min at 0°C. Decelerate the rotor with the brake off.
15. Fractionate each gradient through a spectrophotometer using a 2.5-mm flow cell to monitor the absorbance profile at 260 nm as shown in Fig. 2. A typical A_{260} profile is illustrated in Fig. 3A. Gradients can be fractionated either into about 50 0.75-mL fractions for dot-blotting (Section 3.3.1.) (Fig. 4), or into about 15 2.5-mL fractions for Northern analysis (Section 3.3.2.) (Fig. 5).

3.2. Polysome Gradients: Method 2

In this procedure, cell breakage is achieved by shearing with glass beads. In other respects the method is similar to the spheroplasting method

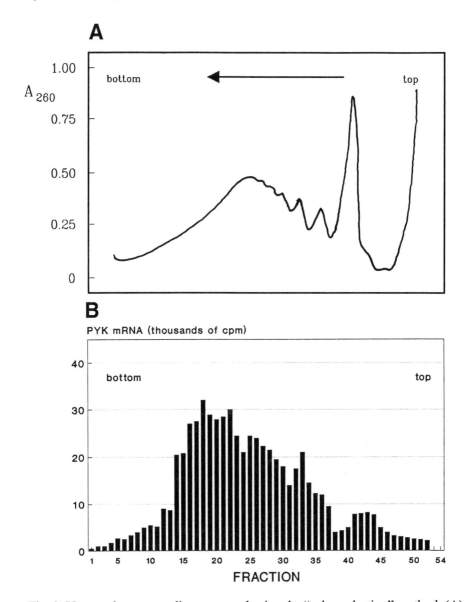

Fig. 3. Yeast polysome gradient prepared using the "spheroplasting" method. (**A**) The absorbance profile at 260 nm of a typical yeast polysome gradient prepared using the spheroplasting technique (Section 3.1.) showing peaks corresponding to the positions at which mRNA carrying 1, 2, 3, 4, or 5 ribosomes sediment on the gradient, and the direction of sedimentation (horizontal arrow). (**B**) The distribution of the pyruvate kinase mRNA across the polysome gradient as determined by cutting out and scintillation counting the individual dots on a blot similar to that shown in Fig. 4.

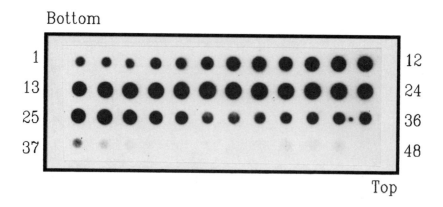

Fig. 4. A dot-blot of fractions from a polysome gradient. The dot-blot was prepared from a polysome gradient prepared using the spheroplasting method (see Fig. 3 and Section 3.3.1.) and probed for the pyruvate kinase mRNA.

described in Section 3.1. Postmitochondrial supernatants are prepared and subjected to sucrose density gradient centrifugation to produce polysome gradients (3) (see Fig. 4). Gradients are then fractionated for mRNA analysis (Section 3.3.). The following procedure (7) is adapted from that of Thireos and coworkers (3).

Follow steps 1–5 as described in Section 3.1.

Perform the following steps in a cold room.

2. Resuspend the cell pellet in 2.5 mL of lysis buffer at 4°C.
3. Transfer the cell suspension to a 15-mL Falcon tube containing 4 g of glass beads. Vortex continuously for 5 min.
4. Quickly transfer the homogenate to a 15-mL Corex tube and centrifuge at 10,000g for 10 min at 4°C.
5. Carefully layer exactly 1.5 mL of the supernatant onto the top of a prethawed and prebalanced 10–45% sucrose density gradient.
6. Centrifuge the gradients using a Beckman SW28 rotor at 25,500 rpm (140,000g) for 170 min at 0°C. Decelerate the rotor with the brake off.
7. Fractionate each gradient through a spectrophotometer using a 2.5-mm flow cell to monitor the absorbance profile at 260 nm, as shown in Fig. 2. A typical A_{260} profile is illustrated in Fig. 5A. Gradients can be fractionated either into about 50 0.75-mL fractions for dot-blotting (Section 3.3.1.) (Fig. 4), or into about 15 2.5-mL fractions for Northern analysis (Section 3.3.2.) (Fig. 5).

3.3. mRNA Analysis Across Polysome Gradients

Having prepared polysome gradients using either of the procedures described earlier, it is possible to determine the distribution of individual mRNAs across these gradients by dot-blotting or Northern analysis. These methods are described here.

3.3.1. Fractionation of Polysome Gradients for Dot-Blotting

1. Collect about 50 0.75-mL fractions from each sucrose gradient into numbered Eppendorf tubes containing 450 µL of 20X SSC and 300 µL of 37X (w/v) formaldehyde.
2. Mix the samples by vortexing, and incubate at 60°C for 15 min.
3. Store the samples, which are now ready for dot-blotting, at −70°C.
4. Dot 10–100 µL of each sample (depending on the abundance of the mRNA to be analyzed) onto a nitrocellulose or nylon membrane using a Hybridot apparatus and wash each sample onto the membrane using 100 µL of 15X SSC.
5. Bake the membrane at 80°C for about 2 h, and perform the prehybridization and hybridization according to standard procedures *(9)*. Ensure that the hybridization is performed under conditions of probe excess (*see* Chapter 28).
6. Following autoradiography to check for background contamination on the membrane (Fig. 4), cut out each dot on the blot and subject each to scintillation counting *(5)* or 2D-Radioimaging if available. The profile of bound radioactivity in the fractions across the gradient represents the distribution of the specific mRNA (Fig. 3B).

3.3.2. Fractionation of Polysome Gradients for Northern Analysis

1. Collect 15 2.5-mL fractions into numbered 15-mL Falcon tubes and immediately freeze on dry ice. Store samples at −70°C.
2. Thaw samples quickly at 50°C. Thaw only four or eight samples at a time since more tubes cannot be handled together quickly and efficiently.
3. Add 300 µL 10% SDS, 300 µL 1*M* Tris-HCl, pH 7.5, and 3 mL phenol:CHCl$_3$ and vortex continuously for 2 min.
4. Centrifuge the samples at 5000*g* for 5 min in a swing-out rotor.
5. Carefully avoiding the interphase, transfer the aqueous phase to a fresh tube containing 3 mL phenol:CHCl$_3$ and vortex for 2 min.
6. Centrifuge the samples at 5000*g* for 5 min in a swing-out rotor.
7. Repeat steps 5 and 6.
8. Transfer the aqueous phase to a fresh tube containing 3 mL CHCl$_3$ and vortex for at least 1 min.

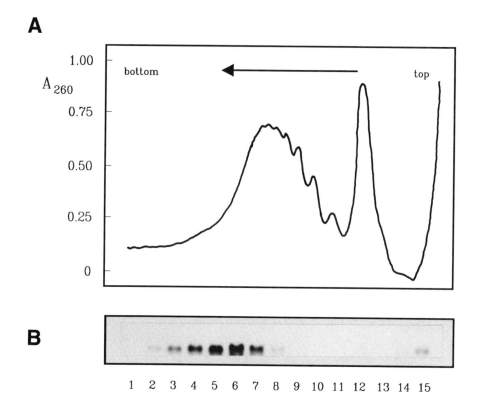
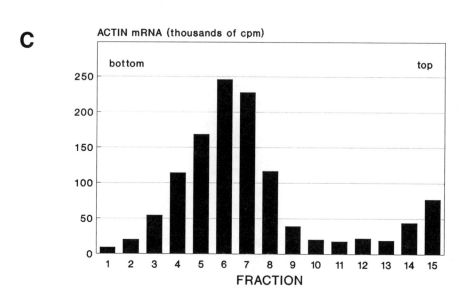

9. Centrifuge the samples at 5000g for 5 min in a swing-out rotor.
10. Remove the $CHCl_3$ and dilute the final aqueous phases three-fold with RNase-free H_2O.
11. Add 3M sodium acetate to a final concentration of 0.2M, add 2 vol of absolute ethanol, vortex, and store at $-20°C$, overnight.
12. Centrifuge the samples at 10,000g for 20 min at 4°C. Pour off the supernatants, drain the tubes, and vacuum desiccate to remove residual ethanol.
13. Redissolve the precipitates carefully in 400 µL T/E. Transfer to Eppendorf tubes, add 35 µL 3M sodium acetate and 900 µL absolute ethanol, mix, and leave at $-20°C$ overnight.
14. Centrifuge the samples at 10,000g for 15 min at 4°C. Pour off the supernatants, drain the tubes, and rinse the precipitates gently with 70% ethanol precooled at $-20°C$. Vacuum desiccate the precipitates to remove residual ethanol.
15. Redissolve the precipitates in 40 µL RNase-free H_2O by warming the samples at 65°C and vortexing. Store at $-70°C$.
16. When the 15 samples representing a single polysome gradient are ready, perform Northern analysis using 4 µL of each sample using standard methods *(9,11,12)* as described in Chapter X. Quantitative Northern analysis of the actin mRNA across a polysome gradient prepared using the "glass bead" method (Section 3.2.) is illustrated in Fig. 5.

3.4. Preparation of Polysomes

The principles of the procedure are similar to those described in Section 3.3. Cell breakage is followed by the preparation and ultracentrifugation of postmitochondrial supernatants through sucrose shelves. The cell lysis methods are adapted from those of Tzamarias and coworkers *(3)*, and the procedures for centrifugal preparation of polysomal pellets are adapted from those of Ross and Kobs *(10)*.

1. Inoculate separate 2.5-mL YPD starter cultures with a single colony of each yeast strain of interest, and grow to stationary phase by incubating at 30°C for 24–48 h with shaking (approx 250 rpm).

Fig. 5. *(preceding page)* Yeast polysome gradient prepared using the "glass bead" method. **(A)** The absorbance profile at 260 nm of a typical yeast polysome gradient prepared using the lass bead technique (Section 3.2.) showing peaks corresponding to the positions at which mRNA carrying 1, 2, 3, 4, 5, or 6 ribosomes sediment on the gradient, and the direction of sedimentation (horizontal arrow). **(B)** Northern analysis of the actin mRNA across the polysome gradient *(see* Section 3.3.2.). **(C)** The distribution of the actin mRNA across the polysome gradient as determined by quantitation of the Northern filter shown in (B).

2. Use about 200 μL of a stationary phase starter culture to inoculate 400 mL of YPD, and incubate at 30°C with shaking at 250 rpm for 14–16 h until the cells reach midexponential growth phase (A_{600} = 0.4–0.6). Use a 2-L flask.
3. Harvest the cells by centrifuging the culture at 5000g for 5 min at 0°C.
4. Resuspend the cells in 5 mL lysis buffer at 0°C. Add 7.8 g glass beads and vortex the mixture continuously for 6 min at 0°C.
5. Add an additional 3 mL of ice-cold lysis buffer and mix the sample.
6. Transfer the homogenate to an Oakridge centrifuge tube and centrifuge at 10,000g for 10 min at 0°C.
7. Take the supernatant, avoiding the cloudy top layer, and layer onto a 1.5-mL 30% sucrose shelf in a precooled Beckman SW55 centrifuge tube at 0°C. Fill these 5-mL tubes almost to the top (approx 4.8 mL), and if necessary top them up with lysis buffer. Centrifuge the samples at 36,000 rpm (280,000g) for 150 min at 0°C using a Beckman SW55 rotor.
8. Remove the supernatant carefully by aspiration using an RNase-free Pasteur pipet, leaving a translucent pellet of polysomal material. (If necessary, the interphase of the sucrose shelf, which contains monosomal material, can be kept for further analysis.)
9. Carefully wash the polysomal pellet twice with 600 μL of lysis buffer, resuspend it in 100 μL of lysis buffer, and centrifuge at 10,000g for 5 min at 0°C to clarify the preparation. Dilute a small portion of the sample for spectrophotometric analysis (Fig. 6), and store the major portion in aliquots at –70°C for future use. Typically, a polysomal yield of approx 600 A_{260} U/mL in a final volume of 100 μL is obtained.
10. At this point, RNA can be purified from the polysomal preparation by phenol extraction (*see* Section 3.3.2.) and subjected to agarose gel electrophoresis (Fig. 7A). Northern analysis *(9,11,12)* of polysomal RNA is presented in Fig. 7B.

4. Notes

It is imperative that precautions be taken to exclude contamination by exogenous ribonucleases (RNases) because these will degrade polysomes. Excess materials should be prepared to prevent shortages, as the preparation of RNase-free materials takes time.

1. Glassware (Pasteur pipets and pipets) is wrapped in aluminum foil and baked at 200°C overnight.
2. Disposable plasticware (tips, Eppendorf tubes, and so on) is taken from new bags that have been set aside for RNA work, and autoclaved.
3. Nondisposable plasticware (for example, 50-mL Oakridge centrifuge tubes and Beckman polyallomer SW28 centrifuge tubes) is soaked in 0.1% (v/v)

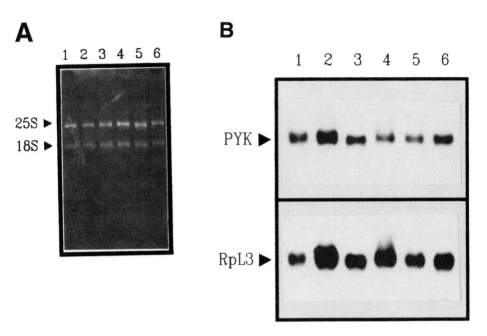

Fig. 6. Absorbance profile of polysomes sedimented though sucrose shelves.

diethylpyrocarbonate for at least 1 h and then rinsed in RNase-free water or incubated at 80°C.
4. Fresh, previously unopened chemicals can be considered to be RNase-free. A set of chemicals are set aside specifically for RNA work to prevent careless contamination by other workers. Spatulas are not used to weigh out these chemicals, because they are a probable source of RNase contamination. Instead, these chemicals are carefully tapped out of the bottles.
5. Solutions that do not contain amines are rendered RNase-free by adding diethylpyrocarbonate to a final concentration of 0.1% (v/v), mixing vigorously, reacting on the bench for at least 1 h, and then autoclaving to inactivate the remaining diethylpyrocarbonate.
6. Solutions containing amines (for example, Tris or EDTA) are not treated with diethylpyrocarbonate. Instead, extra precautions are taken to prevent contamination with RNase. RNase-free chemicals are added to diethylpyrocarbonate-treated distilled water and the pH adjusted using concentrated acid or $10M$ NaOH. The pH is monitored by dropping small samples of the solution onto pH indicator papers to prevent RNase contamination from a pH electrode. Final volumes are then adjusted using volumetric markings on the RNase-free solution bottle, and the solution autoclaved.
7. Gloves should be worn throughout.

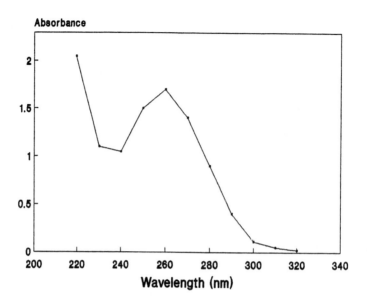

Fig. 7. Analysis of RNA isolated from yeast polysomes. **(A)** Agarose gel electrophoresis of RNA isolated from yeast polysomes. Note the prominent 25S and 18S rRNA bands, and the reduced quantities of 5S and RNA tRNA compared with total RNA preparations (Chapter X, by Brown). **(B)** Northern analysis of the pyruvate kinase and ribosomal protein L3 mRNAs in polysome preparations.

8. Translation rates are extremely sensitive to the physiological status of the cell. The ratio of monosomes:polysomes increases when cells are in late exponential or stationary growth phases, and if the recovery phase is not performed under ideal conditions. Therefore, every attempt must be made to harvest cells in early or midexponential growth phases and to handle the spheroplasts carefully prior to lysis. Hence, temperature shocks are minimized by centrifuging the cells or spheroplasts at room temperature.
9. The efficiency of spheroplasting directly affects the yield of polysomes, and yet the polysome yields are adversely affected by protracted incubation with β-glucuronidase. Therefore, if necessary to achieve efficient spheroplasting, short incubations with relatively high concentrations of β-glucuronidase are preferable to protracted incubations.
10. Some methods include cycloheximide (50 µg/mL) and possibly heparin (0.2 mg/mL) in the lysis buffer to stabilize polysomes *(1,3)*, but this is not a prerequisite for good polysome gradients.
11. In our hands, plasmid-containing cells represent 80–90% of the culture following overnight growth under nonselective conditions. This is the case even for relatively unstable YEp-based plasmids *(6)*.

12. Use only high-quality phenol that is colorless upon melting. Pink or brown phenol should not be used.
13. *Caution: Phenol* is highly corrosive and toxic. Wear gloves and safety spectacles. *Ethidium bromide* is a potential carcinogen. Wear gloves and dispose of appropriately. *Diethylpyrocarbonate* should be handled in a fume cupboard. *Formaldehyde* should be handled in a fume cupboard.

References

1. Hartwell, L. H., Hutchison, H. T., Holland, T. M., and McLaughlin, C. S. (1970) The effect of cycloheximide upon polyribosome stability in two yeast mutants defective respectively in the initiation of peptide chains and in messenger RNA synthesis. *Mol. Gen. Genet.* **106**, 347–361.
2. Maicas, E., Pluthero, F. G., and Friesen, J. D. (1988) The accumulation of three yeast ribosomal proteins under conditions of excess mRNA is determined primarily by fast protein decay. *Mol. Cell Biol.* **8**, 169–175.
3. Tzamarias, D., Roussou, I., and Thireos, G. (1989) Coupling of GCN4 mRNA translational activation with decreased rates of polypeptide chain elongation. *Cell* **57**, 947–954.
4. Baim, S. B., Pietras, D. F., Eustice, D. C., and Sherman F. (1985) A mutation allowing an mRNA secondary structure diminishes translation of *Saccharomyces cerevisiae* iso-1-cytochrome c. *Mol. Cell. Biol.* **5**, 1839–1846.
5. Santiago, T. C., Bettany, A. J. E., Purvis, I. J., and Brown, A. J. P. (1987) Messenger RNA stability in *Saccharomyces cerevisiae*: the influence of translation and poly(A) tail length. *Nucleic Acids Res.* **15**, 2417–2429.
6. Bettany, A. J. E., Moore, P. A., Cafferkey, R., Bell, L. D., Goodey, A. R., Carter, B. L. A., and Brown, A. J. P. (1989) 5'-secondary structure formation, in contrast to a short string of non-preferred codons, inhibits the translation of the pyruvate kinase mRNA in yeast. *Yeast* **5**, 187–198.
7. Sagliocco, F. A., Vega Laso, M. R. V., Zhu, D., Tuite, M. X. F., McCarthy, J. E. G., and Brown, A. J. P. (1993) The influence of 5'-secondary structures upon ribosome binding to mRNA during translation in yeast. *J. Biol. Chem.* **2**, 26,522–26,530.
8. Hutchison, H. T. and Hartwell, L. H. (1967) Macromolecule synthesis in yeast spheroplasts. *J. Bacteriol.* **94**, 1697–1705.
9. Beggs, J. D. (1978) Transformation of yeast by a replicating hybrid plasmid. *Nature* **275**, 104–109.
10. Sambrook, J., Fritsch, E. F., and Maniatis, T. (1989) *Molecular Cloning: A Laboratory Manual*. Cold Spring Harbor Laboratory, Cold Spring Harbor, NY.
11. Ross, J. and Kobs, G. (1986) H4 histone messenger RNA decay in cell-free extracts initiates at or near the 3'-terminus and proceeds 3' to 5'. *J. Mol. Biol.* **188**, 579–593.
12. Lehrach, R. H., Diamond, D., Wozney, J. M., and Boedtker, H. (1977) RNA molecular weight determinations by gel electrophoresis under denaturing conditions, a critical re-examination. *Biochemistry* **16**, 4743–4751.
13. Thomas, P. S. (1980) Hybridization of denatured RNA and small DNA fragments transferred to nitrocellulose. *Proc. Natl. Acad. Sci. USA* **77**, 5201–5205.

CHAPTER 30

Induction of Heat Shock Proteins and Thermotolerance

Peter Piper

1. Introduction

In yeast, as with other organisms, a heat shock causes the induction of the heat shock response. The main consequences of this induction are (at the physiological level) an increased tolerance of high, potentially lethal temperatures, and (at the molecular level) strong induction of a small number of heat shock proteins. The messenger RNAs for the latter proteins are generated by a transcriptional activation of heat-inducible genes. The heat shock response is usually transient, heat shock protein synthesis becoming repressed just a few minutes after an induction by either temperature upshift to stressful temperatures or an upshift followed by a return to normal temperatures *(1)*.

The response to heat shock entails not just the strong induction of genes for heat shock proteins, but the repression of most (but in yeast not quite all) *(4)* of those genes that were being expressed previous to this induction. mRNAs existing prior to the shock are no longer translated, yet they are not always degraded and they can sometimes be stored in inactive form to be translated yet again after the response has been switched off *(1)*.

Gene induction with heat shock often reflects the presence of a heat shock element (HSE) in heat shock gene promoters. This transcriptional activator element has been very extensively studied in *Saccharomyces cerevisiae (2,3)*, and serves as the DNA binding site for a heat shock transcription factor. The latter exhibits some basal activity, yet needs operation of the heat shock trigger before it can efficiently promote RNA initiation *(2)*. HSEs can be used in systems for temperature-regulated heterologous gene expression (*see* Section 3.).

2. Materials

1. Either water baths set to the appropriate temperatures (for large cultures), or a programmable temperature cycler with thermocouple (for small samples of 1 mL or less).
2. Yeast cultures in active growth on any medium (*see* Note 1).

3. Method

3.1. Optimum Induction of Yeast Heat Shock Proteins (see Note 2)

3.1.1. Induction by Temperature

For experimental purposes, heat shock proteins are usually induced by temperature upshift. The optimal temperature for this induction is highly species-dependent, but always close to the maximum that permits cell division. In laboratory *S. cerevisiae* strains, optimal induction is achieved *(3)* by growing cells at relatively low temperatures (18–25°C), then upshifting them to the maximum permitting growth. This is generally around 39°C for *S. cerevisiae* on fermentative media, although it can be somewhat strain dependent. With cultures growing nonfermentatively on gluconeogenic media the maximum temperature for growth of *S. cerevisiae* is usually slightly lower than this value, but a good induction of heat shock proteins still occurs upon shift to 38–40°C *(4)*. For yeasts other than *S. cerevisiae*, the optimum conditions for induction have not yet been so precisely determined. Yeasts vary very widely in their maximal growth temperature, and some are psychrophiles *(5)*. Among well studied species, *Schizosaccharomyces pombe* grows up to about 37°C (displaying marked morphological changes in the phase-contrast microscope at this, compared to lower temperatures), whereas the methylotrophic yeast *Hansenula polymorpha* grows to 49–50°C (Guerra and Piper, unpublished). Probably heat shock gene induction is maximal in these organisms on upshift to temperatures of around these respective values.

3.1.2. Induction Using Chemicals

There are several other strong inducers of heat shock protein synthesis besides thermal stress. Notable among these are ethanol (added to cultures at 6 or 8% [w/v] final concentration) *(6)*, several potentially cyto-

Heat Shock Proteins and Thermotolerance

toxic chemicals (e.g., heavy metal ions such as Cd^{2+}, sodium arsenite, amino acid analogs) *(1)*, and physiological states that may cause the generation of highly reactive free radicals (e.g., refeeding after glucose starvation, recovery from anoxia). These diverse inducers probably share with high temperatures the ability to cause the intracellular accumulation of aberrant or partially denatured protein. Certain heat shock proteins probably help to renature damaged protein, or to sequester such protein prior to its degradation *(1,7)*.

3.2. Induction of Thermotolerance

A mild, nonlethal heat shock (e.g., for *S. cerevisiae*, 25–38°C for 30–40 min) renders most cells much more resistant to short exposure to still higher, normally lethal temperatures. This induced resistance to heat killing (thermotolerance) is the most marked and rapid physiological change associated with induction of the heat shock response. Yeast cells in rapid growth are generally much less thermotolerant than nondividing cultures, thermotolerance induction in *S. cerevisiae* being most rapid in log phase cultures shifted to 45°C *(5)*. This is a temperature well above the optimum for HSE induction (39°C) *(2,3)* and too high for any appreciable synthesis of heat shock proteins. It is therefore thought that the heat shock proteins may not be the major factor determining thermotolerance. Thermotolerance at normally lethal temperatures, the maximum temperature for growth, and even the minimum temperature for growth (cold tolerance) can all be affected by mutational inactivation of *S. cerevisiae* genes for constitutively synthesized proteins related to the heat-induced Hsp70 *(1,7)*. However, these same properties are unaffected by the inactivation of many heat shock genes (*HSP104* excepted) *(7,8)*.

Why is thermotolerance so dependent on physiological state? The large trehalose pool accumulated under certain conditions is almost certainly a factor *(14)*, as also are other intracellular molecules (notably, polyols like glycerol) that influence the interaction between solvent and solute macromolecules. Another part of the answer may lie in the fact that heat shock proteins are often made in situations other than heat shock (*see* Note 2).

3.3. Temperature-Regulated Heterologous Gene Expression

A number of systems have been described for temperature regulation of heterologous gene expression in *S. cerevisiae (2,3,9–12)*. They induce expression in response to either a temperature downshift *(9–11)* or a temperature upshift *(3,12)*. A number of important considerations need to be met before temperature shifts can be used successfully to induce a gene in an industrial-scale fermentation. First, *S. cerevisiae* grows more slowly with reduced biomass yield at temperatures above 34–36°C, so it is usually desirable to grow cultures to high biomass at temperatures slightly lower than this value. Second, although systems using temperature downshift for induction may be straightforward to operate on a small scale, they are not as attractive as those systems that use temperature upshift for larger-scale processes. This is because heat addition to large fermenters is easier than heat extraction.

Most systems for temperature regulation of gene expression in *S. cerevisiae* have to be used with specially developed strains *(9–12)*. However, this is not the case for heterologous genes placed under the control of the HSE promoter element (*see* Section 1.), these being activatable in any genetic background. Such HSE-controlled genes can be activated up to 50-fold by the appropriate heat shock (*see* refs. *[2]* and *[3]* for precise details), their induction occurring with that of genes for heat shock proteins whenever the heat shock response is induced. HSE sequences are also activated up to 15-fold by the simple expedient of adding sublethal concentrations of methanol to *S. cerevisiae* cultures *(13)*. In addition, promoter sequences with HSE elements show authentic regulation when present on high copy number yeast vectors, allowing heat-induction of a protein to the point where it constitutes 30–40% of total cell protein *(15)*.

HSE-directed expression systems are not without potential problems. Most notably:

1. The heat shock response is transient at sublethal temperatures, being switched off within 30–120 min of induction *(1–3)*; and
2. The temperatures used to heat stress yeast may detrimentally affect certain heterologous products.

However, the HSE is a strong transcriptional activator and heterologous gene expression directed by this element may find applications where a limited expression period will suffice, and where authentic rather than aberrant product can accumulate intracellularly at 37–39°C.

4. Notes

1. Stationary phase yeast cells are intrinsically thermotolerant and incapable of any appreciable induction of either heat shock proteins or thermotolerance when heat shocked.

2. *S. cerevisiae* heat-inducible promoters often have other activator elements, in addition to the HSE promoter element, that direct their activation by heat shock. Because of these other elements, specific heat shock proteins are often synthesized in situations other than heat shock (e.g., at the approach to stationary phase, or in sporulating cultures) *(7)*. Therefore, heat shock proteins will often be found in apparently unstressed yeast cells.

References

1. Craig, E. A. (1986) The heat shock response. *CRC Crit. Rev. Biochem.* **18,** 239–280.
2. Sorger, P. K. (1990) Yeast heat shock factor contains separable transient and sustained response transcriptional activators. *Cell* **62,** 793–805.
3. Kirk, N. and Piper, P. W. (1991) Determinants of heat shock element-directed *lacZ* expression in *Saccharomyces cerevisiae*. *Yeast* **7,** 539–546.
4. Piper, P. W., Curran, B., Davies, M. W., Hirst, K., Lockheart, A., and Seward, K. (1988) Catabolite control of the elevation of PGK mRNA levels by heat shock in *Saccharomyces cerevisiae*. *Mol. Microbiol.* **2,** 353–362.
5. Piper, P. W. (1993) Molecular events associated with the acquisition of heat tolerance in the yeast *Saccharomyces cerevisiae*. *FEMS Microbiol. Rev.* **11,** 1–11.
6. Piper, P. W., Talreja, K., Panaretou, B., Moradas-Ferreira, P., Byrne, K., Praekelt, U. M., Meacock, P., Regnacq, M., and Boucherie, H. (1994) Induction of major heat shock proteins of *Saccharomyces cerevisiae*, including plasma membrane Hsp30, by ethanol levels above a critical threshold. *Microbiology* **140,** 3031–3038.
7. Parsell, D. A. and Lindquist, S. (1991) The function of heat shock proteins in stress tolerance: degradation and reactivation of damaged proteins. *Annu. Rev. Genet.* **27,** 437–496.
8. Sanchez, Y. and Lindquist, S. L. (1990) HSP104 required for induced thermotolerance. *Science* **248,** 1112–1115.
9. Brake, A. J., Merryweather, J. P., Coit, D. G., Heberlain, V. A., Maziarz, F. R., Mullenbach, G. T., Urdea, M. S., Valenzuela, P., and Barr, P. J. (1984) α-factor-directed synthesis and secretion of mature foreign proteins in *Saccharomyces cerevisiae*. *Proc. Natl. Acad. Sci. USA* **81,** 4642–4646.
10. Kramer, R. A., DeChiara, T. M., Schaber, M. D., and Hilliker, S. (1984) Regulated expression of a human interferon gene in yeast: Control by phosphate concentration or temperature. *Proc. Natl. Acad. Sci. USA* **81,** 367–370.
11. Sledewski, A. Z., Bell, A., Kelsay, K., and MacKay, V. L. (1988) Construction of temperature-regulated yeast promoters using the MATα2 repression system. *Biotechnology* **6,** 411–416.
12. DaSilva, N. A. and Bailey, J. E. (1989) Construction and characterisation of a temperature-sensitive expression system in yeast. *Biotechnol. Prog.* **5,** 18–26.
13. Kirk, N. and Piper, P. W. (1991) Methanol as a convenient inducer of heat shock element-directed heterologous gene expression in yeast. *Biotechnol. Lett.* **13,** 465–470.
14. Wiemken, A. (1990) Trehalose in yeast, stress protectant rather than reserve carbohydrate. *Ant. van Leeuwenhoek J. Microbiol.* **58,** 209–217.
15. Cheng, L., Hirst, K., and Piper, P. W. (1992) Authentic temperature-regulation of a heat shock gene inserted into yeast on a high copy number vector. Influences of overexpression of HSP90 protein on high temperature growth and thermotolerance. *Biochem. Biophys. Acta* **1132,** 26–34.

CHAPTER 31

Rhodamine B Assay for Estimating Activity of Killer Toxins Permeabilizing Cytoplasmic Membranes

Vladimír Vondrejs and Zdena Palková

1. Introduction

Killer yeasts secrete proteins (killer toxins, zymocins) that kill sensitive yeast strains *(1,2)*. The killer strain is immune (R^+) to the effects or its own toxin, but it can be sensitive to the toxins of another immunity group *(2)*. The killing ability of strains (K^+) is routinely screened *(1)* using buffered rich agar media containing methylene blue (the pH range usually from 4.2 to 4.8) thinly spread with potentially sensitive tester strain. Potential killers are replica plated or heavily streaked across the background lawn. Killer colonies are identified by clear zones usually fringed with blue-stained dead cells. The same assay can be used for distinguishing the sensitivity of selected strains to various killers. A well test *(1)* based on measuring the diameter of inhibition zones was evolved that enables killer toxin activity to be assayed in liquid media added to wells. Also the fraction of colony-forming cells after killer toxin treatment can be considered the measure of killer toxin activity. However, the assays based on the formation of colonies or inhibition zones last usually at least 2 d.

The methylene blue-stained cells and other results indicate that cytoplasmic membranes of cells killed by some killer toxins become permeable after a lag phase for a great number of substances including some dyes *(2)*. For this reason, we decided to examine especially fluorescent dyes *(3)* as they allow a high degree of resolution at low concentrations.

It should be emphasized that the Rhodamine B assay originally evolved in our laboratory for rapidly estimating activity of superkiller toxin K1 in cell-free media after cultivation of superkiller strain *Saccharomyces cerevisiae* T158C *(3)* is presented here in a modified version that is more sensitive, and enables testing of a number of killer toxin samples within a few hours. The procedure may not be directly applicable to all classes of killer toxins, but it illustrates general principles of exploitation of Rhodamine B for estimating relative activities of killer toxins, and other factors, causing permeabilization of cytoplasmic membranes in yeast cells.

2. Materials

1. The following stock solutions in distilled water should be sterilized by autoclaving and stored at room temperature: 5% w/v yeast extract; 5% w/v peptone; 10% w/v glucose; 20% w/v $(NH_4)_2SO_4$; 5% w/v $MgSO_4 \cdot 7 H_2O$; 8% w/v KCl; 3% w/v $CaCl_2 \cdot 2H_2O$; 0.05% w/v $FeCl_3 \cdot 6H_2O$; 0.05% w/v $MnSO_4$; 26 g of glycerol per 253 mL of distilled water; $0.1M$ citric acid; and $0.2M$ Na_2HPO_4.
2. The following stock solutions should be sterilized by filtration and stored in refrigerator at 4°C: 1 mM Rhodamine B, 0.02% w/v biotin, and 0.02% w/v folic acid.
3. McIlvains buffer (pH 4.7) is prepared by mixing of 26 mL of $0.1M$ citric acid with 24 mL of $0.2M$ Na_2HPO_4 and 50 mL of sterilized distilled water.
4. Wickerham's vitamin-solution contains in 100 mL: Ca 40 mg pantothenate, 20 mg inositol; 40 mg nicotinic acid; 20 mg *p*-aminobenzoic acid; 40 mg pyridoxine, HCl; 40 mg thiamine, HCl; 20 mg riboflavin; 1 mL of 0.02% w/v biotin; and 1 mL of 0.05% w/v $FeCl_4 \cdot 6H_2O$. This solution should be sterilized by filtration and stored in a refrigerator at 4°C.
5. Medium J: Mix the following volumes of stock solutions: 50 mL of 5% w/v yeast extract; 50 mL of 5% w/v peptone; 50 mL of 10% w/v glucose; 10 mL of 20% w/v $(NH_4)_2SO_4$; 5 mL of 8% w/v KCl; 5 mL of 5% w/v $MgSO_4 \cdot 7 H_2O$; 5 mL of 3% w/v $CaCl_2 \cdot 2H_2O$; 5 mL of 0.05% w/v $FeCl_3 \cdot 6H_2O$; 5 mL of 0.05% w/v $MnSO_4$; (C-Ph) 50 mL of $0.1M$ citric acid with 24 mL of $0.2M$ Na_2HPO_4; and 265 mL of 26 g of glycerol per 253 mL of distilled water.
6. YEG medium contains in 100 mL: 10 mL of 10% w/v glucose; 10 mL of 5% w/v yeast extract; and 80 mL of sterilized distilled water.
7. Stationary culture of standard sensitive strain, e.g., *S. cerevisiae* S6/1 *(3)* in YEG medium cultivated aerobically on a shaker at 28°C to saturation. The culture of *S. cerevisiae* S6/1 can be stored for several days at 4°C.
8. Killer toxin preparation, e.g., cell-free medium J containing killer toxin K1 (K1 medium J) produced by strain *S. cerevisiae* T158C *(3)*. The cells

should be removed by centrifugation and subsequently filtered using nitrocellulose membrane filter (e.g., Synpor 6, porosity 0.45 µm). K1 medium J can be stored at 4°C for several months.
9. Eppendorf microfuge tubes (1.5 mL) or equivalents.
10. Eppendorf microfuge or equivalent.
11. Fluorescence microscope Bioval PZ0 (Warsaw Poland) or equivalent.
12. Shaker.
13. Thermostatted baths.
14. Spectrophotometer for visible light.

3. Method

1. Inoculate 20 mL of YEG medium in a 100 mL-flask with appropriate amount (e.g., 35 µL) of stationary culture of *S. cerevisiae* S6/1 (*see* Note 1).
2. Incubate the culture on a shaker at 28°C overnight to the final optical density λ = 540 nm) from 0.1–0.5 (*see* Note 2).
3. Determine the cell concentration by haemocytometry (*see* Note 3).
4. Centrifuge aliquots (two parallels for each toxin dilution) of the culture containing 1×10^6 cells for 15 s (about 15,000 g), allow the centrifuge to slow without braking and pour of the supernatants (*see* Note 4).
5. Resuspend the pellets each in 1 mL of appropriately diluted sample of killer toxin K1 in medium J (*see* Note 5).
6. Add 0.1 mL of 10% w/v glucose and 0.1 mL of 1 mM Rhodamine B to each sample, mix the samples, and incubate them without shaking for 2 h at 2°C in the dark (*see* Note 6).
7. Centrifuge the samples (as in step 4), wash the pellets with distilled water, and resuspend harvested cells in small amounts of distilled water for determining fractions of stained cells using fluorescence microscope (*see* Note 7).
8. Specific details for estimating fractions of dead (Rhodamine B-stained) cells (F) after killer toxin treatment vary, depending on which microscope is used. When, e.g., a mercury lamp in connection with filter BG 12/2 (fluorescence microscope, Bioval PZO) is used as the excitation source, blue ghosts of living cells can be easily distinguished from Rhodamine B-stained cells (red fluorescence) (*see* Note 8).
9. Present your data by plotting the fractions F = N_C/N_T (N_C is the number of stained cells and N_T is the total number of cells in the counted sample after killer toxin treatment) against the relative concentrations (C) of killer toxin in samples (*see* Note 9).
10. A lethal dose (LD_{50}) expressed as the relative concentration of killer toxin that causes the death in 50% of cells can be easily estimated (Fig. 1) in order to characterize the relative killing activity of the killer toxin preparation under standard conditions (*see* Notes 10 and 11).

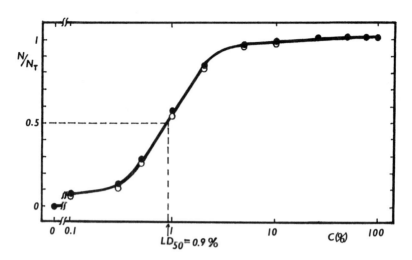

Fig. 1. Rhodamine B assay: An example of determination of a lethal dose (LD_{50}). Cell-free K1 medium J was prepared by centrifugation and filtration of *S. cerevisiae* T158C/IP3 culture incubated on a shaker for 16 h at 20°C. N is the number of stained cells, $N_T = 400$ is a total number of cells counted in each sample, C (%) is a relative concentration of examined K1 medium J in samples, $C = 100\%$ for undiluted K1 medium J, $(LD_{50}) = 0.9\%$ is a concentration of K1 medium J that causes the death of 50% of cells under conditions of Rhodamine B assay, and -○-●- are the results of two parallel experiments.

4. Notes

1. Some modifications of the procedure concerning, namely, the selection of standard sensitive strain and optimal conditions (namely, pH and temperature) for estimating activity of different killer toxins may be required. A suitable standard strain usually has to be found by trial and error. The following criteria should be taken into account:
   a. It should be the most sensitive one among strains examined.
   b. It should not form cell-clusters or cell-aggregates under conditions of assays.
   c. Clones with remarkably different sensitivities and resistant cells should not appear in cultures of standard sensitive strains.
   d. The fraction of stained cells in control samples untreated with toxin should be negligible.
   e. Stable prototrophic strains highly sensitive to killer toxins of several classes are the best candidates for usage as standard strains in the Rhodamine B-assay.

2. As stationary cultures of sensitive strains are usually not susceptible to killer toxins, a crucial point of the second step is to obtain an exponential culture of the sensitive strain. For estimating the appropriate amount of inoculum (step 1) and time of cultivation (step 2), it is important to know the doubling time (T) of the strain, e.g., for *S. cerevisiae* S6/1 growing aerobically in YEG medium at 28°C t = 100 min.
3. The determination of cell number does not need to be very accurate because it has been shown that small differences in cell concentrations (step 4) do not remarkably change the results.
4. Manipulation of samples before the addition of killer toxins should be as fast as possible. Conical Eppendorf tubes sterilized by boiling in distilled water appear to be the most convenient centrifugation cuvets for this purpose. Centrifugation of 15 s in an Eppendorf centrifuge at room temperature is sufficient.
5. In addition to zero concentration of killer toxin and undiluted K1 medium J (relative concentration 100%), at least four concentrations in the range of increasing part of sigmoidal dose-response curve should be selected for estimating LD_{50}. When a sample of purified killer toxin is available, the concentration of killer protein in samples can be determined, and LD_{50} can be expressed as the concentration of killer protein that causes death in 50% of cells.
6. The addition of glucose solution is important, especially when killer toxin activities in cell-free media after cultivation of killer strains are estimated, because it is known that the effect of killer toxins is usually dependent on the cytoplasmic membranes in sensitive cells being in an energized state. The susceptibility of sensitive cells to killer toxins in media that are not glucose-supplemented is lower than in presence of glucose *(3)*. Gentle mixing of samples and relatively low temperature of incubation (20°C) are recommended, particularly when some fragile and temperature-sensitive killer toxins are assayed. The exploitation of 2*M* glucose and 2 m*M* Rhodamine B can increase the sensitivity of the assay if necessary.
7. Washing the cells after staining decreases the background fluorescence, stops additional staining, and eases the observation of cells in the fluorescence microscope.
8. The advantage of staining with a fluorescence dye is that determination of cell number in a fluorescence microscope can be replaced by flow cytometer counting. The counting of dead cells should be as fast as possible because the fluorescence of killed cells suspended in water is slowly decreasing. The microscopic examination of 400 randomly selected cells in each sample appears to be satisfactory for approximately estimating the relative killer toxin activity. Washing the cells (step 7) can be replaced by dilution of stained samples (1/1000) when flow cytometer counting is used.

9. The corrected fraction $F' = (N_C - N_O)/N_T$ has to be plotted against relative concentration of killer toxin in samples, when the fraction N_O/N_T (N_O is the number of stained cells at zero concentration of toxin) is not negligible.
10. The rapid assay described earlier is convenient, not only for estimating the relative killer toxin activities in killer toxin samples, but also for determining killer protein concentrations in cell-free media when a purified killer toxin is available for calibration.
11. The same procedure has been used repeatedly in our laboratory for quantitatively evaluating the effects of various factors on the activity of certain killer toxins by comparing the results obtained under standard and modified conditions. It is obvious that the determination of fractions of stained cells at one concentration of killer toxin under various conditions can replace more complex determination of LD_{50} when exact quantitative analysis is not required. In addition, the procedure can be used for evaluating the relative sensitivity of various strains to some killer toxins.
12. A similar fluorescent staining method for assaying killer toxin has recently been published *(4)*.

References

1. Woods, D. R. and Bevan, E. A. (1968) Studies on the nature of the killer factor produced by *Saccharomyces cerevisiae*. *J. Gen. Microbiol.* **51**, 115–126.
2. Tipper, D. J. and Bostian, K. A. (1984) Double-stranded ribonucleic acid killer systems in yeasts. *Microbiol. Rev.* **48**, 125–156.
3. Spacek, R. and Vondrejs, V. (1986) Rapid method for estimation of killer toxin activity in yeasts. *Biotechnol. Lett.* **8**, 701–706.
4. Kurzweilova, H. and Sigler, K. (1993) Fluorescent staining with bromocresol purple: A rapid method for determining yeast cell dead count developed as an assay of killer toxin activity. *Yeast* **9**, 1207–1212.

CHAPTER 32

Nystatin-Rhodamine B Assay for Estimating Activity of Killer Toxin from *Kluyveromyces lactis*

Zdena Palková and Vladimír Vondrejs

1. Introduction

Killer toxin activity is commonly determined by the well test *(1)*, which ensures the inhibition zone on a substratum of sensitive strain cells. The Rhodamine B assay described in Chapter 31 is based on estimation of the fraction of killed cells, which stain with the fluorescent dye Rhodamine B. Nevertheless, this test cannot be used to assess the effect of *Kluyveromyces lactis* killer toxin, since it does not render sensitive cells permeable to this stain but merely causes a G1 block in the cell cycle *(2)*. It was, however, expected that killing yeast cells other than by the toxin, e.g., with nystatin, would also make them stainable by Rhodamine B. A clear-cut killing effect of nystatin was only observed in exponentially growing cells, whereas stationary cells were much less susceptible *(3,4)*. At the exponential growth phase, nystatin-treated *Saccharomyces cerevisiae* cells stain with Rhodamine B, if they have not previously been exposed to killer toxin produced by *Kluyveromyces lactis*. This finding constitutes the principle of a method for estimating the activity of this toxin *(5)*. The fraction of unstained cells increases in dependence on concentration of killer toxin in medium. As the lowered sensitivity of exponentially growing *S. cerevisiae* culture to nystatin after exposure to this toxin is owing to its inhibitory effect, it would probably be possible to determine in similar manner the activity of any inhibitor that suppresses the growth of cells in the exponential phase. The main disadvantage of the original version of the nystatin-Rhodamine B assay *(5)* is that the protective effect against nystatin caused by treating exponentially growing cells with toxin, develops rather slowly. An improved nystatin-Rhodamine B assay presented here is based on the observation that the susceptibility to nystatin in the absence of killer toxin is fully developed in all cells after 3 h of cultivation of diluted stationary culture at 28°C, whereas in the presence of appropriately concentrated samples of killer toxins all stationary phase cells are protected. Determination of the intermediate concentrations of killer toxin capable of protecting 50% of cells against nystatin treatment (PD_{50}) can be used to characterize the relative activity of a killer toxin preparation.

2. Materials

1. The following stock solutions in distilled water should be sterilized by autoclaving and stored at room temperature: 5% w/v yeast extract; and 10% w/v glucose.
2. YEG medium is prepared by mixing the following volumes of stock solutions: 50 mL of 5% w/v yeast extract; 50 mL of 10% w/v glucose; and 400 mL of sterilized distilled water.
3. Stock solution of 100 µ*M* Rhodamine B: should be sterilized by filtration and stored in refrigerator at 4°C.
4. Nystatin-Rhodamine B solution: Prepare by dissolving nystatin in the stock solution of Rhodamine B to the final 4 µM nystatin concentration. It must be freshly prepared before each experiment.
5. Stationary culture of *S. cerevisiae* S6/1 in YEG medium cultivated aerobically on a shaker at 28°C to stationary phase; the culture can be stored for several days at 4°C. The strain S6/1 was selected here because it is a very sensitive one and for this reason it is used in many laboratories; however, any sensitive laboratory strain can be used as well.
6. Killer toxin preparation, e.g., cell-free YEG medium containing killer toxin (KK) produced by *Kluyveromyces lactis* IF01267 *(5)* (KK medium YEG);* this can be stored at 4°C for several weeks.
7. Eppendorf conical centrifugation cuvets or equivalents.
8. Eppendorf microfuge or equivalent.
9. Fluorescence microscope (Bioval) or equivalent.
10. Shaker.
11. Thermostatted baths.
12. Spectrophotometer.

* The cells were removed by centrifugation and subsequent filtration using nitrocellulose membrane filter (e.g., Synpor 6, porosity 0.45 µm).

3. Method

1. Inoculate 20 mL of YEG medium in a 100-mL flask with 20 μL of stationary culture of *S. cerevisiae* S6/1 (*see* Note 1).
2. Incubate the culture on a shaker for 16 h at 28°C (*see* Note 2).
3. Dilute the culture with YEG medium to a final absorbance ($A_{1cm, \lambda = 540\,nm}$) of approx 0.1 (*see* Note 3).
4. Pipet 9-mL aliquots of the diluted culture each into a 100-mL flask containing 1 mL of appropriately diluted KK medium YEG (*see* Note 4).
5. Incubate the samples on a shaker for 3 h at 28°C (*see* Note 5).
6. Withdraw 1 mL of cell suspension from each sample and transfer it to a conical Eppendorf tube (*see* Note 6).
7. Centrifuge the samples in an Eppendorf centrifuge for 15 s (approx 15,000g) at room temperature, allowing the centrifuge to slow without breaking and pour off the supernatants (*see* Note 7).
8. Resuspend each pellet in 1 mL of 4 μM nystatin-100 μM Rhodamine B stock solution (*see* Note 8).
9. Incubate the samples statically for 1 h at 28°C in the dark (*see* Note 9).
10. Determine the fractions of stained cells using a fluorescence microscope. The details of this procedure are described in Chapter 31, on the Rhodamine B assay (*see* Note 10).
11. Plot the fractions of stained cells (F) against concentrations (C) (Fig. 1) (*see* Note 11).
12. A protection dose (PD_{50}) expressed as the relative concentration of killer toxin that protects 50% of cells against the permeabilization by nystatin is used to characterize the relative killing activity of killer toxin preparation (*see* Note 12).

4. Notes

1. Inocula prepared under the same conditions were always used in our experiments.
2. It was shown that an interval of incubation between 12 and 17 h does not make any difference in the results.
3. This cell concentration is convenient for subsequently counting the cells in fluorescence microscope.
4. Final relative concentrations (C) of KK medium YEG in 10-mL samples should be calculated.
5. 3 h appear to the shortest possible interval of incubation with KK medium YEG. The prolongation of interval causes a slow increase in protection (Fig. 1).
6. Withdraw several parallel 1-mL samples if more exact determination of (PD_{50}) (step 12) is required.

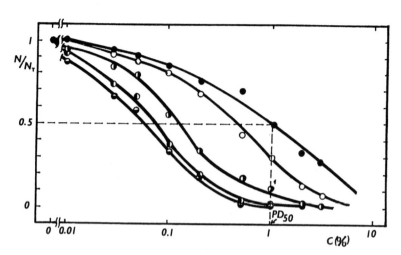

Fig. 1. Nystatin-Rhodamine B assay: An example of determination of the protection dose (PD_{50}). Cell-free KK medium YEG was prepared by centrifugation and filtration of *K. lactis* IFO1267 culture incubated on a shaker for 24 h at 28°C. N, number of stained cells; N_T, 300 (total number of cells counted in each sample); C (%), (relative concentration of examined KK medium YEG in samples); C, 100% (for undiluted KK medium YEG); (PD_{50}), 1% (concentration of KK medium YEG that protects 50% of cells against the permeabiliziation by nystatin under conditions of the standard Nystatin-Rhodamine B assay [-●-]). Results obtained under modified conditions, after 4, (-○-), 5 (-◐-), 6 (-◑-), or 7 (-○-) h of incubation of sensitive cells (*S. cerevisiae* S6/1) with killer toxin samples.

7. Manipulation of samples should be as fast as possible (*see* Note 6).
8. Vigorous mixing of samples is permitted (and recommended) in this case because the killer toxin from *Kluyveromyces lactis* is not labile.
9. Incubation at 28°C is permitted in this case because the killer toxin from *K. lactis* is not sensitive to this temperature.
10. Determination of the fractions of stained cells can be replaced by flow-cytometer counting. Dilution of stained samples (1:1000) is recommended when the flow cytometer is exploited. When the background fluorescence from the medium makes the observation of cells in microscope complicated, it is possible to wash the pellets with distilled water and resuspend them in the same volume of water.
11. According to our experience, usually no corrections of F' for the fraction of stained cells at zero concentration of toxin are necessary. When the concentration of killer toxin in the sample to be examined is too low,

the sensitivity of the assay may be increased by modification of step 3. The final concentration of K medium YEG can be increased by the addition of larger volumes of undiluted medium YEG. When the samples with very low activities are compared it is possible, in addition, to prolong the incubation at 28°C (step 4).
12. As in the case of the Rhodamine B assay, the Nystatin-Rhodamine B assay can be exploited also for evaluating the relative sensitivity of various strains to the killer toxin, for evaluating effects of various factors on activity of the killer toxin and for determining killer protein concentration (*see* Chapter 31, Notes 10 and 11). It is very likely that the same or slightly modified procedure can be applied also to estimations of activities of both different killer toxins and other factors that suppress the growth of treated cells; however, we have not yet experimented with these applications.

References

1. Woods, D. R. and Bevan, E. A. (1968) Studies on the nature of the killer factors produced by *Saccharomyces cerevisiae. J. Gen. Microbiol.* **51,** 115–126.
2. Sugisaki, Y., Gunge, N., Sakaguchi, K., Yamasaki, M., and Tamura, G. (1983) *Kluyveromyces lactis* killer toxin inhibits adenylate cyclase of sensitive yeast cells. *Nature* **304,** 464–466.
3. Moat, A. G., Peters, N., and Srb, A. M. (1959) Selection and isolation of auxotrophic yeast mutants with the aid of antibiotics. *J. Bacteriol.* **77,** 673–677.
4. Saul, D. J., Walton, E. F., Sudbery, P. E., and Carter, B. L. A. (1985) *Saccharomyces cerevisiae* whi2 mutants in stationary phase retain the properties of exponentially growing cells. *J. Gen. Microbiol.* **131,** 2245–2251.
5. Palková and Cvrčková, F. (1988) Method for estimating activity of killer toxin from *Kluyveromyces lactis. Folia Biologica (Praha)* **34,** 277–281.

CHAPTER 33

Application of Killer Toxins in Stepwise Selection of Hybrids and Cybrids Obtained by Induced Protoplast Fusion

Vladimír Vondrejs, Zdena Palková, and Zuzana Zemanová

1. Introduction

The most current method for selecting hybrid yeast clones obtained by means of protoplast fusion or transformation procedures is based on the auxotrophy of manipulated strains. Unfortunately, preparation of auxotrophic mutants without damage to the rest of the genome is time consuming, and, sometimes, as in the case of polyploid strains, practically impossible. For these reasons, namely in connection with improvement of industrial strains, selection techniques in which mutagenesis of manipulated parental strain is not required are attractive. Elimination of the industrial parent strain can be based, for example, on its natural sensitivity to some antibiotic or toxic compound, if the donor parental cell confers on the hybrid the ability to grow in the presence of such an agent.

A relevant widespread natural property in yeast (including industrial yeast strains) is, for example, sensitivity to various killer toxins usually produced by related killer strains (1,2). Killer strains occur relatively frequently in yeast of various species. They produce killer toxins (zymocins)—proteins killing sensitive cells. The killer phenotype was found to be associated with cytoplasic double–stranded (ds) RNA in *Saccharomyces* sp. or with dsDNA in *Kluyveromyces* sp. The genetic determinant of killer characters in many other killer strains has not been established clearly. Nuclear genes are very likely candidates for this function, at least in some cases. In addition, it has been shown that many nuclear and cytoplasmic genes take part in maintenance and expression of the killer phenotype (production of killer toxin K^+, and immunity to it, R^+).

The selection techniques developed in our laboratory are, in principle, similar to those used in bacteria in which sensitive strains are eliminated by the action of some bacteriophages or bacteriocins. A standard selection technique based on the application of killer toxin was developed in our laboratory in 1983 *(3)* using model systems of induced protoplast fusion of superkiller strains T158C (his$^-$, K^+, R^+) with various sensitive strains of *S. cerevisiae* (K^-, R^-). The method was also applied to the construction of new killer brewing strains *(4–6)*. After induced protoplast fusion, the auxotrophic killer parental strain and cybrids (phenotype his$^-$, K^+, R^+) were eliminated during regeneration of protoplasts in minimal medium. The second parental strain (K^-, R^-) was eliminated by addition of cell-free medium containing killer toxin K1 produced in an independent experiment by *S. cerevisiae* T158C. Hybrid strains and cybrids of the same phenotype (his$^+$, K^+, R^+) formed large colonies under these conditions on the background of the large number of small colonies of different phenotypes.

In an attempt to make the procedure more simple and efficient, improved versions of standard selection techniques were evolved *(7–9)*. The main goal of the so-called stepwise selection technique, which is presented here, is that it does not require the preparation of killer toxins. The principle of the technique is based on the fact that the auxotrophic parental killer strain can itself produce the killer toxin to eliminate sensitive protoplasts, if the first step of selection and reversion proceeds in complex medium. After several days of incubation at 20°C, the cells growing in the top layer are released from agar, centrifuged, washed, and resuspended in minimal medium. If the parental killer strain used in the experiment is a stable auxotrophic mutant (e.g., with a deletion in the marker gene, or doubly auxotrophic), then it is completely eliminated during the second step of selection on minimal medium at 28°C.

A disadvantage of this technique is that, because of the stepwise selection procedure, the original hybrid or cybrid clones cannot be recovered.

2. Materials

1. Parental strains: Killer strain, e.g., *S. cerevisiae* GRF18/ZH (α, *leu2-3, leu2-112, his3-11, his3-15*, K$^+$, R$^+$) and sensitive strain *S. cerevisiae* S6/1 (α, *LEU2, HIS3*, K$^-$, R$^-$).
2. The following stock solutions in distilled water should be sterilized by autoclaving and stored at room temperature: 5% w/v yeast extract; 5% w/v peptone; 10% w/v glucose; 20% w/v $(NH_4)_2SO_4$; 5% w/v $MgSO_4 \cdot 7H_2O$; 8% w/v KCl; 3% w/v $CaCl_2 \cdot 2H_2O$; 0.05% $FeCl_3 \cdot 6H_2O$; 0.05% w/v $MnSO_4$; 26 g glycerol per 253 mL of distilled water; 0.1M citric acid; 0.2M Na_2HPO_4; 0.8M KCl, 0.6M KCl, 1M $CaCl_2$, 0.3M $CaCl_2$, 4% w/v agar in 1.2M KCl, 3% w/v agar in 0.6M KCl (top agar), 4% w/v agar, 60 mM EDTA-NaOH (pH 8.0), and 33% w/v polyethylene glycol (4000).
3. The following stock solutions in distilled water should be sterilized by filtration and stored in the refrigerator at 4°C: 0.02% w/v biotin (B) and 0.02% w/v folic acid (F).
4. McIlwain's buffer (pH 4.7) is prepared by mixing 26 mL of 0.1M citric acid with 24 mL of 0.2M Na_2HPO_4 and 50 mL of sterilized distilled water.
5. Wickerham's vitamin-solution contains in 100 mL: 40 mg Ca pantothenate; 20 mg inositol; 40 mg nicotinic acid; 20 mg *p*-aminobenzoic acid; 40 mg pyridoxine HCl; 40 mg thiamine HCl; 20 mg riboflavin; 1 mL of 0.02% w/v biotin, and 1 mL of 0.02% w/v folic acid. This solution should be sterilized by filtration and stored in the refrigerator at 4°C.
6. ME-EDTA solution contains 9.9 mL of 60 mM EDTA-NaOH, pH 8.0, and 0.1 mL β-mercaptoethanol.
7. Snail digestive juice (SDJ): Dissolve 100 mg of lyophilized digestive juice from hepatopancreas of *Helix pomatia* in 10 mL of 0.8M KCl containing 10 µL of β-mercaptoethanol. Freshly prepared (SDJ) has been used in all of our experiments; however, commercially prepared SDJ or, e.g., Novozyme, is also effective in protoplasting (*see* Chapter 5).
8. PEG solution: Add 1 mL of 1M $CaCl_2$ to 9 mL of 33% w/v polyethylene glycol (4000).
9. YEG medium contains in 100 mL: 10 mL of 10% w/v glucose; 10 mL of 5% w/v yeast extract; and 80 mL of sterilized distilled water.
10. Minimal medium: Mix the following volumes of stock solutions: 50 mL of 10% w/v glucose; 10 mL of 20% w/v $(NH_4)_2SO_4$; 5 mL of 8% w/v KCl; 5 mL of 5% w/v $MgSO_4 \cdot 7H_2O$; 5 mL of 3% w/v $CaCl_2 \cdot 2H_2O$; 5 mL of 0.05% $FeCl_3 \cdot 6H_2O$; 5 mL of 0.05% w/v $MnSO_4$; 50 mL of 0.1M citric acid; 0.2M Na_2HPO_4; 265 mL of 26 g glycerol per 253 mL of distilled water; 0.5 mL of Wickerham's vitamin-solution; and 100 mL of sterilized distilled water.

11. Minimal agar medium: Mix the following volumes of stock solutions: 50 mL of 10% w/v glucose; 10 mL of 20% w/v $(NH_4)_2SO_4$; 5 mL of 8% w/v KCl; 5 mL of 5% w/v $MgSO_4 \cdot 7H_2O$; 5 mL of 3% w/v $CaCl_2 \cdot 2H_2O$; 5 mL of 0.05% $FeCl_3 \cdot 6H_2O$; 5 mL of 0.05% w/v $MnSO_4$; 50 mL of $0.1M$ citric acid; $0.2M$ Na_2HPO_4; 0.5 mL of Wickerham's vitamin-solution; 250 mL of 4% w/v agar, and 120 mL of sterilized distilled water.
12. Complex agar medium is prepared by mixing the following volumes of stock solutions: 50 mL of 5% w/v yeast extract; 50 mL of 5% w/v peptone; 50 mL of 10% w/v glucose; 10 mL of 20% w/v $(NH_4)_2SO_4$; 5 mL of 5% w/v $MgSO_4 \cdot 7H_2O$; 5 mL of 3% w/v $CaCl_2 \cdot 2H_2O$; 5 mL of 0.05% $FeCl_3 \cdot 6H_2O$; 5 mL of 0.05% w/v $MnSO_4$; 50 mL of $0.1M$ citric acid; $0.2M$ Na_2HPO_4; 250 mL of 4% w/v agar in $1.2M$ KCl, and 20 mL of sterilized distilled water.
13. Centrifuge for low speed centrifugation (2000g, 10 min) in swing-out rotor.
14. Shaker.
15. Thermostatted baths.
16. Spectrophotometer.
17. Minimal agar medium plates.
18. Complex agar medium plates.

3. Methods

3.1. Protoplasting of Parental Cells (see Note 1)

1. Centrifuge 20 mL of exponentially growing culture of each parental strain (e.g., *S. cerevisiae* GRF18/ZH and S6/1) containing approx 10^7 cells counted using a hemocytometer/mL of YEG medium.
2. Resuspend the pellets each in 10 mL of ME-EDTA.
3. Incubate the cell suspensions for 30 min at 30°C.
4. Centrifuge the samples and wash the pellets with 10 mM of $0.8M$ KCl.
5. Resuspend the harvested cells in 6 mL of SDJ.
6. Incubate the samples or 60 min at 30°C with occasional shaking.
7. Centrifuge the samples and wash the pellets twice with 10 mL of $0.8M$ KCl, once with 10 mL of $0.6M$ KCl, and once with 10 mL of $0.3M$ $CaCl_2$.
8. Resuspend the pellets each in 1 mL of $0.3M$ $CaCl_2$.
9. Determine the concentration of protoplasts in samples by counting in a hemocytometer.

3.2. Induced Fusion of Protoplasts (see Note 2)

1. Prepare a mixed sample of parental protoplasts containing 5×10^7 protoplasts of each parental strain.

2. Centrifuge the sample, pour off the supernatant, and resuspend the pellet in a small amount of $0.3M$ $CaCl_2$ remaining at the bottom of centrifugation cuvet.
3. Add 2 mL of PEG, resuspend the pellet by gentle mixing, and leave the protoplast suspension for 20 min at room temperature.
4. Divide the sample into 0.3-mL aliquots in 10-mL tubes.
5. Pipet 5 mL of melted top agar medium held at 45°C to each tube, and immediately pour on complex agar medium plates preincubated at 37°C.

3.3 Stepwise Selection Technique

1. Incubate the plates for 4 d at 20°C (*see* Note 3).
2. Scrape the top agar layers off from all plates into a beaker and release the regenerated cells by squeezing the agar against the wall in 60 mL of distilled water (*see* Note 4).
3. Transfer the cell suspension after sedimentation of agar under gravity to a centrifugation tube, harvest the cells by centrifugation, and wash the pellets twice with distilled water (*see* Note 5).
4. Resuspend the pellet in an appropriate amount of sterilized distilled water to the final absorbance (λ = 540 nm) of 0.1 (*see* Note 6).
5. Transfer 1 mL of cell suspension into 25 mL of minimal medium J in a 250-mL flask and incubate the culture in a shaker at 28°C to the final cell concentration approx 10^7 cell/mL (approx for 3 d) (*see* Note 7).
6. Check the cell-concentration, withdraw a sample, and dilute it with sterilized distilled water to approx 5×10^2 cells/mL (*see* Note 8).
7. Spread four minimal medium plates, each with 0.2 mL of diluted cell suspension, and incubate the plates for 2 d at 28°C (*see* Note 9).
8. Assay the killer phenotype (K^+) of colonies by replica-plating or streaking them onto the lawn of sensitive strain *S. cerevisiae* S6/1 growing on minimal medium plates (*see* Note 10).
9. Incubate the plates for 2 d at 20°C (*see* Note 11).
10. Identify colonies of killer hybrid or cybrid clones by clear zones (*see* Note 12).

4. Notes

1. This protocol is a modified version of one described by Vondrejs et al. *(3)*. It is presented here as an example because it has been used repeatedly in our laboratory. It can be replaced by any different protoplasting procedure.
2. This protocol is also a modified version of one described earlier *(3)*. Any different procedure, including electrofusion, can be used for obtaining used protoplasts.

3. The first step of selection proceeds during incubation at 20°C because regenerating killer cells produce, under these conditions, enough killer toxin to eliminate the sensitive parental cells (K⁻R⁻). The recommended temperature of incubation (20°C) is relatively low because the killer toxin K1 is temperature sensitive. In the case of other killer systems, higher temperatures may be employed. The efficiency of killing can be increased when $0.6M$ KCl (top agar) and $1.2M$ KCl (complex agar medium) is replaced by, respectively, $1M$ sorbitol and $2M$ sorbitol.
4. The amount of cells released from one top agar is usually sufficient.
5. Large volumes of water should be used for washing the cells in order to remove histidine and leucine from the sample because the second selection step is based on auxotrophy of the killer parental strain and of cybrid clones containing nuclei from the killer parent.
6. It is obvious that the concentration of cells may be varied when it is necessary for some reason.
7. A relatively low concentration of cells (step 5) is required in order to make the second step of selection efficient enough. However, account should also be taken of the need for the number or cells in the inoculum to be large enough to contain the killer hybrid and cybrid cells being selected. This selection step is efficient only when a parental killer strain is a nonreverting auxotrophic mutant. Otherwise killer-revertants would grow under these conditions on minimal medium plates.
8. The extensive dilution recommended in step 6 is required only if replica plating technique is going to be used in step 10.
9. It is obvious that the interval of incubation may be prolonged in order to obtain larger colonies.
10. Different sensitive strains and/or assay conditions may be required in connection with other killer systems.
11. The low temperature of incubation (20°C) is required only in connection with temperature-sensitive killer toxins.
12. Hybrids and cybrids (containing nuclei from the sensitive parental prototrophic strain) selected under these conditions may be distinguished, e.g., by determination of cell ploidy or by recovering the original phenotype of auxotrophic killer parent after benomyl treatment. In addition, a centrifugation technique *(10)* can be used for separating diploid hybrids and haploid cybrids.

It should be mentioned that, in some cases, the killer phenotype can be exploited for construction of hybrid strains also as a temporary marker merely for the purposes of selection, and can be eliminated when no longer desired. Simple methods for curing killer strains belonging to groups K1, K2, and K3 of the killer character by elevated temperature, cycloheximide, or 5-flourouracil have been described *(11–15)*. In connection with curing, the method of selection may be used repeatedly, even in multiple "cascade" fusion of protoplast *(3)*. The hybrids formed on the first fusion may be cured of the killer character and thus become sensitive again to a particular killer factor. In the following fusion, an additional killer cell may be "added."

The killer selection techniques are advantageous also for selecting killer strains that arise by other procedures (e.g., by mating, rare mating, cytoduction, or transformation).

Individual killer toxins, however, are often labile and protease-sensitive proteins with a relatively narrow range of optimal conditions for specific killing action. The dependence of the effect on the metabolic status of sensitive cells and strain variations in sensitivity have to be taken into account. It can be expected that the exploitation of some different killer systems may require modification of the procedure described earlier in order to reach an adequate efficiency of selection. In addition, success in transfer of killer character between strains cannot be universally predicted because the present knowledge of interactions among the components involved is poor. Some 30 or more nuclear genes have been shown to be involved in the maintenance or expression of killer character in K1 killer strain of *S. cerevisiae*. The basis for the killing phenomenon in the non-*Saccharomyces* yeasts may be entirely different from *Saccharomyces* sp. In some cases, there is evidence or a possible chromosomal location of the killer toxin determinant. For these reasons, it is likely that stable killer cybrids will not arise and may even be hybrids when some parental strains are fused. Further limitations on selection of parental strains are also imposed by a highly strain-specific action spectrum of some killer toxins.

On the other hand, it should be emphasized that selection procedures based on killer toxins are universal enough, because sensitivity to a killer toxin is a current natural property of many yeast strains, including industrial yeasts and yeast-like organisms. In addition, the cytoplasmic determinants of killer character can be transmitted easily from strain to strain, e.g., by sexual hybridization, fusion of protoplasts with miniprotoplasts that are devoid of nuclei, or by transformation of protoplasts using isolated virus-like particles in the case of *Saccharomyces* sp., or by the dsDNA killer genome of *Kluyveromyces* sp.

References

1. Bevan, E. A. and Makower, M. (1963) The physiological basis of the killer character in yeast. *Proceedings of the XIth International Congress on Genetics* **1**, 202–203.
2. Tipper, D. J. and Bostian, K. (1984) Double-stranded ribonucleic acid killer systems in yeast. *Microbiol. Rev.* **48**, 125–156.
3. Vondrejs, V., Pšenička, I., Kupcová, L., Dostálová, R., Janderová, B., and Bendová, O. (1983) The use of a killer factor in the selection of hybrid yeast strains. *Folia Biologica (Praha)* **29**, 372–384.
4. Bendová, O., Kupcová, L., Janderová, B., Vondrejs, V., and Vernerová, J. (1983) Ein beitrag zur brauerei-hefehybridisierung. *Monats. Brauwiss.* **36**, 167–171.
5. Janderová, B., Davaasurengijn, T., and Bendová, O. (1986) Hybrid strains of brewers yeast obtained by protoplast fusion. *Folia Microbiol.* **31**, 339–343.
6. Janderová, B., Davaasurengijn, T., Vondrejs, V., and Bendová, O. (1986) A new killer brewing yeast capable of degrading dextrin and starch. *J. Basic Microbiol.* **26**, 627–631.
7. Vondrejs, V., Cvrčková, F., Janatová, I., Janderová, B., and Spacek, R. (1988) Applications of killer toxins to selection techniques. *Prog. Biotechnol.* **4**, 133–143.
8. Vondrejs, V. (1987) A killer system in yeasts: applications in genetics and industry. *Microbiol. Sci.* **4**, 313–316.
9. Vondrejs, V. and Palková, Z. (1988) Progress in applications of killer toxins to selection techniques. *Yeast* **4**, (special issue), S 189.
10. Vagölgyi, C., Kucsera, J., and Ferenczy, L. (1988) A physical method for separating *Saccharomyces cerevisiae* cells according to their ploidy. *Can. J. Microbiol.* **34**, 1102–1104.
11. Fink, G. R. and Styles, C. A. (1972) Curing of a killer factor in *Saccharomyces cerevisiae*. *Proc. Natl. Acad. Sci. USA* **69**, 2846–2849.
12. Wickner, R. B. (1974) "Killer character" of *Saccharomyces cerevisiae:* curing by growth at elevated temperature. *J. Bacteriol.* **117**, 1356,1357.
13. Mitchell, D. J., Herring, A. J., and Bevan, E. A. (1976) The genetic control of dsRNA virus-like particles associated with *Saccharomyces cerevisiae* killer yeast. *Heredity* **37**, 129–134.
14. Young, T. W. and Yagiu, M. (1978) A comparison of the killer character in different yeasts and its classification. *Antonie van Leeuwenhoek J. Microbiol. Serol.* **44**, 59–77.
15. Nesterova, G. F., Semykina, L. V., and Filatov, A. A. (1981) A study of the killer plasmid mutants isolated by treatment with 5-fluorouracil. *Genetika* **17**, 391–398.

CHAPTER 34

Killer Plaque Technique for Selecting Hybrids and Cybrids Obtained by Induced Protoplast Fusion

Zdena Palková and Vladimír Vondrejs

1. Introduction

The "killer plaque" technique *(1)* appears to be the simplest of all versions of the selection procedure *(2)*, based on application of killer toxins, that have been examined in our laboratory (*see* Chapter 33). The procedure is based on the observation that the selection of killer hybrids, cybrids, or transformants that produce killer toxin K1 or *Kluyveromyces lactis* killer toxin does not require application of killer toxin. The toxin production from the cells of a single colony is large enough to make a small zone (plaque) in the lawn of regenerated cells of sensitive parental strain in a minimal agar medium (Fig. 1). The parental auxotrophic killer strain can neither grow nor produce plaques under these conditions.

2. Materials

1. Parental strains: Killer strain, e.g., *Saccharomyces cerevisiae* GRF18/ZH (α, *leu2-3, leu2-112, his3-11, his3-15*, K^+, R^+) and sensitive strain, e.g., *Saccharomyces cerevisiae* S6/1 (α, *LEU2, HIS3*, K^-, R^-).
2. The following stock solutions should be sterilized by autoclaving and stored at room temperature: 5% w/v yeast extract; 10% w/v glucose; 20% w/v $(NH_4)_2SO_4$; 5% w/v $MgSO_4 \cdot 7H_2O$; 3% w/v $CaCl_2 \cdot 2H_2O$; 0.05% w/v $FeCl_3 \cdot 6H_2O$; 0.05% w/v $MnSO_4$; 26 g of glycerol per 253 mL of distilled water; 0.1M citric acid; 0.2M Na_2HPO_4; 0.8M KCl, 0.6M KCl, 1M $CaCl_2$, 0.3M $CaCl_2$, 4% w/v agar in 1.2M KCl, 3% w/v agar in 0.6M KCl (top agar); 60 mM EDTA-NaOH, pH 8.0, and 33% w/v polyethylene glycol (4000).

From: *Methods in Molecular Biology, Vol. 53: Yeast Protocols*
Edited by: I. Evans Humana Press Inc., Totowa, NJ

Fig. 1. Demonstration of "killer plaques." Petri dish and an enlarged section from it. Colonies of hybrid and cybrid clones (prototroph, K^+, R^+) of *S. cerevisiae* GRF18/ZH and *S. cerevisiae* S6/1 obtained by induced protoplast fusion are forming plaques in the lawn of sensitive cells of *S. cerevisiae* S6/1.

3. The following stock solutions in distilled water should be sterilized by filtration and stored in the refrigerator at 4°C: 0.02% w/v biotin and 0.02% w/v folic acid.
4. McIlvain's buffer (pH 4.7) is prepared by mixing 26 mL of $0.1M$ citric acid with 24 mL of $0.2M$ Na_2HPO_4 and 50 mL of sterilized distilled water.
5. Wickerham's vitamin-solution contains in 100 mL: 40 mg Ca pantothenate; 20 mg inositol; 40 mg nicotinic acid; 20 mg *p*-aminobenzoic acid; 40 mg pyridoxine, HCl; 40 mg thiamine, HCl; 20 mg riboflavin; 1 mL of 0.02% w/v biotin; and 1 mL of 0.02% w/v folic acid. This solution should be sterilized by filtration and stored in the refrigerator at 4°C.
6. ME-EDTA solution contains 9.9 mL of 60 mM EDTA-NaOH, pH 8.0, and 0.1 mL β-mercaptoethanol.

Killer Plaque Technique for Protoplast Fusion

7. Snail digestive juice (SDJ) is prepared by dissolving 100 mg of lyophilized digestive juice from hepatopancreas of *Helix pomatia* in 10 mL of $0.8M$ KCl containing 10 µL of β-mercaptoethanol. Freshly prepared SDJ has been used in all of our experiments, however, commercially prepared SDJ, or, e.g., Novozym is also effective in protoplasting (*see* Chapters 5 and 18).
8. PEG solution: Add 1 mL of $1M$ $CaCl_2$ to 9 mL of 33% w/v polyethylene glycol (4000).
9. YEG medium contains in 100 mL: 10 mL of 10% w/v glucose; 10 mL of 5% w/v yeast extract; and 80 mL of sterilized distilled water.
10. Minimal agar medium: Mix the following volumes of stock solutions: 50 mL of 10% w/v glucose; 10 mL of 20% w/v $(NH_4)_2SO_4$; 5 mL of 5% w/v $MgSO_4 \cdot 7H_2O$; 5 mL of 3% w/v $CaCl_2 \cdot 2H_2O$; 5 mL of 0.05% w/v $FeCl_3 \cdot 6H_2O$; 5 mL of 0.05% w/v $MnSO_4$; 50 mL of $0.1M$ citric acid; $0.2M$ Na_2HPO_4; 0.5 mL of Wickerham's vitamin-solution; 250 mL of 4% agar in $1.2M$ KCl and 120 mL of sterilized distilled water.
11. Centrifuge for low speed centrifugation ($2000g$, 10 min) in a swing-out rotor.
12. Shaker.
13. Thermostatted baths.
14. Spectrophotometer.
15. Minimal agar medium plates.

3. Methods

3.1. Protoplasting and Induced Protoplast Fusion

The procedures for protoplasting and induced fusion of protoplasts are the same as those described in Chapter 33 (Sections 3.1. and 3.2.). The only difference is in the last step of fusion procedure (Section 3.2.). Minimal agar medium plates are used instead of complex agar medium plates, in order to eliminate prototrophic cells.

3.2. "Killer Plaque" Technique

1. Incubate the minimal medium plates with protoplasts inside the top agar for 4 or 5 d at 20°C (*see* Note 1).
2. Pick up the largest microcolony from inside of each plaque (Fig. 1) and streak it onto a minimal medium plate.
3. Incubate the plate for 24 h at 28°C.
4. Assay the killer phenotype of colonies by replica-plating or streaking them onto lawns of sensitive cells of *S. cerevisiae* S6/1 growing on minimal medium plate.
5. Incubate the plates for 2 d at 20°C.
6. Confirm identity of colonies of killer hybrid or cybrid clones by clear zones (*see* Fig. 1).

4. Notes

1. The procedure described above has been so far tested in our laboratory using model systems of intraspecific induced protoplast fusion of sensitive strain *S. cerevisiae* S6/1 with K1-superkiller strains of *S. cerevisiae* (Fig. 1). In addition, it has been verified that the same procedure can be used after intergeneric fusion of *K lactis* with *S. cerevisiae* S6/1. It is obvious that this technique can be exploited only when the sensitive strain is susceptible enough to killer toxin produced by one growing colony of killer hybrid or cybrid under prescribed conditions. A useful test of the possibility that the procedure may be applied successfully is a preliminary experiment in which the formation of plaques is inspected in a mixed culture of killer (1×10^2 cells/plate) and sensitive (5×10^5/plate) parental strains, on a minimal agar medium plate, supplemented with amino acids and other ingredients required for the growth of the auxotrophic killer strain. The efficiency of killing can be increased when $1.2M$ KCl in minimal agar medium and $0.6M$ KCl in top agar is replaced by $2M$ sorbitol resp. $1M$ sorbitol.

    It is obvious that nonreverting auxotrophic mutants of parental killer strain have to be used, otherwise the revertants will form plaques on unsupplemented minimal agar medium plates, and the selection technique will be inefficient.
2. For general notes on advantages and limitations of killer selection techniques, *see* Chapter 33.

References

1. Vondrejs, V. and Palková, Z. (1988) Progress in applications of killer toxins to selection techniques. *Yeast* **4** (special issue), S 189.
2. Vondrejs, V., Pšenička, I. Kupcová, L., Dostálová, R., Janderová, B., and Bendová, O. (1983) The use of a killer factor in the selection of hybrid yeast strains. *Folia biologica (Praha)* **29,** 372–384.

CHAPTER 35

Genotoxicity Testing in *Schizosaccharomyces pombe*

Pamela McAthey

1. Introduction

During the last 20 yr, the established molecular relationship between mutagenicity and carcinogenicity has been exploited in the development of numerous microbial tests designed to detect mutagenic activity and hence predict carcinogenic activity in industrial and environmental chemicals. The best documented and most widely used of these systems, the Ames test, monitors reversion to prototrophy in histidine-requiring strains of the bacterium, *Salmonella typhimurium*. In order to supply the necessary eukaryotic enzymes to convert promutagens to active mutagenic species, this bioassay requires the addition of a rodent-derived microsomal fraction, the so-called S9 mix. It has long been argued, however, that as eukaryotic test organisms, such as yeast, provide an endogenous supply of metabolic activation, increasing their use would, in turn, reduce the demand for experimental animals in genotoxicity testing.

Although the majority of genotoxicity testing in yeast has been concentrated on one species *(Saccharomyces cerevisiae)*, the fission yeast, *Schizosaccharomyces pombe*, has a number of advantages as an experimental organism. The physiology, biochemistry, and genetics of this latter yeast are well characterized and it has become the model system for analyzing the molecular biology of the eukaryotic cell cycle. Comparative studies have established that it is equally as sensitive to chemical mutagens as *S. cerevisiae* but, as it has only three linkage groups, subsequent genetic analysis can be facilitated. Genotoxicity testing may be

From: *Methods in Molecular Biology, Vol. 53: Yeast Protocols*
Edited by: I. Evans Humana Press Inc., Totowa, NJ

carried out in either batch cultures or continuous cultures of *Sch. pombe* and one protocol of each type is described in this chapter.

The most commonly used system monitors forward mutation in batch cultures of *ade6* or *ade7* auxotrophs. Because of their particular biochemical blockage, such strains accumulate a pigmented derivative of the intermediate aminoimidazole ribonucleotide and hence produce deep red colonies. The induction of mutation in one of the five genes that precede this step in the adenine biosynthetic pathway will result in doubly defective mutants producing white colonies. Testing for forward mutation in this system, therefore, involves treating the auxotroph with the test chemical and then visually screening for white colonies on a medium containing a low concentration of adenine.

One disadvantage of genotoxicity testing in batch cultures lies in the necessity of exposing the experimental organism to an acute dose of the test chemical. Chronic exposures, however, which are more likely to be responsible for environmentally induced cancers, can be administered in continuous cultures. In these cultures, forward mutation is usually monitored by screening for induced resistance to a range of antibiotics. Resistance to some antibiotics (e.g., trichodermin, anisomycin, and cycloheximide) results from mutation in nuclear genes, whereas resistance to others (e.g., chloramphenicol and erythromycin) results from mitochondrial mutation. It is useful to measure the induction of both types simultaneously, as certain carcinogens appear to exhibit a greater mutagenic effect on mitochondrial DNA than on nuclear DNA.

2. Materials

1. The wild-type and auxotrophic mutant strains of *Sch. pombe* used in genotoxicity testing may be obtained from the National Collection of Yeast Cultures, AFRC Institute of Food Research, Norwich Laboratory (Norwich, UK). Refer to Note 1 for storage details.
2. Yeast Extract Liquid (YEL) is used as a complete growth medium. This contains per liter: 5 g yeast extract; 30 g glucose. YEL is sterilized by autoclaving and may then be stored at 4°C prior to use.
3. Edinburgh Minimal Medium Number 2 (EMM2) is used as a defined growth medium. In continuous cultures the growth of the cells is limited by the concentration of glucose (*see* Note 7). This medium contains per liter: 5 g glucose, 5 g NH_4Cl, 300 mg NaH_2PO_4, 1 g CH_3COONa, 1 g KCl, 500 mg $MgCl_2$, 10 g Na_2SO_4, 10 mg $CaCl_2$, 10 mg inositol, 10 mg nicotinic acid, 1 mg calcium pantothenate, 10 µg biotin, 500 mg H_3BO_3,

400 µg MnSO$_4$H$_2$O, 400 µg ZnSO$_4$ · 7H$_2$O, 200 µg FeCl$_3$ · 6H$_2$O, 160 µg H$_2$MoO$_4$ · H$_2$O, 100 µg KI, 40 µg CuSO$_4$ · 5H$_2$O, 1 mg citric acid. EMM2 is sterilized by autoclaving (*see* Note 6) and may then be stored at 4°C prior to use.

It is convenient to keep the components, especially the salts, vitamins, and trace elements, as concentrated stock solutions (salts at 50 times working strength, and vitamins and trace elements at 1000 times working strength). Store these at 4°C over a few drops of ethanol and examine periodically for signs of microbial contamination.

4. For pouring Yeast Extract Agar (YEA) and Edinbugh Minimal Medium Number 2 Agar (EMM2A) plates, YEL and EMM2 are, respectively, solidified by the addition, prior to sterilization, of 15 g agar/L. Plates may be stored in sealed polythene bags at 4°C, but it is recommended that they are used within a few days of preparation.
5. For culturing auxotrophic mutants, EMM2 or EMM2A is supplemented, prior to sterilization, with 50 mg of appropriate amino acids and/or 50 mg of appropriate bases/L.
6. For the isolation of nuclear mutants, molten sterilized YEA is supplemented with 2 mg trichodermin (Leo Pharmaceuticals Ltd., Ballerup, Denmark), 25 mg anisomycin (Sigma, St. Louis, MO) or 100 mg cycloheximide (Sigma) per liter. These antibiotics are filter sterilized prior to their addition to the growth medium.
7. For the isolation of mitochondrial mutants, molten sterilized YEA, containing 30 g/L glycerol in place of glucose, is supplemented with 1 g chloramphenicol (Sigma) or 1.5 g erythromycin (Sigma) per liter. These antibiotics are filter sterilized prior to their addition to the growth medium.
8. Giese's Salts Vitamins Buffer (GSVB) is used for washing cells and for making dilutions. This contains per liter: 3 g (NH$_4$)$_2$SO$_4$, 700 mg MgSO$_4$ · 6H$_2$O, 500 mg NaCl, 400 mg Ca(NO$_3$)$_2$, 6.8 g KH$_2$PO$_4$, 10 mg inositol, 10 mg nicotinic acid, 1 mg calcium pantothenate, 10 µg biotin. GSVB is sterilized by autoclaving and may then be stored at 4°C prior to use.

It is convenient to keep the components as concentrated stock solutions (salts at 10 times working strength and vitamins at 1000 times working strength). Store these at 4°C over a few drops of ethanol and examine periodically for signs of microbial contamination.

9. The chemicals under test are preferably dissolved in distilled water or, if insoluble in water, in a minimum volume of dimethyl sulfoxide and then made up to the required concentration in distilled water. Because of the instability of some chemicals, solutions generally are prepared just before use. All chemicals must be handled as recommended by the current codes of practice for working with chemical carcinogens.

10. Ethyl methanesulfonate at a concentration of 120 mg/L in distilled water is used as a positive control of genotoxicity.
11. Continuous cultures of *Sch. pombe* are grown in a chemostat, an apparatus that consists essentially of four parts: a growth chamber with a built-in stirrer, a medium supply system, a gaseous supply system, and an overflow system (*see* Fig. 1). Although chemostats are available commercially, perfectly adequate chemostats can be constructed from standard laboratory glassware *(1,2)*.

3. Methods
3.1. Genotoxicity Testing in Batch Cultures

1. Inoculate YEL with a deep red colony of an *ade6* or *ade7* auxotrophic strain of *Sch. pombe* and incubate at 32°C for 2 d. A stationary phase culture will be obtained with an approximate cell density of 1×10^7 cells/cm^3.
2. Harvest the cells by centrifugation at 4000g for 10 min. Wash the cells twice in GSVB (i.e., resuspend in GSVB to produce a homogeneous suspension, harvest again, and then repeat). Resuspend the washed cells to their original volume in GSVB.
3. Treatment is carried out in disposable screw-cap bottles or tubes. Set up reaction mixtures consisting of 1 mL washed cells and 0.1 mL test compound (or appropriate control) and incubate at 32°C. To avoid cell sedimentation, incubation should preferably be carried out on an orbital shaker. The duration of the treatment will vary with the cellular toxicity and genotoxic potency of the chemical being tested. A period of between 10 and 18 h should be used in preliminary experiments when these characteristics are not known.
4. A minimum of five concentrations of the chemical should be tested (*see* Note 12).
5. A number of controls should be included in each assay. A positive control is used to ensure that the experimental procedure carried out is capable of detecting the activity of a known chemical mutagen, whereas a negative control is used to enable the spontaneous mutation frequency to be determined. If the test chemical has been dissolved in a solvent other than water, a solvent control should also be run to assess the genotoxicity of the solvent itself.

    Replace the 0.1 mL test chemical in the reaction mixture with the same volume of ethyl methanesulfonate for a positive control, with GSVB for a negative control and with the appropriate solvent (at its in-use concentration) for a solvent control.

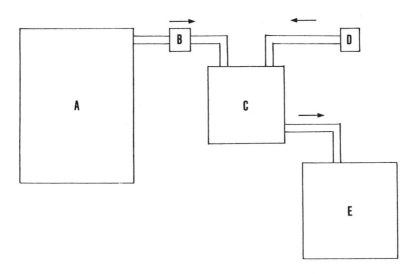

Fig. 1. Diagrammatic representation of a chemostat. Fresh medium is pumped from the medium reservoir (**A**) by a peristaltic pump (**B**) into the growth chamber (**C**). The cells in the growth chamber are kept in suspension by a built-in stirrer and are aerated by an air pump (**D**). For every volume of fresh medium entering the growth chamber, an equal volume of spent medium and cells leaves the growth chamber via the overflow system and is collected in the effluent reservoir (**E**).

6. Terminate the reaction in each treatment bottle or tube by harvesting the cells and then washing them at least three times in ice-cold GSVB. Resuspend the washed cells to their original volume in GSVB.
7. Plate out 0.1 cm^3 aliquots of 10^{-2}, 10^{-3}, and 10^{-4} dilutions of the treated cells on YEA plates and incubate at 32° for 3–5 d when the colonies should be clearly visible (*see* Note 3). For strongly mutagenic agents, 10 plates should be used for each concentration of the chemical, whereas for weakly mutagenic ones as many as 100 plates per concentration may be required (*see* Note 13).
8. Score both the number of white colonies (*see* Note 4) and the total number of colonies (red plus white). As ideally plates should contain 30–300 of the type of colonies being counted, the 10^{-2} or 10^{-3} dilution will probably be required for the former count and the 10^{-3} or 10^{-4} dilution for the latter count. Storage of the plates at 4°C for 1–2 d prior to counting usually improves the pigmentation of the red colonies and hence facilitates this step (*see* Note 5).

9. Calculate the number of mutant cells (i.e., white colony producers) per mL after treatment as follows:

average number of white colonies per plate
× (1/dilution factor) × 10

Similarly, calculate the total number of surviving viable cells (i.e., red colony producers plus white colony producers) per mL after treatment as follows:

average number of colonies per plate
× (1/dilution factor) × 10

10. Express the mutation frequency at each chemical concentration and for each control as number of mutants per 10^6 survivors of treatment.
11. Finally, the data obtained in the assay must be statistically evaluated. The observation of a positive correlation between increasing concentration of test chemical and mutant yield (i.e., a dose-related response) is a good indication of genotoxicity. Test compounds that elicit at least a 1.5-fold increase in mutation compared with the negative (and, if relevant, solvent) controls in the assay are considered to be of biological significance, provided this value is in excess of the minimum detectable by statistical analysis. Probability values are calculated from one-sided tests and the critical value at which the null hypothesis is rejected is usually set at 0.05 or 0.01.

3.2. Genotoxicity Testing in Continuous Cultures

1. Inoculate EMM2 with a colony of a wild-type strain of *Sch. pombe* and incubate at 32°C for 2 d. An exponential phase culture will be obtained.
2. Add the required volume of EMM2 (containing the test chemical where appropriate) to the growth chamber of the chemostat and inoculate with 2% v/v exponential phase cells. Switch on the stirrer and the gaseous supply system. Incubate at 32°C overnight to permit the cell density to increase to approx 1×10^7 cells/cm^3 (*see* Note 2).
3. Connect the medium supply system, setting the peristaltic pump to deliver fresh EMM2 (containing the test chemical where appropriate) at the required flowrate (*see* Note 8). The main parameters of continuous cultures are the dilution rate (D) and the cell generation time (t_d) (*see* Note 9). These can be determined as follows: $D = F/v$, where F is the flowrate of the fresh medium (measured in mL/h) and v is the working volume of the culture in the growth chamber (measured in cm^3); $t_d = 0.693/\mu$, where μ is the specific growth rate of the culture (measured as the rate of increase in cell mass or cell numbers/h). At steady-state $\mu = D$. Although, for most pur-

poses, a dilution rate of 0.1/h (equal to a cell generation time of 7 h) is recommended, steady-state may be achieved over a range of dilution rates (*see* Note 10).
4. Allow time for the system to settle down. Initially, cell numbers will oscillate. However, steady-state, as judged by the onset of a stable cell density, should be achieved within 1–2 d.
5. Take samples from the steady-state culture at least twice daily for a minimum of 8 d. The volume removed on each occasion should be as small as possible, sufficient only to provide the cells required for further processing (*see* Note 11). Samples may be stored at 4°C for short periods of time, but it is recommended that they are processed immediately.
6. Harvest the cells by centrifugation at 4000g for 10 min. Wash the cells twice in GSVB and then resuspend them in GSVB to a concentration of approx 1×10^8 cells/cm^3.
7. Plate out 0.1 mL aliquots of undiluted resuspended cells on to sets of YEA plates containing appropriate antibiotics (*see* Notes 14–16). At least three plates of each type used should be inoculated at each sample time.
8. Plate out 0.1 mL aliquots of 10^{-4} and 10^{-5} dilutions of the resuspended cells on to unsupplemented YEA. At least three plates should be inoculated with each dilution at each sample time.
9. Incubate all plates at 32°C for 3–4 d except for those containing the antibiotics chloramphenicol and erythromycin, which should be incubated at 32°C for 10–12 d.
10. Score the number of colonies on each plate. As ideally plates should contain 30–300 colonies for viable count estimations, only use the counts from the more appropriate of the two dilutions on unsupplemented YEA, in the subsequent calculations. In some circumstances, colony counts on the antibiotic containing plates may fall below the ideal minimum value (*see* Note 17).
11. Calculate the number of mutant cells (i.e., those resistant to a particular antibiotic) per mL at each sample time as follows:

$$\text{average number of colonies per YEA plate containing that antibiotic} \times 10$$

Repeat this calculation for each antibiotic used. Similarly calculate the number of viable cells/mL at each sample time as follows:

$$\text{average number of colonies per unsupplemented YEA plate} \times (1/\text{dilution factor}) \times 10$$

12. Express the frequency of each type of antibiotic resistant mutant at each sample time as the number of mutants per 10^6 viable cells.

13. The increase in mutant numbers can then be followed by plotting mutation frequencies against the sample times. Calculate the least squares linear regression lines and fit these in the standard manner.
14. Repeat steps 1–13 for each concentration of test chemical to be assayed, for the negative control (where no additions are made to EMM2), for the positive control (where ethyl methanesulfonate is added to EMM2 in place of the test chemical) and, where relevant, for the solvent control (where the appropriate solvent, at its in-use concentration, is added to EMM2 in place of the test chemical).
15. Finally, the data obtained in the assay must be statistically evaluated. Comparison of the rate of accumulation of mutants resistant to a particular antibiotic in the presence and in the absence of a test chemical permits a quantification of the mutagenic activity of that chemical. If the rate of accumulation of induced mutants is found to be twice or more than that of the spontaneous mutants (and, if relevant, than that of the mutants in the solvent control), then the chemical is considered to be mutagenic in *Sch. pombe*. As in the case of genotoxicity testing in batch cultures, any increase in mutation over the negative control(s) is only considered to be significant if this is in excess of the minimum value detectable by statistical analysis.

4. Notes

1. For short-term storage of *Sch. pombe*, streak plates should be prepared. Inoculate YEA plates with material from single colonies of the appropriate yeast strains and incubate at 32°C for 3–5 d when the colonies should be clearly visible. The plates may then be stored at 4°C for several weeks. Depending on usage, strains should be subcultured every 3–4 wk.

    For long-term storage of *Sch. pombe* in a culture library, strains should be dried on to white silica gel crystals. Bijoux bottles three-quarters filled with crystals may be conveniently sterilized in a drying oven at 180°C for 2 h. Grow the yeast strains on YEA slopes at 32°C for 3 d. Wash off the cells in small volumes of sterile 50 g/L skimmed milk and then slowly add 1 cm^3 aliquots into the bijoux bottles. Store the bottles with their caps loosely screwed on for one week at room temperature. Tighten the caps and then continue storage at 4°C. *Sch. pombe* will remain viable for at least 5 yr under these conditions. To rehydrate the cells, transfer a few crystals into 5 cm^3 YEL and incubate at 32°C. Growth should occur within 3 d. Provided aseptic techniques are used, the bottles may be resealed after the removal of some of the crystals, and storage continued.
2. With experience, the cell density of *Sch. pombe* in liquid medium or buffer may be roughly gaged by eye. Workers unfamiliar with this organism, however, should check cell densities using a hemocytometer.

3. Cell enumeration on solid medium relies on the premise that individual colonies are formed by the growth and multiplication of single cells. Inaccurate results will therefore be obtained when more than one cell initiates colony formation. Where chemicals are suspected of causing the clumping of cells, the treated culture should be examined microscopically and care should be taken to break up any clumps observed (e.g., by whirli-mixing) prior to plating.
4. Sectored colonies (i.e., colonies with segments of red and white pigmentation) are often observed when *ade6* and *ade7* strains are grown on YEA. As such colonies arise as a result of mutation, they should be classified as white in mutational analysis.
5. The characteristic deep red pigmentation of *ade6* and *ade7* strains is usually readily visible on YEA. Some batches of yeast extract, however, contain sufficiently high concentrations of adenine to repress the adenine biosynthetic pathway and hence preclude the development of the red pigmentation. As an alternative, EMM2 may be used to distinguish the red and white colonies, provided its adenine concentration is not in excess of 20 mg/L.
6. Caramelization of glucose sometimes occurs when large volumes of EMM2 are autoclaved. If this is a problem, the glucose may either be sterilized separately in solution or be added *carefully* as a solid to hot (but not boiling!) EMM2.
7. EMM2 can be modified to enable nutrients other than glucose to control the growth of continuous cultures of *Sch. pombe*. However, because the kinetics of mutant accumulation become significantly altered under some growth limitation regimes (for example, where a required amino acid is used *[3]*), it is important to carry out preliminary experiments to assess the effect of any new limiting nutrient used on the rate of accumulation of spontaneous mutants.
8. Ideally, a chemostat containing sufficient medium for a whole experiment should be sterilized as a single unit. In practice, however, it is often more convenient to sterilize the component parts separately and then reassemble them before use. The medium reservoir may be topped up during the course of an experiment but, as this is a potential source of contamination, care should be taken to ensure that any additions are made under aseptic conditions.
9. Although growth rates in chemostats are relatively independent of external conditions, care should be taken to maintain the temperature of the growth chamber at 32°C. The most practical arrangement is to site the whole apparatus in a hot room. However, if such a facility is not available, then the growth chamber may be immersed in a water bath or placed on a heating block. Where the latter is used, it is useful to have a stirrer attachment

so that the temperature of the chamber and the agitation of its contents may be controlled by the same component. It may not always be necessary to continually monitor the pH or the oxygen concentration in the growth chamber. However, if these parameters are found to be affected by the chemical under test, then pH and oxygen electrodes should be included in the chemostat design, and any required adjustments made periodically to the ionic composition of the medium and/or to the aeration rate. Depending on the sophistication of the system, these adjustments can be made either manually or automatically.

10. Steady-state may be achieved in continuous cultures of *Sch. pombe* grown at dilution rates ranging from 0.22–0.07/h (equal to generation times of 3.2–10.0 h).

11. Depending on the chemostat design, samples may be taken from either the growth chamber via a sampling port or from the overflow system. If the former arrangement is used, care should be taken to avoid introducing contamination into the growth chamber.

12. As a general guide, the concentrations of test chemicals used should be evenly spaced and yield between 100 and 5% cell survival. Some chemicals, however, are cytotoxic above a sharply defined threshold and exhibit mutagenicity only within a very narrow concentration range. It is therefore important to carry out pilot experiments to determine the appropriate range of concentrations to use with a particular test chemical.

13. Because of the large number of plates required to establish the genotoxic potential of weak mutagens in the colony pigmentation assay with *ade6* or *ade7* strains, some workers prefer to use a selective system. Forward mutation may be monitored by screening for antibiotic resistance (as described under the heading of genotoxicity testing in continuous cultures) or reverse mutation monitored by screening for prototrophic reversion in auxotrophic strains plated on defined medium deficient in the required growth factors. Assays employing forward mutation are generally preferable in genotoxicity testing as they are sensitive to a range of genetic alterations, whereas those using reverse mutation usually rely on specific changes and hence are prone to mutagen specificity.

14. As explained in the introduction, in genotoxicity testing in continuous cultures, it is advisable to monitor mutation in both nuclear DNA (by screening for resistance to trichodermin, anisomycin, or cycloheximide) and mitochondrial DNA (by screening for resistance to chloramphenicol or erythromycin). Furthermore, as the degree of mutation induction varies with the individual antibiotic resistance scored as well as the inducing agent used *(2)*, relying on a single screening system is risky, as this may yield anomalous data.

15. Selective systems other than scoring for resistance to the five antibiotics described in this chapter may be used in genotoxicity testing in continuous cultures. However, before any assay can be used in mutational studies on continuous cultures, preliminary experiments have to be carried out to ensure that it fulfills three main criteria:
    a. The mutants obtained must be amenable to effective screening;
    b. Spontaneous and induced mutation frequencies must be sufficiently high to permit the recovery of mutants at cell concentrations that can be maintained in chemostats; and
    c. Mutant cells must not have a selective growth advantage or disadvantage over the non-mutant cells *(2)*.
16. Antibiotics insoluble in water should be dissolved in a minimum volume of ethanol or acetone and then made up to the required concentration in distilled water.
17. Where the average colony count on a selective medium is lower than 30, the number of these plates used at each sample time should be correspondingly increased.

References

1. Kubitschek, H. E. (1970) in *Introduction to Research with Continuous Cultures*. Prentice-Hall, New Jersey.
2. McAthey, P., and Patel, B. (1987) Testing the Industrial Environment for Carcinogens/Mutagens using *Schizosaccharomyces pombe*, in *Industrial Microbiological Testing* (Hopton, J. W. and Hill, E. C., ed.), Blackwell, Oxford, pp. 137–149.
3. McAthey, P. and Kilbey, B. J. (1978) Mutation in Continuous Cultures of *Schizosaccharomyces pombe* II. Effect of amino acid starvation on mutational response and DNA concentration. *Mutat. Res.* **50,** 175–180.

CHAPTER 36

Purification and Quantification of *Saccharomyces cerevisiae* Cytochrome P450

Ian Stansfield and Steven L. Kelly

1. Introduction

Cytochrome P450s (cyt.P450s) are a class of hemoprotein enzymes found in a wide variety of organisms, where they play an important role in endogenous metabolism and the metabolism of xenobiotic compounds. In eukaryotes these enzymes are membrane-bound, in most cases located in the endoplasmic reticulum. Higher eukaryotes often have complex systems containing many cyt.P450s. However, vegetatively growing *Saccharomyces cerevisiae* contains only one major cyt.P450, the lanosterol 14α demethylase *(1)*, and thus provides an ideal model system for the study of microsomal eukaryote cyt.P450s. Methods will be described for the quantification of yeast cyt.P450s, and the purification of the main yeast cyt.P450. This latter method was first described by Yoshida and Aoyama *(1)*, and King et al. *(2)*. Recently, the mRNA of a second *S. cerevisiae* cyt.P450 gene has been detected among sporulation-specific transcripts *(3)*.

The methods to be described may also be applied, with some modification, to the preparation of heterologously expressed cyt.P450 in *S. cerevisiae (4)*.

Quantification of cyt.P450 is achieved spectrophotometrically after first reducing the enzyme preparation with a chemical reductant followed

From: *Methods in Molecular Biology, Vol. 53: Yeast Protocols*
Edited by: I. Evans Humana Press Inc., Totowa, NJ

355

by binding of carbon monoxide. The reduction of the iron atom at the center of the heme moiety coupled with the binding of a molecule of carbon monoxide in the fifth ligand position of the iron causes the cyt.P450 to absorb light strongly at or near 450 nm.

Cytochrome P450 can be partially purified from yeast by first preparing microsomes, membrane vesicles consisting of disrupted endoplasmic reticulum. Following breakage of the yeast cell wall, microsomes are prepared by a series of differential centrifugation steps.

Using the microsomal fraction as a starting material, the enzyme can be further purified. Eukaryote cyt.P450s are membrane-bound enzymes and their purification requires extensive use of detergents to maintain the proteins in a soluble form. A series of three chromatographic resins are used in the purification protocol. Of these, the hydrophobic affinity resin amino-octyl Sepharose is the most important. This is thought possibly to bind cyt.P450 through interaction with the hydrophobic domains of the protein responsible for membrane binding.

Methods (*see* Sections 3.1., 3.2., and 3.3.) are presented detailing cyt.P450 quantification, microsome preparation, and enzyme purification, respectively.

2. Materials

1. Sodium dithionite (solid).
2. Carbon monoxide (*see* Note 1).
3. Scanning spectrophotometer.
4. Glass beads (0.3–0.4 mm diameter).
5. A bead mill such as the Braun MSK bead mill.
6. YEPD: Yeast growth medium containing 1% w/v yeast extract, 2% w/v bactopeptone, and 5% w/v glucose.
7. Buffer A: 100 mM phosphate buffer pH 7.2, 20% v/v glycerol, 1 mM EDTA, 1 mM dithiothreitol (DTT) (*see* Note 2).
8. Phenylmethylsulfonyl fluoride (PMSF): A 100-mM solution in ethanol, store at −20°C (*see* Note 3).
9. A hand-held Potter homogenizer.
10. Sodium cholate.
11. Nitrogen gas.
12. Ammonium sulfate (analytical grade).
13. 3M Ammonium hydroxide solution.
14. Buffer B: 20-mM phosphate buffer pH 7.2, 1 mM EDTA, 1 mM DTT, 20% v/v glycerol and 0.5% w/v sodium cholate (*see* Note 2).

Purification of Cytochrome P450

15. Buffer C: 10-mM phosphate buffer pH 7.0, 1 mM EDTA, 1 mM DTT, 20% v/v glycerol, and 0.3% sodium cholate (*see* Note 2).
16. Buffer D: 10-mM phosphate buffer pH 7.0, 1 mM EDTA, 20% v/v glycerol, 0.3% w/v sodium cholate, and 0.1% v/v Emulgen 911.
17. Amino-octyl agarose (Sigma, St. Louis, MO) or amino-octyl Sepharose (Pharmacia, Uppsala, Sweden), equilibrated before use with 10–20 column volumes of buffer D.
18. Emulgen 911 (Kao-Atlas, Tokyo) a nonionic detergent (*see* Note 4).
19. Buffer E: 10-mM phosphate buffer, pH 7.0.
20. Hydroxy-apatite, equilibrated before use with buffer E (*see* Note 5).
21. Buffer F: 30-mM phosphate buffer, pH 7.0, 0.1% v/v Emulgen.
22. Buffer G: 75-mM phosphate buffer, pH 7.0, 0.1 %v/v Emulgen.
23. Sephadex G-25, equilibrated before use with buffer H.
24. Buffer H: 10-mM phosphate buffer, pH 7.0, 0.2% v/v Emulgen.
25. Carboxy-methyl Sephadex (mesh size C-50), equilibrated before use with buffer H (*see* Note 6).
26. Buffer 1: 50-mM phosphate buffer, pH 7.0, 0.2% v/v Emulgen.

3. Methods
3.1. Cytochrome P450 Quantification

1. To the preparation of cyt.P450 (whether crude microsomes or an aqueous solution of pure protein), add a few grains of solid sodium dithionite on the tip of a small spatula to 1 mL of the preparation. The actual amount added is not critical (*see* Note 7).
2. While the enzyme preparation is being reduced by the sodium dithionite, turn the carbon monoxide supply on to purge any air from the supply lines. Feed the gas through a pressure regulator to a syringe needle. Perform all manipulations in a fume hood.
3. Use a scanning spectrophotometer (*see* Note 8) to record the baseline absorption of the enzyme preparation between 390 and 500 nm. Then take the cuvet and pass carbon monoxide through the syringe needle into the liquid at the approximate rate of 1 bubble/s for 60 s.
4. Rescan the absorption between 390 and 500 nm. The presence of cyt.P450 is revealed by a peak in the absorption at or around 450 nm (*see* Fig. 1 and Note 9).
5. The concentration of cyt.P450 can be calculated by subtracting the absorption at 490 nm from that at the peak of absorption, 450 nm. 490 nm is the isosbestic point, at which cyt.P450 and CO-cyt.P450 complexes absorb light to an equal extent. An mM^{-1} cm^{-1} extinction coefficient of 91 is used with the $A_{450-490}$ difference to calculate the concentration of the enzyme.

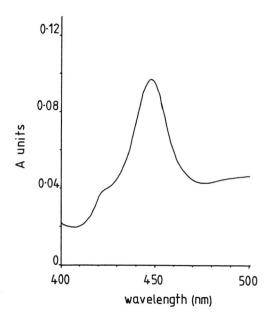

Fig. 1. A CO-reduced difference spectrum of *S. cerevisiae* cyt.P450 from a microsomal sample. A very slight amount of cyt.P420 is discernible, apparent from the slight inflexion at 420 nm.

3.2. Microsome Preparation

All manipulations should be performed at 4°C.

1. Grow 2 L of yeast culture in YEPD medium (1 L of medium per 2.5 L flask) at 30°C and 100 rpm on an orbital shaker until the culture reaches late logarithmic phase (1×10^8 cells/mL) (*see* Note 10).
2. Harvest the cells by centrifuging the culture at 1500g for 10 min. Resuspend the cell pellet in 20 mL buffer A, adding PMSF to 0.5 mM.
3. Break open the cells using either a bead mill or a French Press, following the manufacturer's instructions (*see* Note 11).
4. Take the cell lysate and centrifuge for 10 min at 1500g at 4°C. Retain the supernatant. The pellet will comprise unbroken cells, large fragments of cell wall, and glass beads.
5. Take the supernatant from step 4 and centrifuge at 12,000g for 10 min. This centrifugation step sediments mitochondria and smaller pieces of cell debris. Retain the supernatant.
6. Centrifuge the supernatant from step 5 at 100,000g for 90 min. This will sediment the microsomes (the endoplasmic reticulum membrane fraction

Purification of Cytochrome P450 359

formed into vesicles by the cell breakage procedure). The pellet will appear faintly pink/red in color (*see* Note 12).
7. Remove the pellet from the tube with a clean spatula and place it in a prechilled hand-held homogenizer. Add 2 mL of ice-cold buffer A and resuspend the pellet thoroughly using the homogenizer (*see* Note 13). The microsome preparation can be stored frozen at $-70°C$ for at least 1 yr.

3.3. Cytochrome P450 Purification

1. Adjust microsomes suspended in buffer A to a total protein concentration of 30 mg/mL.
2. Add solid sodium cholate to a final concentration of 1% w/v, and PMSF to a final concentration of 0.5 mM. Stir the solution for 1 h at 4°C under a stream of nitrogen gas (*see* Note 14).
3. Centrifuge the solubilized preparation at 100,000g for 1 h to remove unsolubilized membrane fraction and proteins. Retain the supernatant, which contains a crude preparation of solubilized cyt.P450.
4. To this supernatant, add, slowly with stirring, solid ammonium sulfate (finely powdered to aid even dissolution) to a final concentration of 35% w/v. Monitor the pH constantly, ensuring the pH is maintained at pH 7.0 by the careful addition of 3M ammonium hydroxide (*see* Note 15). Sediment the precipitated protein by centrifuging at 10,000g for 10 min.
5. Retain the supernatant and add more ammonium sulfate to a final concentration of 65% w/v. Again monitor the pH and maintain at pH 7.0 (*see* Note 15).
6. Sediment at 10,000g and retain the precipitated protein pellet. This fraction of protein precipitating between 35 and 65% w/v ammonium sulfate has been found to contain the greater proportion of yeast cyt.P450.
7. Resuspend this protein precipitate in a large volume of buffer B (100 mL/ [g protein]). Dialyze the resuspended protein solution overnight against three changes of buffer C (each "change" is 2 L/[g protein] in volume) (*see* Note 16).
8. Remove insoluble material by centrifugation at 100,000g for 30 min. Adjust the sodium cholate concentration to 0.5% w/v before loading onto a column packed with amino-octyl agarose resin equilibrated with buffer C (*see* Note 17).
9. The amino-octyl chromatography (*see* Note 18) should be performed at a flowrate of approx 10 mL/h (cm^2 resin cross section area). Use a size of column of approx 10 mL resin/g of total microsomal protein used as starting material The column diameter should be approx 1.5 cm.
10. Protein bound to the amino-octyl agarose can be seen as a faintly pink band at the top of the column. Wash the column with buffer C until the A_{280} of the eluate returns to baseline levels. Elute the protein by washing the resin with buffer D.

11. Monitor the column eluate at 280 nm. Assay fractions containing protein for cyt.P450 content using the CO reduced-difference spectrum.
12. Pool the fractions containing cyt.P450 and load onto a hydroxy-apatite column (see Note 19) equilibrated with buffer E. After loading the cyt.P450 fractions, wash the column with buffer E until the A_{280} returns to a baseline value. Then wash the column with buffer F. Continue the wash again until the absorbance returns to a baseline value (see Note 20). Elute the cyt.P450-containing fractions with buffer G.
13. Desalt the eluted cyt.P450, now in a 75-mM phosphate buffer, on a Sephadex G-25 column equilibrated with buffer H.
14. Apply the desalted protein fraction containing the cyt.P450 to a carboxymethyl Sephadex column equilibrated with buffer H (see Note 19). Wash the column-bound protein with buffer H until the A_{280} returns to the baseline value (see Note 20). The cyt.P450 is eluted with buffer I. The elution process can be observed as the red band representing bound cytochromes moves down the column (see Note 21).
15. Assay the purity of the cyt.P450 using SDS-polyacrylamide gel electrophoresis. Typical yields (i.e., recovery of cyt.P450 initially present in the microsomal fraction) of between 10 and 15% have been obtained, although some as high as 40% have been reported. Purity can be conveniently estimated by determining the specific content of cyt.P450 (nmol/mg total protein); given that the molecular weight of cyt.P450 protein is 54 kDa (see Note 22), and since:

$$\text{moles} = \text{weight/relative molecular mass}$$
$$18.5 \times 10^{-9} \text{ mol cyt.P450} = 1 \times 10^{-3} \text{g}/54{,}000$$

Thus 1 mg of cyt.P450 contains 18.5 nmol of cyt.P450 protein. The number of nmol of cyt.P450 can be assayed using the CO-reduced difference spectrum. From this figure, and a knowledge of the total protein content, the percentage purity of a cyt.P450 preparation can be calculated (see Note 23).

4. Notes

1. Carbon monoxide can be purchased bottled or generated in the laboratory by the action of concentrated sulfuric acid on sodium formate (see any basic chemistry synthesis text). Carbon monoxide is an extremely toxic gas. All manipulations involving it should be performed in a fume hood. Normal precautions for dealing with a concentrated acid should be employed if synthesis of the gas is attempted.
2. DTT has a limited lifespan in solution. It should be added after autoclaving and immediately before use. Basic phosphate buffers described containing glycerol and EDTA can be stored for >6 mo after autoclaving.

3. PMSF, a serine protease inhibitor, is, first, very toxic and, second, unstable in aqueous solution. It should be made up in ethanol, stored at −20°C, and added to the buffer A immediately before use as a cell breakage buffer.
4. Emulgen 911, the nonionic detergent routinely employed in cyt.P450 purification, may prove difficult to obtain. Some researchers have used Renex 690 and Lubrol PX with success in the purification of mammalian cyt.P450s. The latter detergent also has the advantage that it is not a UV chromophore.
5. Many types of hydroxyapatite are commercially available, some offering better performance than others. A mixture of cellulose and hydroxyapatite (e.g., Bio-Rad [Hercules, CA] Bio-Gel HT) was found to give good results, allowing good flowrates and giving resistance to "packing." Approximately 10 mL of hydroxyapatite resin is used for every gram of microsomal protein used as a starting material. The column should have a diameter of approx 1.5 cm, and is run at a similar flowrate to the amino-octyl agarose column.
6. The carboxymethyl Sephadex column contains 3–4 mL of resin/g of starting microsomal protein, with a column diameter of approx 0.75 cm. The resin is translucent when hydrated, and bound cyt.P450 can be seen as a red band in the resin. The cyt.P450 binds to this cation exchange resin at pH 7.0 owing to the alkaline isoelectric point of yeast lanosterol demethylase cyt.P450.
7. The addition of sodium dithionite ($Na_2O_4S_2$; less commonly known as sodium hydrosulfite) to cyt.P450 preparations during quantification of the enzyme is usually described in the literature as a few grains. The actual amount added is not critical. In our experience, a few grains approximates 5 mg. The chemical must be stored dry. It oxidizes slowly in the presence of air, the rate of oxidation being increased by the presence of water. The purpose of the addition of the reductant is to reduce the ferric (Fe^{3+}) iron atom at the center of the heme moiety to the ferrous (Fe^{2+}) form.
8. During spectrophotometric quantification of cyt.P450, the absorption maximum of reduced CO-bound enzyme is not always seen at 450 nm. Pure yeast enzyme absorbs maximally at 448 nm (447 nm has been quoted in other work; *1*).

Furthermore, the absorption maximum can be affected by contaminating enzymes. The presence of cytochrome oxidase, caused by contamination with the mitochondrial fraction of a cell, causes a red shift in the maximum absorption by as much as 10 nm. Figure 2 shows an example of gross cytochrome oxidase contamination of a microsomal preparation, illustrating the characteristic cytochrome oxidase CO-reduced difference spectrum. It is also possible to observe, on occasion, a peak at 420 nm in the CO-reduced difference spectrum (Fig. 3). This so-called cyt.P420 is thought to

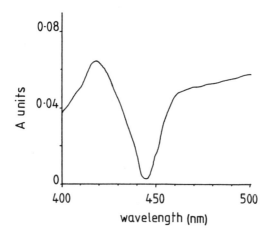

Fig. 2. A CO-reduced difference spectrum of a microsomal fraction of *S. cerevisiae* contaminated with the mitochondrial cell fraction. The spectrum obtained is attributable to cytochrome oxidase. Less severe cytochrome oxidase contamination can result simply in a red shift of the cyt.P450 absorption maximum of between +5 and +10 nm.

be a breakdown product of cyt.P450, and can result when a cyt.P450 preparation is not maintained at 4°C. Cyt.P420 is often present also when the yeast culture from which the enzyme extract was prepared is harvested in the stationary phase of growth. That the 420 nm peak may also represent other hemoproteins cannot be excluded.

For quantification of cyt. P450 levels in either whole cells or microsomes, the spectrophotometer will need to be equipped with a turbid sample cuvet holder. This allows the placement of a turbid sample near the light detector of the instrument, thus minimizing signal loss owing to light scattering. Transparent, as opposed to translucent, cyt.P450-containing fractions can be assayed using a normal cuvet holder.

It has been observed that the spectrophotometer will respond in a linear fashion to a concentration of microsomes defined by a protein concentration of 30 mg/mL. Beyond this value the sample becomes too turbid. However, individual experimenters should determine the linearity of their own spectrophotometers with respect to turbid samples.

The quantification of cyt.P450 can also be carried out on a suspension of whole yeast cells. A turbid suspension of yeast (1×10^9 cells/mL) is treated with sodium dithionite and carbon monoxide as described earlier. Care must be taken to shake the cuvet immediately before the spectropho-

Purification of Cytochrome P450

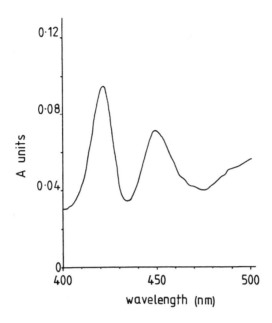

Fig. 3. A CO-reduced difference spectrum of the microsomal fraction of a *S. cerevisiae* culture harvested during early stationary phase. An increased amount of cyt.P420 is apparent from the peak of absorbance at 420 nm.

tometer reading is made. This avoids settling of the cell suspension with a consequent false reading.

9. An "ideal" cyt.P450 spectrum is shown in Fig. 1, and spectra of this appearance can be obtained using microsome preparations. However, it is not unusual to see some cyt.P420 in the preparation even if great care is taken with temperature and pH maintenance.
10. Cyt.P450 levels have been found to be induced during late stationary phase when the yeast are grown in low oxygen tensions and high glucose concentrations, hence the use of the growth conditions described.
11. For the breakage of yeast the use of bead mill such as the Braun MSK or a similar model is optional, but recommended for breakage of large amounts of yeast cells (>10 g wet wt). Alternatively, the French press can be used for cell breakage, following the manufacturer's instructions. It has been found that good cell breakage is achieved with the Braun apparatus if the breakage vessel is filled with 1/3 volume glass beads, 1/3 vol cell slurry and 1/3 vol air. The breakage vessels of both the bead mill and French press should be prechilled. If neither of these pieces of equipment is available, glass beads can be added to the cell slurry to just below the meniscus

and the preparation vortexed for 2–3 minutes using a bench-top vortex. This latter manual method is not recommended far large volumes of yeast suspension.

Good cell breakage is obviously essential in order to obtain a good yield of microsomes/cyt.P450. The proportion of breakage should approach 70%. This can be assayed by microscope examination. Broken yeast cells (termed ghosts) are distinguishable from intact cells, appearing fainter and more transparent.

12. Microsomal pellets obtained should be pink/red in color. This color is thought to be a result of the presence of cytochromes such as cyt.P450 and cyt.b5 and therefore provides a good indication as to the suitability of the culture conditions used in producing a high microsomal content of cyt.P450.

13. Once the microsomal pellet is obtained, thorough resuspension with the hand-held homogenizer is essential, especially if the preparation is to be used for further purification. Inadequate resuspension will result in a very poor release from the membrane of the cyt.P450 during sodium cholate solubilization (Section 3.3.2.).

14. It is important for efficient solubilization of the membrane that a microsomal protein concentration of 30 mg/mL is not exceeded. Also, as stressed before, the microsomal pellet must be thoroughly homogenized to aid sodium cholate-induced disruption of the membrane. The sodium cholate, an ionic detergent, services to release a limited fraction of microsomal proteins from the membrane, including yeast cyt.P450.

15. During addition of the ammonium sulfate, it is essential that the pH is maintained at pH 7. Deviation of the pH by as little as 1 pH unit from pH 7 can cause a conversion of cyt.P450 to the P420 form.

16. It is important that removal of ammonium sulfate from the cyt.P450 preparation by dialysis proceed to completion. The dialysis buffer should therefore be changed during dialysis to maintain the concentration gradient. The presence of any remaining salt in the form of ammonium sulfate will reduce and possibly prevent binding of protein to the first column, amino-octyl agarose.

The dialysis procedure is fairly slow. Since the cyt.P450 has been observed to be salt-labile, the use of a more rapid desalting process, such as G-25 chromatography, is obviously advantageous. If a large column is available, G-25 chromatography may be used to remove the ammonium sulfate, remembering that the column volume must be at least four times that of the sample to be loaded. The G-25 Sephadex resin should be equilibrated with buffer C ready for loading onto the amino-octyl agarose column.

17. Equilibrate the amino-octyl agarose resin before use with 10–20 column volumes of buffer C. Check equilibration by ensuring the column eluate pH is within 0.2–0.3 pH U of that of buffer C (pH 7.0). If a conductivity meter is available, it should also be employed to determine when the column is equilibrated: if equilibrated, the conductivity of buffer eluting from the column should be roughly equal to the conductivity of buffer C.
18. During the experimental procedures leading up to the application of the crude cyt.P450 preparation to the amino-octyl agarose resin, some cyt.P420 is inevitably generated. It has been noted, however, that this degraded form of the enzyme does not bind to the column (authors' unpublished observation). This chromatographic step thus achieves a useful "cleaning up" of the cyt.P450. Emulgen 911 is used to elute cyt.P450 from the column; the detergent disrupts the hydrophobic interactions between the resin and hydrophobic regions of bound proteins.

The protocol for the three column chromatography stages uses a step elution procedure for eluting the proteins from the resins. The use of a salt gradient in column chromatography normally provides better resolution. However, efforts to utilize gradient elutions in this purification protocol have been found not to be as effective as the step-wise approach.
19. It has been noted (authors' unpublished observation) that yeast cyt.P450 is fairly labile while bound to the hydroxy-apatite and carboxy-methyl Sephadex resins. The reason for this is not clear. However, better yields have been obtained when the cyt.P450 remained on the resin for as short a time as possible.
20. Monitoring the column eluate at 280 nm has the major drawback that the Emulgen 911 employed in this protocol absorbs light strongly at this wavelength. Some researchers thus monitor the eluate at 420 nm. Heme absorbs light at this wavelength and this method therefore provides quite a specific indicator of heme proteins, including cyt.P450.
21. Once purified the cyt.P450 can be stored at –70°C for at least 1 yr.
22. Calculation of the percentage purity of the cyt.P450 preparation requires a knowledge of the molecular weight of the enzyme. In our hands, this has been found to be 54 kDa, although a weight of 58 kDa has been reported *(1)*. These discrepancies are thought to be owing to the aberrant migration of this membrane-bound protein on SDS-PAGE.
23. Different yeast strains have been found to produce varying levels of cyt.P450. It has been shown that strain NCYC 754, a brewing strain, is a "good producer" of cyt.P450, and is routinely used in our studies. A typical cyt.P450 content obtained with the above strain is 0.1 nmol/mg microsomal protein. It should be noted, however, that this strain is unsuitable for heterologous cyt.P450 expression studies, having no auxotrophic mutations for use in plasmid selection.

References

1. Yoshida, Y. and Aoyama, Y. (1984) Yeast cytochrome P450 catalysing lanosterol 14α demethylation: I. Purification and spectral properties. *J. Biol. Chem.* **259**, 1655–1660.
2. King, D. J., Azari, M. R., and Wiseman, A. (1984) Studies on the properties of highly purified cytochrome P448 and its dependent benzo[a]pyrene hydroxylase from *Saccharomyces cerevisiae*. *Xenobiotica* **14**, 187–206.
3. Briza, P., Breitenbach, M., Ellinger, A., and Segall, J. (1991) Isolation of two developmentally regulated genes involved in spore maturation in *Saccharomyces cerevisiae*. *Genes Dev.* **4**, 1775–1789.
4. Ching, M. S., Lennard, M. S., Tucker, G. T., Woods, H. F., Kelly, D. E., and Kelly, S. L. (1991) Expression of human cytochrome P450 1A1 in the yeast *Saccharomyces cerevisiae*. *Biochem. Pharmacol.* **42**, 753–758.

CHAPTER 37

Calorimetry of Whole Yeast Cells

Linda J. Ashby and Anthony E. Beezer

1. Introduction

Microcalorimetry provides a quick and accurate nonspecific technique for the direct measurement of metabolic activity of biological cells either under growth or nongrowth conditions. The method has added advantages such as good reproducibility, no requirement for optically transparent media, and ease of relating results to biomass or metabolic rate. Two applications that relate to the use of flow calorimetry will be discussed, but with certain modifications batch-type calorimetry can be used *(1)*.

The direct measurement of metabolic activity by flow microcalorimetry is actually quite simple. By choosing the medium employed either the growth or nongrowth condition for metabolic activity can be forced. Under both conditions, the stirred culture is situated outside the calorimeter in a fermentor and is pumped through the flow cell (usually the outflow from the cell is returned to the fermentor, i.e., a closed loop is established; alternatively, the medium could be flowed to waste following passage through the microcalorimetric flow cell), which is in good thermal contact, via semiconductor thermopiles, with a large heat sink that normally consists of a metal block. The flow cell is usually a gold spiral tube (operational volume approximately 0.6 mL).

Heat flow is usually detected by a "thermopile wall" situated between the flow cell and heat sink. The temperature difference across the thermopile gives rise to a voltage signal that may be displayed on a power meter. In addition, most modern microcalorimeters operate by the twin

From: *Methods in Molecular Biology, Vol. 53: Yeast Protocols*
Edited by: I. Evans Humana Press Inc., Totowa, NJ

principle where one vessel contains the reaction system and the other an inactive material. This allows for a differential signal to be recorded, and, consequently, disturbances (such as small thermal fluctuations in the thermostat temperature) are not detected.

Microcalorimeters are routinely used for precise determinations of millijoule and even microjoule quantities of heat with a power sensitivity of the order of a microwatt (μW). Accurate calibration of the instrument is therefore extremely important and can be achieved by the use of electrical or chemical calibration procedures (2,3).

Electrical calibrations, using the relationship:

$$\text{power supplied} = I^2 R t$$

can be carried out for all calorimeters. The current, I, is usually in the order of milliamps and is variable depending on the sensitivity required. The resistance is fixed (e.g., at 50.0 ohms) and therefore the power supplied (μW) can be calculated and related to the deflection from the baseline achieved during electrical calibration of the calorimeter (Fig. 1). Full-scale deflection in μW is hence calculable. Theoretical aspects of flow microcalorimetry have been described by Beezer and Tyrrell (4), who have considered all types of calorimetry. For a continuous flow-type system, the average heat output rate (power), dq/dt, relative to the baseline for a first-order reaction may be expressed as follows:

$$dq/dt = -RC \Delta H [1-\exp(-k\tau)] \exp(-kt)$$

where R is the flowrate, C is the concentration of the reactant, ΔH is the reaction enthalpy, k is the first order rate constant, τ is the average residence time of the reaction mixture in the sample cell; and t is the time from initiation of the reaction.

The above equation can be used to obtain analytical, thermodynamic, and kinetic information and therefore demonstrates the usefulness of calorimetry.

The metabolism of microorganisms under nongrowing conditions is observed in media lacking one or more nutrients required for growth. Respiring cells exhibit zero order behavior in the presence, of, for example, excess glucose thus giving rise to a constant heat output rate over the timespan for which glucose remains in excess.

The observed calorimetric deflection from a baseline (Fig. 2) is determined by inoculum size and metabolic "state" of the biological cells. As the deflection is directly proportional to total enzymic concentration, the method may therefore be used as a cell counting procedure.

Calorimetry of Whole Yeast Cells

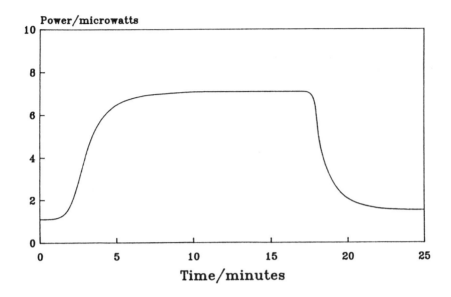

Fig. 1. Electrically induced calibration curve.

The growth of microorganisms can be studied in media that contain all the salts, vitamins, and minerals that are essential and specific for their growth.

After an initial lag period, the energy owing to growth, observed calorimetrically, usually rises exponentially to a peak. At this time, the carbon source has become limiting and this results in a subsequent decline in energy until a new baseline is finally achieved, at which point there is no further increase in cell numbers (Fig. 3).

Large variations in inoculum size, i.e., cells numbers, do not alter the shape of the power-time curve, i.e., peak height is constant; however, the time to achieve maximum power increases greatly with a decrease in inoculum size.

An increase in glucose concentration increases both the peak height and the time to achieve maximum power by broadening the growth phase period.

In summary, the energy measured via the calorimeter is the sum of the energies of all processes occurring within the system, i.e., it follows the progress and process of metabolism. Calorimetry can thus be used to determine reaction enthalpies and reaction kinetics of biological systems, providing that extremely fast rates are not involved. (The time constant of the instrument is about 120 s and therefore by a fast reaction here is meant a reaction with a lifetime of less than approx 360 s.)

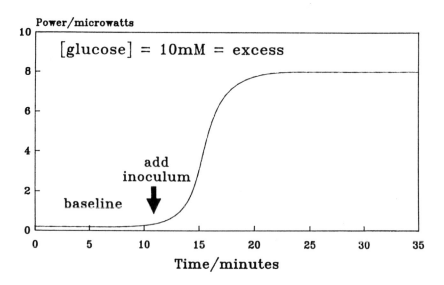

Fig. 2. Typical p-t curve for glucose metabolism by *Saccharomyces cerevisiae*.

Finally, as there is no requirement for optical transparency, calorimetry may be employed for the study of all manner of biological material, for example, bacteria, yeasts, blood cells, isolated animal cells, and, as demonstrated more recently, whole tissues or organs. Applications to media and strain identification and to drug bioassay are discussed in Notes 1 and 2, respectively. References *(5,6)* describe two recent yeast studies employing microcalorimetric approaches.

2. Materials

The fermentors are not "special," nor are they commercially available—they are, in our laboratory, simply convenient three-necked flat-bottomed flasks. All reagents and materials should be of the purest grade available.

1. Hycol (a commercial sterilant/disinfectant) or similar disinfectant for the cleaning of all equipment that comes into contact with biological cells (excluding the calorimeter).
2. Ampules of microorganisms stored under liquid nitrogen (*see* Note 1). Ampules should be stored totally submerged in liquid nitrogen and under this condition should remain viable for at least 5 yr.
3. 0.1*M* HCl.
4. 0.1*M* NaOH.

Calorimetry of Whole Yeast Cells

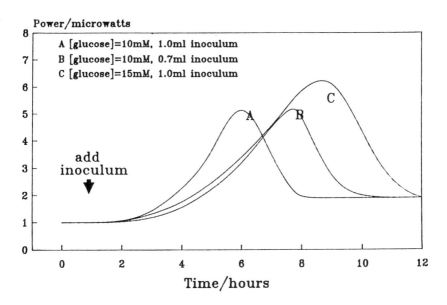

Fig. 3. Typical p-t curves for growth of *Saccharomyces cerevisiae*.

5. Sterilized deionized water.
6. Nitrogen cylinder for anaerobic incubations or oxygen cylinder for aerobic incubations.
7. Nongrowth medium (*see* Note 3): 14.91 g potassium hydrogen phthalate, 0.36 g KOH, 1.80 g D-glucose (10 mM). Make up to 1.0 L with deionized water. Adjust to pH 4.5 using KOH.
8. The following growth medium is specific for most strains of *Saccharomyces cerevisiae* and is usually made up and stored as five separate solutions for simplicity (*see* Note 4).
   a. Solution A, salts (sterilize by autoclave):

	g/L deionized water
$(NH_4)_2HPO_4$	40.0
KH_2PO_4	20.0
Citric acid, H_2O	73.5
Trisodium citrate, $2H_2O$	191.2

   b. Solution B, salts (sterilize by autoclave):

	g/L deionized water
$MgSO_4 \cdot 7H_2O$	10.0
$CaCl_2 \cdot 2H_2O$	1.68

c. Solution C, trace elements (sterilize by autoclave):

	g/L deionized water
H_3BO_3	0.10
KI	0.10
$CuSO·5H_2O$	0.045
$FeSO_4·7H_2O$	0.25
$ZnSO_4·7H_2O$	0.04
$MnSO_4·4H_2O$	0.04
$(NH_4)_6Mo_7O_{24}·4H_2O$	0.02

d. Solution D, vitamins (sterilize by membrane filtration): Protect from light—store in brown glass bottle.

	mg/100.0 mL deionized water
Thiamine hydrochloride	10.0
Pyridoxine hydrochloride	10.0
Calcium pantothenate	10.0
Nicotinic acid	10.0
Meso-inositol	100.0
p-Aminobenzoic acid	1.0
Riboflavin	0.5
Biotin	0.2

Glucose solution (sterilize by autoclave). Make 56.25 g glucose up to 250.0 mL with deionized water.

All solutions can be stored in a refrigerator for up to 4 mo but should be checked regularly for contamination.

Growth medium (250.0 mL); Solution A, 25.0 mL; Solution D, 2.5 mL; Solution C, 0.25 mL; Solution B, 25.0 mL; glucose solution (15 mM); H_2O, 194.0 mL.

The medium pH is 5.1 and as this pH is not detrimental to the proliferation of *S. cerevisiae* no adjustment is necessary.

The following quantities of antibiotics should be added to the growth medium for long-term incubations (*see* Note 10). Stock solutions of antibiotics can be stored at $-20°C$. Penicillin G, 1000 U/mL; ampicillin, 100 µg/mL; streptomycin sulfate, 20 µg/mL.

3. Method

The methods described employ the use of controlled (i.e., recovered from storage in liquid nitrogen and reproducible to approx ± 1% in con-

Calorimetry of Whole Yeast Cells 373

trolled experiments) organisms (*see* Note 5) for the study of glucose metabolism by, and growth of, *S. cerevisiae*.

1. Set the temperature of the calorimeter; for example, the yeast *S. cerevisiae* requires a working temperature of 30°C. Give sufficient time for the calorimetric apparatus to equilibrate to any new temperature as this process may take up to 2 d.
2. Initiate cleaning and sterilization of the calorimeter by placing the inlet lead into a flask of $0.1 M$ HCl and the outlet lead to waste. Switch on the peristaltic pump for approx 10 min.
3. Repeat with deionized water, then $0.1 M$ NaOH for a time of 10 min each.
4. Pump sterilized deionized water through the flow lines and continue pumping while other preparations are carried out.

   For glucose metabolism studies; dispense 50.0 mL (volumes stated here are illustrative not prescriptive) nongrowth medium into a three-necked (inlet, outlet, and inoculation port) respiratory flask containing a magnetic stirring bar (without ridges).

   Alternatively, for growth studies, dispense 150.0-mL growth medium into a multinecked incubation vessel (containing magnetic stirring bar) using aseptic techniques. Fit a reflux condenser and a nitrogen line to the incubation vessel, if anaerobic conditions are required (*see* Note 13).
6. Place the incubation vessel in a thermostatted water bath (30°C) external to the calorimeter in order to equilibrate.
7. Pump a separate solution of either nongrowth or growth medium (i.e., other than that contained in the incubation vessel) through the calorimetric flow lines in order that all residual HCl, NaOH, and H_2O is thoroughly washed out and replaced with medium.
8. When the incubation vessel has had sufficient time to equilibrate, leave the pump on and place the inlet lead into the solution in the incubation vessel. If a slug of air is allowed to be drawn through the inlet lead before placing it in the incubation medium, the "slug" will reappear in the outlet lead after being pumped around the whole calorimetric system—the reappearance of the air slug will signal the correct time to place the outlet lead into the incubation vessel and therefore set up a flow loop of an exact and known volume of buffer.
9. Set the amplifier to the correct sensitivity, the choice of which will be dependent on
   a. Type of metabolism studied (e.g., greater sensitivity is required when studying nongrowth metabolism in comparison to that of growth); and,
   b. Inoculum size.
10. Turn on the chart recorder.

11. When a baseline is achieved, indicating thermal equilibration of solution flowing through the calorimeter, prepare to inoculate the yeast cells into the medium.
12. Remove 1 ampule of yeast cells from liquid nitrogen storage (*see* Note 5).
13. Place in a water bath set at 40°C and start a stop-clock (i.e., at $t = 0$); thaw the cells for 3 min.
14. Agitate the ampule vigorously for 1 min using a whirlimixer.
15. Use a pipet (e.g., Gilson, Paris) with a sterile tip to dispense an accurate volume of cells (0.5–1.0 mL assuming an inoculum density of approx 10^8 colony forming U/mL cell suspension) into the medium at $t = 5$ min. Timing the process ensures the reproducibility of the method, which is essential when such a sensitive technique is used for biological work.

4. Notes

1. Evaluation of media and strain identification: Microbial strain identification is possible, using microcalorimetry, by use of controlled media as power-time curves of microorganisms in batch culture are very detailed, i.e., strain-specific. This is especially true when media containing more than one carbon source are utilized. In a mixed carbon source medium, the power-time curve will reflect the order, and rate, of adaptation to utilization of the energy source.

    Media can be controlled by either the use of a minimal/defined medium in which all required components are specified in their required quantities, or by the use of *one batch* of enriched medium throughout the studies.

    Batch-to-batch variations in enriched media are extremely large as some of its components, and more specifically peptone and yeast extracts, are undefined in themselves and therefore have limited reproducibility.

    The controlled media can be used to inspect the growth characteristics of genetically distinct strains of organisms; for example, Perry et al. *(7)* found marked differences in power-time curves derived by flow microcalorimetry for brewing, baking, and distilling grains of *S. cerevisiae* in defined media containing:
    a. Glucose;
    b. Glucose and maltose; or,
    c. Maltose as the carbon source(s).
    This led to the characterization of representative strains of *S. cerevisiae* used in the food industry.

    If, alternatively, the microorganisms are controlled—by the use of liquid nitrogen stored inocula—the components of the media employed for their growth can be examined.

Commercial substrates such as molasses (used in the industrial production of baker's yeast) and ethanol often show marked batch-to-batch variability, despite an acceptable chemical analysis—thus trace components must be responsible for determining the capacity to maintain satisfactory growth. Flow microcalorimetry has been used to identify industrially poor and acceptable samples of molasses *(8)* and it is probable that this technique could be used to evaluate components of other complex media.

2. Drug bioassay: Microcalorimetry has been successfully employed in determining the comparative biological activity of groups of metabolic modifiers, e.g., antifungals and antibacterial antibiotics, against a wide range of biological microorganisms *(9–11)*.

As described previously, microcalorimetry can be used to follow the energetic processes of growth and nongrowth metabolism. The presence of an active antimicrobial agent would distort the power-time curve and the extent of this distortion gives a measure of the biological activity of the modifier.

Standard liquid nitrogen stored inocula are used in order to achieve reproducibility, in bioassay data of better than ±3% (in comparison to that of ±10% by the agar plate diffusion method, for example).

The incubation vessel is charged with either nongrowth or growth medium (as described earlier) together with the antimicrobial agent of the desired concentration. Once a baseline is achieved, the appropriate biological cells are inoculated into the composite solution. The power-time curves are compared to a control in which no antimicrobial agent is present. The *relative* biological activities of a group or series of compounds may thus be obtained.

Alternatively the metabolic modifier can be added to the incubation vessel at some point during the growth/respiration cycle of the culture.

Metabolic modifiers suppress substrate metabolism (for biological cells under nongrowth conditions), often giving rise to a decay curve, as observed calorimetrically.

In other instances, the enthalpy of metabolism is reduced to a plateau, in parallel to that of the control power-time curve, as shown in Fig. 4. Differing modes of action are therefore discernible by microcalorimetry.

Empirically based biological activity parameters for metabolic modifiers are obtainable from drug-modified power-time curves. Useful parameters include:
a. Time required for the power output to fall to a certain value;
b. Power output at a fixed time from time of inoculation of the medium; and,
c. Percentage change from control response.

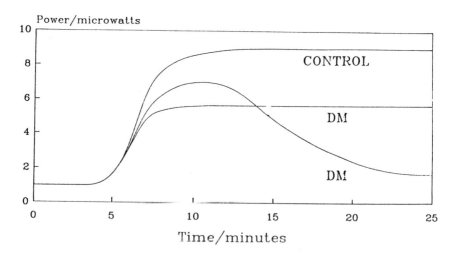

Fig. 4. Control and drug modified (DM) p-t curves for glucose metabolism.

Drug-modified *growth* curves possess two notable characteristics. First, the peak-time delayed, and second, the peak height is somewhat depressed, thereby giving rise to a flatter, broader curve (Fig. 5). In addition to the biological activity parameters described earlier, when the growth of biological organisms is studied, parameters relating to the peak-height and peak-time may be used.

Very often metabolic modifiers are hydrophobic in character and therefore have limited solubility in aqueous solution. In such cases, the compounds can be solubilized by the use of a "carrier" solvent, such as dimethylsulfoxide or methanol, providing that the solvents are used at low concentrations to ensure nonlethal damage of the biological cells.

Additional problems owing to adhesion, sedimentation, and so on, may arise, owing to the presence of the antimicrobial agent, and these should be identified and overcome (*see* Notes 5–13).

3. The medium employed when studying nongrowth metabolism is essentially a glucose buffer, in which glucose is present in excess, and should be of suitable pH for the organisms employed. This medium is suitable for studies of the yeast *S. cerevisiae* and should be made up fresh on a daily basis.
4. Special attention should be paid to aseptic techniques when making up and handling growth media as lengthy incubations are more susceptible to contamination by bacteria.
5. Batches of nonsynchronous cells that are freshly prepared on a daily basis will not only contain cells of varying ages but will be intrinsically different owing to chance mutations, slight changes in media employed for growth, and temperature and pH fluctuations during growth. Power output is strongly

Calorimetry of Whole Yeast Cells

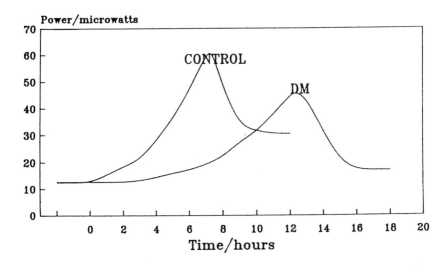

Fig. 5. Control and drug modified (DM) p-t curves for microbial growth.

influenced by cell history and therefore in order to obtain good reproducibility in power-time curves, and comparability between experiments, a single microbial strain should be grown under standard defined conditions in one batch, with a number of subcultures, and finally stored in liquid nitrogen.

Cryogenic storage is suitable for both yeasts and bacteria. The cells are normally suspended in a solution of quarter-strength Ringers' containing 10% v/v DMSO *(12)* in order to protect against freeze damage that would be otherwise incurred at liquid nitrogen temperatures.

The optimum inoculum is that which, in the chosen medium, yields an output equivalent to around 60–80% of the full scale response; some pilot experiments may be necessary to determine this.

6. Changes in pH during calorimetric incubations can give rise to anomalous thermal effects owing to protonation and deprotonation enthalpies. In order to overcome these problems, buffered solutions with adequate buffer capacity are always used and a pH specific to the study organism should be chosen; for example, media of pH 4.5 should be used when studying the metabolism of *S. cerevisiae*.

7. Notionally, all cell types can be studied by batch microcalorimetry, but this is not the case when flow-lines are utilized. The Teflon lines that are used to pump the culture medium through the calorimeter are very small in diameter (1.0 mm) and are therefore susceptible to hyphal forms, e.g., *Candida albicans*, should therefore be avoided. The yeast form of *Candida albicans* is, however, acceptable.

A special strain of nonaggregating *S. cerevisiae*, NCYC 239, is suitable for use in flow microcalorimeters. *Eschericha coli* and fungal spores give rise to very fine homogeneous suspension cultures and are therefore suitable for flow calorimetric studies. However, those cells which grow in mycelial forms, such as the fungus *Septoria nodorum*, should only be considered for batch-type work.

8. Sedimentation of calorimetric suspension cultures should be avoided by the use of stirred cultures.
9. Partial/preferential adhesion between biological cells and the walls of the incubation vessel (a), or flow cell (b), or even the flow lines (a), may occur, thereby causing the measured power owing to the metabolizing cells to be (a) greater or (b) less than expected. This problem can be identified by performing a cell count on the culture medium before and after an incubation experiment. Teflon leads may be exchanged for metal leads (e.g., steel) and any glassware causing adhesion can be coated with silicone (e.g., Repelcote). Modified flow through cells can also be constructed.
10. In particular, when studies requiring lengthy incubations (i.e., more than 2 h) are carried out, provisions for control of pH, sterilization of equipment, and prevention of loss of medium owing to evaporation/condensation must be made.

Sterilization of media and equipment in order to suppress unwanted bacterial growth can be achieved by autoclaving. For those media components that are denatured at high temperatures, e.g., vitamins, sterilization by membrane filtration is used. This usually entails passing aliquots of the solution through a millipore filter that has been autoclaved.

Of course, a whole calorimeter cannot be sterilized in an autoclave, but the leads which transport the culture suspension around the calorimeter can be sterilized by pumping sterilized deionized water through them, after thoroughly cleaning the leads/lines with acid and base (0.1M HCl, 0.1M NaOH, respectively).

During growth studies of microorganisms, it should be established that the growth energy of only the test organism is being monitored. The growth of other possible contaminant organisms that may feed from the rich growth medium can be prevented by adding specific antibiotics to the growth medium. For instance, when studying the growth of *S. cerevisiae*, a mixture of penicillin, ampicillin, and streptomycin can be added to the growth medium to suppress bacterial growth.

Excess evaporation of medium from the incubation vessel can affect medium concentration. To prevent this, a double-surface reflux condenser can be attached to the multinecked incubation flask.

11. Studying the growth characteristics of microorganisms, with a time scale of 8–15 h (dependent on study organism), will give rise to a great increase in cell numbers. The resulting viscous culture may cause intermittent blockages in the flow lines—this is easily identifiable as "peaks" in the power-time curves or by energy profiles which do not appear to be smooth and continuous. Evolution of gaseous CO_2 may also occur with dense cultures—again "peaks" will be observed (see Note 12). The problem can be overcome by increasing the pump speed (to a limited extent), by decreasing inoculum size, and by limiting carbon source concentration, hence limiting the extent of microbial growth.

    Optimum pump speed is dependent on the factors outlined above and pump rates described in the literature are in the range 40.0—80.0 mL/h.

    Slower pump speeds should be used whenever possible, provided that problems with aeration do not arise, as this ensures temperature equilibration of the suspension medium prior to its reaching the reaction cell. In addition, more moderate speeds, e.g., 40 mL/h for *S. cerevisiae*, are acceptable for nongrowth metabolic studies that are both short in duration and nongrowth in character, and therefore pose very little threat of causing blockages.

12. Very sharp "spikes" present in power-time curves indicate the presence of air/carbon dioxide bubbles in the flow lines. If all unions/joints of the flow setup are found to be secure, then the pH of the medium should be checked, as adverse pH conditions can often lead to the production of *excess* carbon dioxide by microorganisms that remains insoluble, once the saturation solubility of carbon dioxide by the medium has been exceeded.

13 For some calorimetric experiments, there will be a need to create an aerobic system in which a sufficient and constant supply of oxygen must be maintained or, conversely, a need for an anaerobic environment where the system must be completely purged of oxygen.

    An anaerobic system can be obtained within a charged incubation vessel by first bubbling nitrogen through the incubation medium for approximately 15 min and then raising the gas-line outlet to a position just above the surface of the medium. This ensures that an atmosphere of nitrogen is maintained over the surface of the medium throughout the duration of the experiment. The incubation vessel should be fitted with a reflux condenser as a means for gas exhaustion.

    In situations where *strict* anaerobic conditions are required, solutions may be degassed by conventional methods. In addition, the Teflon leads normally employed in flow microcalorimetry, and which are gas permeable, may be exchanged for metal leads. Constant aeration can be achieved in batch-type calorimetric work by simply bubbling a constant pressure of air/oxygen through the medium situated in the static cell. This is particu-

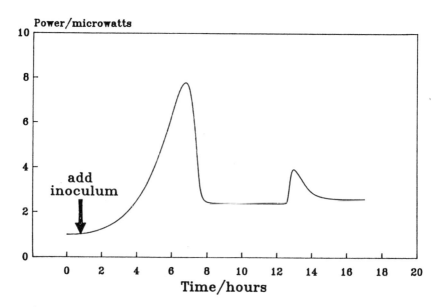

Fig. 6. P-t curve for *K. fragilis* grown in oxygen sparged medium.

larly simple when using aeration/perfusion cells as they are permanently fitted with suitable lines for gas flow into the reaction cell. However, a test of aerobiosis is required at the calorimetric cell, i.e., it should be ensured that oxygen starvation does not occur in the leads to and from the fermentor (by, e.g., carefully simulating the microcalorimetric experiment using an oxygen electrode assembly in place of the cell).

In addition, when aerobic flow microcalorimetry is made use of, it should be ensured that there is no starvation of oxygen at the measuring site that is at a considerable distance from the fermentor. This can be achieved by the use of a segmental flow aeration vessel, as designed by Eriksson and Wadson *(13)*. The apparatus enables alternative aliquots of the reaction medium and air/oxygen to be pumped continuously around the flow system, thereby ensuring sufficient aeration throughout the system.

The influence of oxygen on derived power-time curves is striking. For example, Fig. 6 depicts the p-t curve for the anaerobic growth of *Kluyveromyces fragilis* in glucose-limited medium. When air is substituted for nitrogen gas, a p-t curve is obtained in which three, as opposed to two, distinct energetic regions are discernible (Fig. 7) as oxygenation results in the appearance of the second of these peaks. It has been established that the first peak corresponds to aerobic utilization of glucose and the second anaerobic fermentation. Further to this it was found to be possible to produce the second peak at will by changing the gas supply from nitrogen to air *(14)*.

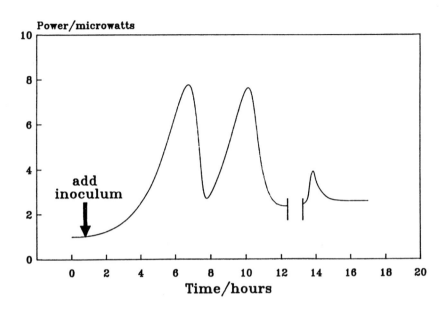

Fig. 7. P-t curve for *K. fragilis* grown under aerated conditions.

References

1. Wadson, I. (1987) Calorimetric Techniques, in *Thermal and Energetic Studies of Cellular Biological Systems* (James, A. M., ed.), I.O.P., Bristol, pp.45–60.
2. Wadson, I. (1980) Some Problems in Calorimetric Measurements on Cellular Systems, in *Biological Microcalorimetry* (Beezer, A. E., ed.), Academic, London, pp. 247–274.
3. Chem, A-T. and Wadson, I. (1982) A test and calibration process for micro calorimeters used as thermal power meters. *J. Biochem. Biophys. Meth.* **6**, 297–306.
4. Beezer, A. E. and Tyrrell, H. J. V. (1972) Applications of Flow Microcalorimetry to Biological Problems. Part I Theoretical Aspects. *Sci. Tools* **19**, 13–16.
5. Ölz, R., Larsson, K., Adler, L., and Gustafsson, L. (1993) *J.Bacteriol.* **175**, 2205.
6. Vásquez-Juárez, R., Andlid, T., and Gustafsson, L. (1994) Colloids and Surfaces B: *Biointerfaceas* **2**, 199.
7. Perry, B. F., Beezer, A. E., and Miles, R. J. (1983) Characterization of Commercial Yeast Strains by Flow Microcalorimetry. *J. Appl. Bacteriol.* **54**, 183–189.
8. Perry, B. F., Beezer, A. E., and Miles, R. J. (1981) Microcalorimetry as a Tool for Evaluation of Complex Media: molasses. *Microbios.* **37**, 163–172.
9. Beezer, A. E., Chowdhry, B. Z., Newell, R. D., and Tyrrell, H. J. V. (1977) Bioassay of Antifungal Antibiotics by Flow Microcalorimetry. *Anal. Chem.* **49**, 1781–1784.
10. Beezer, A. E. and Chowdhry, B. Z. (1981) Flow microcalorimetric bioassay of polyene antibiotics: interaction with growing *Saccharomyces cerevisiae*. *Experientia* **37**, 828–830.

11. Joslin, N. (1984) Bioassay and Interactions of Antibiotics with Sensitive Cells. A microcalorimetric Study. PhD Thesis, University of London.
12. Beezer, A. E., Newell, R. D., and Tyrrell, H. J. V. (1976) Application of flow microcalorimetry to analytical problems: The preparation, storage and assay of frozen inocula of *Saccharomyces cerevisiae*. *J. Appl. Bacteriol.* **41,** 197–207.
13. Eriksson, R. and Wadson, I. (1971) Design and Testing of a Flow Microcalorimeter for Studies of Aerobic Bacterial Growth, in *Proceedings First European Biophysical Congress*, Baden (Broada, E., Locker, A., and Springer-Lederer, H., eds.), vol. 4, Verlag der Wiener Medizinischen Akademie, Austria, p. 319.
14. Beezer, A. E., Newell, R. D., and Tyrrell, H. J. V. (1979) Characterization and metabolic studies of *Saccharomyces cerevisiae* and *Kluyveromyces fragilis* by flow microcalorimetry. *J. Microbiol.* **45,** 55–63.

CHAPTER 38

Light Microscopy Methods

Eva Streiblová and Jiří Hašek

1. Introduction

The molecular biology of yeasts places increasing emphasis on the genetic and biochemical complexity of the cell in the contest of its spatial organization. This development has led to a renaissance of yeast cytology. Classical cytological methods are being replaced by new techniques partially adopted from related fields of cell biology. The remarkable refinement of indirect immunofluorescence methods has been a great advance, which has opened a new chapter in the precise localization of enzymes and other gene products, including cytoskeletal and karyoskeletal components. Exploiting these techniques, the yeast cell can be approached at almost all levels of its structural organization.

A complete review of light microscopy methods for yeasts is beyond the scope of this and its accompanying chapter. The impact of light microscopic techniques and an extended survey of cytological methods have been presented elsewhere (1). This chapter includes simple light microscopic staining procedures to define essential cellular components, and Chapter 39 covers:

1. Useful fluorescence techniques using fluorochrome and brighteners;
2. Procedures using fluorochrome-labeled antibodies to cytoskeleton proteins; and
3. A well-established technique of F-actin staining by fluorochrome-labeled phalloidin.

We focus attention on procedures that are routinely used in our laboratory.

Basically, two approaches are possible to obtain a clear picture of cellular detail using light microscopy. One way is to increase the contrast of living cells by optical means using phase-contrast, dark-field, or Nomarski differential interference optics. For background information, *see* ref. *(2)*.

The second possibility is to use bright-field microscopy and to increase contrast by using dyes that selectively stain various structures or substances within the cells. The two approaches are:

1. Staining of living and unfixed cells; and
2. Staining made after application of fixatives.

For general information concerning dyes, *see* ref. *3*.

2. Materials
2.1. Immobilization of Cells

1. Egg white supernatant: Stir egg white with an equal volume of distilled water in a cylinder. When the floccules of insoluble globulins settle, the soluble albumin fraction is decanted and a drop of the supernatant is used to coat coverslips or slides.
2. Haupt's adhesive: Dissolve 1 g of pure gelatine in 100 mL of distilled water at 30°C. When dissolved, add 2 g of phenol crystals and 15 mL of glycerol to prepare the adhesive.
3. Poly(L-lysine): Dissolve 1 mg/mL of poly(L-lysine) in water.
4. Concanavalin A (Con A): Dissolve 1 mg/mL Con A in water.

2.2. Vital Staining and Staining of Unfixed Cells

1. Janus green B: Dissolve 5 mg of Janus green in 100 mL of 1% (w/v) glucose. Unstable. Prepare the staining solution immediately before use and aerate for 30 min.
2. Neutral red: Dissolve 1 mg/mL stock solution of neutral red in water. Keep in the dark and filter before use.
3. Lomofungin: Prepare stock solution 2 mg/mL of lomofungin (Upjohn, Kalamazoo, MI) in dimethylsulfoxide, store in dark at 4°C, and filter before use.
4. Lugol agent: Dissolve 1 g of iodine and 2 g of potassium iodide in approx 2 mL of H_2O and make up to 100 mL with H_2O when fully dissolved. Store the reagent in the dark at 4°C and filter before use.
5. Methylene blue: Dissolve 10 g of methylene blue first in 100 mL of H_2O, then add 1 mL of 1% (w/v) aqueous KOH and 30 mL of absolute ethanol. Stir the solution several hours at elevated temperature (approx 50°C) until fully dissolved. Store in a tightly sealed bottle at 4°C and filter before use.

Light Microscopy Methods

6. Sudan Black B: Dissolve 0.5 g of Sudan black B in 100 mL of warm 70% (v/v) ethanol, mix by shaking thoroughly, and leave to stand at 60°C overnight. Prepare 1 d in advance and filter before use.

2.3. Staining of Nuclei in Fixed Cells

1. Acetic acid-formaldehyde-ethanol fixative: Mix 50 mL of 90% (v/v) ethanol with 5 mL of glacial acetic acid, 5 mL of formaldehyde, and 40 mL of water. Keep in the dark at 4°C in a tightly sealed bottle.
2. Modified Helly's fixative: Dissolve 5 g of mercuric chloride and 3 g of potassium dichromate in 100 mL of H_2O at elevated temperature. Keep in the dark at 4°C in a tightly sealed bottle. Immediately before use, add formaldehyde to 5% (w/v).
3. Schaudin's fixative (half strength): Mix 20 mL of saturated solution of mercuric chloride with 10 mL of absolute ethanol and 30 mL of H_2O. Then add 1 mL of glacial acetic acid.
4. Acid fuchsin: Dissolve acid fuchsin in 1% (w/v) acetic acid (1:40,000 w/v), mix by shaking thoroughly, and leave to stand 2 d at room temperature. Filter before use. Store at 4°C in a tightly sealed bottle.
5. Gurr's buffer: Prepare $0.067M$ phosphate buffer, pH 6.8.
6. Giemsa solution: Mix 0.35 mL of Giemsa stain (Gurr or R 66) with 10 mL of Gurr's buffer ($0.067M$). Prepare immediately before use.
7. Aceto-orcein solution: Dissolve one part of synthetic orcein in 100 parts of 60% (w/v) acetic acid, store at 4°C, and filter before use.

3. Methods
3.1. Immobilization of Cells

It is recommended to make yeast cells adhere firmly to coverslips or slides before subjecting the specimens to fixation and staining procedures. Generally, one of the following methods is convenient (*see* Note 1).

1. Place a small drop of yeast suspension onto the center of a clean slide and spread by means of a thin glass rod or the edge of a coverslip. Allow the smears to air dry for 20 min. Do not speed the drying process by heating the slide.
2. Coat a slide with a thin film of the supernatant obtained from fresh egg white (*see* Section 2.1.). Place a drop of the supernatant onto the slide and smear evenly with the tip of a finger. Spread a yeast suspension.
3. Coat a slide with Haupt's adhesive (*see* Section 2.2.) by immersing it in the reagent kept in a conventional staining jar. Spread a yeast suspension.
4. Coat slides with poly(L-lysine) (*see* Section 2.3.) and leave to stand in a moist chamber for 30 min. Remove excess liquid, then wash with water and mount yeast cells to be examined.
5. Coat slides with Con A and air dry at room temperature before spreading a yeast suspension.

3.2. Vital Staining and Staining of Unfixed Cells

Staining with dyes such as methylene blue, benzidine blue, Nile blue, rose Bengal, neutral red, and Janus green, which have no toxic effect at concentrations ranging from 0.1–100 µg/mL, may, in many cases, permit going beyond a simple description of what is colored in the yeast cells.

3.2.1. Vital Staining of Mitochondria

1. Spin down 1 mL of actively growing yeast culture (5×10^6 cells).
2. Place a drop of cells from the pellet on a microscopic slide.
3. Mix with an equal volume of freshly prepared Janus green solution without putting the coverslip.
4. Observe immediately with a water-immersion objective (×40 or ×70).

Janus green stains mitochondria in living cells greenish-blue owing to the action of the mitochondria-bound cytochrome oxidase system that maintains the dye in a colored form. In the cytoplasm the dye is in the reduced, colorless state.

3.2.2. Vital Staining of Vacuoles

1. Mix a drop of the suspension of yeast cells (10^9/mL) with an equal volume of the staining solution of neutral red on a microscope slide.
2. Apply a coverslip.
3. Observe immediately under the microscope.

Neutral red is actively taken up and internalized by the yeast vacuolar system. The vacuoles stain reddish-purple. The staining may be viewed up to 15–30 min. Then the stain in the vacuoles tends to fade, leaks out of the vacuoles, and stains the cytoplasm brownish-pink.

3.2.3. Vital Staining of the Nucleus

1. Spin down 10 mL of the yeast cell suspension (10^9/mL; 500g; 5 min).
2. Wash once with distilled water.
3. Resuspend in a solution of lomofungin (dilute the stock solution 1:1 [v/v] with distilled water).
4. Mount a drop onto a slide.
5. Apply a coverslip.
6. Examine immediately under the microscope.

The antibiotic lomofungin belongs to the group of phenazine dyes. It reveals deep red chromosome-like structures in a variety of yeast species in vivo during mitosis and meiosis *(4)*.

3.2.4. Detection of Glycogen

1. Prepare smears by mounting a drop of yeast suspension (10^9/mL) on a microscopic slide.
2. Spread the suspension into a thin film with the edge of a coverslip.
3. Make the cells adhere by air-drying or by mixing a drop of egg white with the cells while spreading them on the slide.
4. Add a drop of Lugol solution.
5. Apply a coverslip.
6. Examine under the microscope.

Glycogen deposits in the cytoplasm of yeast cells, composed of polymerized glucose molecules, and thus an important reserve of energy, are invisible in unstained cells. Lugol-stained cells display scattered reddish-brown precipitates of glycogen. Cytoplasmic proteins typically stain yellow. Stained glycogen deposits temporarily disappear after a brief heating of the mounts to about 70°C.

3.2.5. Detection of Volutin

1. Prepare and air dry smears prepared by spreading a yeast suspension (10^9/mL) into a thin film with the edge of the cover slip.
2. Stain for 30 min with methylene blue solution in a conventional staining dish.
3. Drain off the excess stain and blot.
4. Place the specimen into a dish with 1% (w/v) H_2SO_4 and decolorize for about 1 min.
5. Watch for the blue-black precipitates to appear.
6. Once precipitates become obvious, stop bleaching by rinsing the slides with water.
7. Apply a coverslip and examine.

Volutin is a polymeric compound used as reserve by yeast cells. Phosphoric acid groups make the substance readily stainable with metachromatic methylene blue, which forms aggregates upon the polymers. Vacuoles with volutin take on a dark blue to black coloration.

3.2.6. Detection of Lipids

1. Prepare smears of yeast cells (*see* Section 3.1.1.).
2. Apply a drop of Sudan black B and stain for 30 min.
3. Drain off the excess stain and blot dry.
4. Rinse with 50% (v/v) aqueous ethanol.
5. Watch for dark globules to appear.
6. Then stop the decolorization by rinsing the specimen with water.
7. Dry in air.
8. Apply a coverslip and examine.

Sudan stains act by simple diffusion and accumulation of the dyes in the interior of lipid droplets. Sudan black has the advantage producing greater contrast than Sudan II and IV. Lipid granules are bluish-grey in contrast to the pale pink cytoplasm.

3.3. Staining of Nuclei in Fixed Cells (see Notes 2 and 3)

3.3.1. Fixation

1. Acetic acid-formaldehyde-alcohol fixation: Make air-dried smears and immerse the slide in the fixative kept in a closed staining jar. Allow a minimum time of 50 min for fixation at room temperature prior to further treatment.
2. Helly's fixation: Make smears and fix with the fluid for 10–20 min, rinse several times with 70% (v/v) ethanol, and store the slides for 2 d (at the most) prior to further treatment.
3. Schaudin's fixation: Fix the specimen with the fluid for 10 min, rinse with ethanol, and store in 70% (v/v) ethanol, prior to staining.

3.3.2. Acid Fuchsin Staining

1. Fix cells according to Helly or Schaudin (*see* Section 3.3.1.).
2. Remove from 70% (v/v) ethanol.
3. Rinse several times with 1% (w/v) acetic acid.
4. Immerse slides for 1–5 min in the acid fuchsin solution kept in conventional staining dishes.
5. Rinse briefly with 1% (w/v) acetic acid.
6. Decolorize in 70–90% (v/v) ethanol, watching the development of deep staining of the nucleolus and spindle, by making temporary mounts with a coverslip and observing with a microscope.
7. Stop the differentiation by rinsing the specimen with 1% (w/v) acetic acid, drain excess fluid, and allow the cells to flatten.
8. Apply a coverslip and examine (i.e., mounting in the residual fluid from step 7).
9. Use a green filter and optimal illumination.

Acid fuchsin acts by creating differences in light absorption between yeast structures. Gentle flattening of the specimen and appropriate magnification are indispensable to resolve detail of the yeast nucleus. Preparations may undergo rapid decolorization when rinsed with tap water. Rinsing with slightly acidulated water for a few seconds and/or differentiation in ethanol help to discern the nucleolus and the spindle.

3.3.3. HCl-Giemsa Staining

1. Fix with one of the fixatives described above (*see* Section 3.3.1.).
2. Remove from ethanol (or formaldehyde).
3. Transfer to 1% (w/v) NaCl at 60°C for 1 h.
4. Hydrolyze for 8–10 min with $1 M$ HCl at 60°C.
5. Rinse with water at 37°C.
6. Rinse in Gurr's buffer.
7. Stain with Giemsa solution for 1–2 h in a glass-covered staining dish, or in a Coplin jar.
8. Check the quality of staining using a ×40 or ×70 water-immersion objective.
9. Bleach overstained specimens with acidulated water in a Petri dish (10 mL of distilled water and a loopful of glacial acetic acid).
10. Watch for staining of the chromosomes.
11. Mount the finished specimen in a solution containing two drops of Giemsa stain per 12 mL of Gurr's buffer.

Giemsa staining is the procedure of choice for the study of chromosome in mitosis and meiosis. A strong pressure on the coverslip is recommended to resolve the arrangement of chromosomes. For the study of meiosis, acid hydrolysis ($1 N$ HCl at 60°C) may be replaced by treatment with 100 µg/mL RNase in 50 mM Tris-HCl, pH 7.4, 5 mM EDTA, at 60°C for 2 h, prior to Giemsa staining.

3.3.4. Aceto-Orcein Staining

1. Fix with one of the fixatives described earlier (*see* Section 3.3.1.).
2. Hydrolyze for 8–10 min with $1 M$ HCl at 60°C.
3. Rinse with water.
4. Immerse for a few minutes in 60% (w/v) acetic acid.
5. Apply the aceto-orcein solution for 20–30 min.
6. Watch for staining of the chromatinic part of the nucleus.
7. Mount the finished specimen using a drop of the same stain and observe.

HCl-aceto-orcein staining is an alternative for chromosome staining. It often gives good results when individual chromosomes are hardly visible with Giemsa.

4. Notes

1. To prevent the finished specimen from dehydrating, seal the edges of the cover glass with wax, nail polish, or petroleum jelly. Place petroleum jelly in a syringe, heat the needle of the syringe in a flame to melt the jelly, and apply to the ridges of the cover glass.

2. Fixation is an intimate part of the staining procedure. It attaches the cells firmly to the surface of the slide and retains the structure of the nucleus in as natural a state as possible. Fixation procedures are essentially methods of controlled protein denaturation. The procedures covered in this section designed for yeast cells were originally by Robinow *(5)*.
3. Complementary staining of fixed nuclei with acid fuchsin (for nucleolus and mitotic spindle) and with HCl-Giemsa or HCl-aceto-orcein (for chromosomal components) is needed to examine the cytological basis of mitosis and meiosis in different yeast species.

References

1. Streiblová, E. (1988) Cytological methods, in *Yeast—A Practical Approach* (Campbell, I. and Duffus, J. H., eds.), IRL, Oxford, UK, pp. 9–49.
2. Lacey, J., ed. (1989) *Light Microscopy in Biology.* IRL, Oxford, UK.
3. Clark, G., ed. (1980) *Staining Procedures.* Williams & Wilkins, Baltimore, MD.
4. Kopecká, M. and Gabriel, M. (1978) Staining the nuclei in cells and protoplasts of living yeasts, moulds and green algae with the antibiotic Lomofungin. *Arch. Microbiol.* **119,** 305–311.
5. Robinow, C. F. (1975) The preparation of yeasts for light microscopy, in *Methods in Cell Biology*, vol. XI (Prescott, D. M., ed.) Academic, New York, pp. 2–22.

CHAPTER 39

Fluorescence Microscopy Methods

Jiří Hašek and Eva Streiblová

1. Introduction

The last decade has seen an extraordinary improvement in fluorescence microscopy instruments. These have been mainly in the development of epi-illumination systems, including dichroic beam splitters, matching excitation and barrier filters, and special objectives (*see* ref. *1* for background information). Routinely we use a JenaLumar fluorescence microscope (Zeiss, Jena, Germany) equipped with oil immersion wide field ×100 objective with an adjustable iris diaphragm. According to our experience, the best documentation is obtained when the photographs are taken on Kodak Tri X-Pan film processed in Kodak developer.

Checking autofluorescence is a prerequisite for any experiments involving application of fluorescence substances to yeast cells. This phenomenon is significant in many yeast species, the common pattern being a faint diffuse blue fluorescence of the cytoplasm and a yellow fluorescence of the granules. One way to reduce the the autofluorescence is to use narrow-band cut-off suppression filters. Another possibility is to experiment with the conditions of the cultivation of the yeast under study.

Berberin sulfate, aurophospin, coriphospin, thioflarin S, and neutral red are fluorochromes that impart brilliant fluorescence patterns to yeast cells. Although their potential has not been fully realized so far, they are of considerable interest for yeast cytology. Useful information on fluorochromes currently available are given in ref. *2*.

Tinopal LPW (Sigma, St. Louis, MO), Uvitex BOPT (Ciba-Geigy, Basel), Photine LV (Hickson and Welch, Nijmegen, Holland), Leucophor (Sandoz, Basel) and Fluolite RP (ICI, Manchester, UK) are brighteners of potential interest for yeast cytology. Originally used as detergents in textile and paper treatments, they give a strong bright fluorescence associated mainly with the cell wall. The most popular brightener is Calcofluor White (American Cyanamid Co., Princeton, NJ); Calcofluor White M2R; Fluorescent brightener 28.

Fluorochromes and brighteners used at low concentrations ranging from 0.001–10 µg/mL are essentially nontoxic, water-soluble vital stains. When added to the nutrient medium, some of them transmit fluorescence markers to subsequent generations of yeast cells.

Fluorochrome-labeled antibodies are used to localize various macromolecules at the cellular level. The most suitable is the indirect immunofluorescence procedure since it provides high sensitivity of labeling. In the first step, a primary antibody specifically reacts with intracellular epitopes. In the second step, the epitope-bound primary antibody is recognized and visualized by a secondary, fluorochrome-labeled, antibody (conjugate). This approach is widely used to follow the dynamics of various proteins in yeast cells, especially of the components of the cytoskeleton *(3)*.

Indirect immunofluorescence techniques utilize various commercially available antibodies with previously characterized specificity. For conjugate preparation, fluorescein isothiocyanate and tetramethylrhodamine isothiocyanate are most widely used as discreet labels. Both fluorochromes have a suitable spectral resolution. Tagged fluorescein gives a strong green-yellow fluorescence. Rhodamine emits a bright red-orange fluorescence. To date, secondary antibodies labeled with new fluorochromes, such as Phycoerythrin B, Texas red, and Bodipi, are available on the market.

The most critical step for the application of immunofluorescence techniques in yeasts is the penetration of antibodies through the cell wall. To overcome this general problem, the following approaches may be used:

1. To work with protoplasts;
2. To digest the cell wall of fixed cells with various lytic enzymes; or
3. To break the cell wall of fixed cell mechanically using glass beads.

Fluorescence Microscopy Methods

Each of the mentioned approaches has its own advantages and/or disadvantages. The choice depends mainly on the purpose of the study. We suggest cultivated protoplasts as a suitable experimental system *(4,5)* for screening and localization of yet unknown proteins using various antibodies. On the other hand, treatment of fixed cells with glass beads is useful when cells with resistant cell walls are studied (e.g., nocodazole-affected, *cdc 24* or *ras* mutant cells of *S. cerevisiae*).

Phalloidin (M_W 800)—a toxin obtained from the mushroom *Amanita phalloides*—has a high affinity for polymeric F-actin but not for monomeric G-actin. Fluorochrome-tagged phalloidin *(6)* is therefore a unique compound that is generally used to monitor the distribution of F-actin in various eukaryotic cells. Moreover, in yeast cells, this labeling can be performed directly without cell wall removal. Since the phalloidin-staining pattern coincides with actin localization by immunofluorescence; labeling with fluorochrome-tagged phalloidin is generally preferred in most studies on yeast *(7)*. Our method of choice, which is a modified version of the procedure proposed by Heath *(8)*, works well with wild-type cells and various mutants.

2. Materials

2.1. Fluorochrome Staining Different Cell Components

1. Carnoy's fixative: Mix absolute ethanol with glacial acetic acid (1:3) and store at 4°C in a tightly sealed bottle.
2. Acridine orange (excitation 400–490 nm, emission 510–530): Prepare 1 mg/mL stock solution of acridine orange in $0.2M$ acetate buffer, pH 4.5. Keep the stain in a dark bottle at 4°C and filter before use; working concentrations range from 0.01–1 mg/mL.
3. $0.2M$ acetate buffer, pH 4.5.
4. Mithramycin (excitation 400–500 nm, emission 490–530 nm): dissolve 0.6 mg/mL of mithramycin in aqueous 22.5 mM $MgCl_2$.
5. DAPI (4',6-diamidino-2-phenylindole) (excitation 340–370 nm, emission 420–530 nm): Store as 1 mg/mL stock solution of DAPI in H_2O at –20°C until needed; working concentrations of DAPI are indicated in the protocols. With fixed protoplasts, dissolve 1 µg/mL DAPI in the following solution: 2 mM Tris-HCl, pH 7.6, $0.25M$ sucrose, 1 mM EDTA, 1 mM $MgCl_2$, 0.1 mM $ZnSO_4$ 0.4 mM Ca Cl_2, 1.5% (v/v) 2 mercaptoethanol; store at –20°C.
6. Use aqueous rhodamine 123, berberine sulfate, or coriphospin solutions at concentrations in the range 0.25–20 µg/mL.

7. Calcofluor (excitation 340–370 nm, emission 420–530 nm): Prepare 1 mg/mL stock solution of Calcofluor (American Cyanamid Co.) in H_2O and keep in dark at 4°C (lasts for 6 mo). Filter before use; working concentrations range from 0.01–1 µg/mL.
8. Primulin (excitation 400–500 nm, emission 490-530 nm): Prepare stock solution of primulin in H_2O as in step 7.

2.2. Staining with Fluorochrome-Labeled Antibodies

1. Prepare PEM buffer as a fourfold concentrated stock solution: $0.4M$ PIPES, 20 mM EGTA, 20 mM $MgCl_2$, pH 6.9 (KOH). Store at 4°C up to 1 mo.
2. Prepare 150 mM EGTA, pH 7 (KOH).
3. 7.4% (w/v) formaldehyde: Dissolve 2.96 g of paraformaldehyde in 10 mL of H_2O at 60°C in the presence of 10 µL of $1N$ KOH; mix the solution with 20 mL of PEM buffer stock solution, filter through a membrane filter (0.45 µm), and adjust the volume to 40 mL. Store at 4°C up to 1 wk.
4. Prepare $0.1M$ potassium phosphate buffer, pH 6.5.
5. Prepare 2% (w/v) bovine serum albumin (BSA) in PEM buffer, filter, and store in aliquots at −20°C.
6. Prepare 1% (v/v) Triton X-100 in PEM buffer; store at 4°C (up to 1 wk).
7. Snail digestive juice: Clarify fresh crude snail gastric juice from *Helix pomatia* by centrifugation and store in aliquots at −20°C. Prepare the digestive mixture immediately before use. It contains 1 vol of snail juice and 4 vols of $0.1M$ potassium phosphate, pH 6.5. Related enzyme preparations are commercially available (e.g., β-glucuronidase from *H. pomatia*, Boehringer Mannheim, Mannheim, Germany; Zymolyase-20T, Seikagaku Corp.).
8. Phosphate buffered saline (PBS): Dissolve 8 g NaCl, 0.2 g KCl, 1.14 g $Na_2HPO_4 \cdot H_2O$, 0.2 g KH_2PO_4 in 1 L of deionized H_2O, pH 7.4 (NaOH).
9. Phenylmethyl sulfonylfluoride (PMSF): Prepare stock solution of 87.5 mg PMSF in 5 mL methanol. Store at 4°C. **Caution:** PMSF is highly toxic.
10. Pepstatin A: Prepare stock solution 1 mg/mL in methanol. Store at −20°C.
11. Aprotinin: Prepare stock solution 1 mg/mL in H_2O.
12. Leupeptin: Prepare stock solution 1 mg/mL in H_2O. Store inhibitors 11 and 12 in aliquots at −70°C. Add inhibitors 9, 10, 11, and 12 to cell suspension in ratio 1:100 vol:vol in case of each inhibitor.
13. TU-01 mouse monoclonal antibody (Sanbio, Serva, Heidelberg, Germany) raised against porcine brain d-tubulin *(9)* is a primary antibody of IgG_1 subclass. Use as an ascitic fluid fiftyfold diluted with PEM buffer.
14. DM 1A mouse monoclonal antibody against α-tubulin (Sigma).

15. MA-01 mouse monoclonal antibody recognizing porcine brain MAP 2 *(10)* is a primary antibody of IgG$_1$ subclass purified from the ascitic fluid by DEAE-chromatography following ammonium sulfate precipitation. Use at final concentration 10 µg/mL.
16. Dilute swine anti-mouse antibody conjugated with fluorescein isothiocyanate (SwAM/FITC) (Sevac, Prague, Czech Republic) 1:10 (v/v) with 1% (w/v) BSA in PEM buffer or PBS and membrane-filtered. Store all antibodies in aliquots at –20°C until used.
17. Mounting medium: Prepare 0.1% (w/v) *p*-phenylenediamine (Sigma) in 100 mM Tris-HCl, 100 mM NaCl, 5 mM MgCl$_2$, pH 9.5 and 0.5 µg/mL of DAPI (*see* Section 2.1., step 5). The solution deteriorates. Prepare just before use.

2.3. Staining of Actin with Fluorochrome-Labeled Phalloidin

1. PRE buffer: 0.06M PIPES, 30 mM EGTA, 5 mM MgCl$_2$, pH 7 (KOH); store at 4°C up to 2 wk.
2. PME buffer: 0.06M PIPES, 15 mM EGTA, 5 mM MgCl$_2$, pH 7 (KOH); store at 4°C up to 2 wk.
3. FIX buffer: 0.06M PIPES, 15 mM EGTA 5 mM MgCl$_2$, 7.4% (w/v) formaldehyde—dilute 38% (w/v) formaldehyde, pH 7 (KOH); prepare immediately before use.
4. Rh-palloidin: Dissolve 400 µg/mL of TRITC-labeled phalloidin (Sigma P-5157, Mol. Probes R-415) in methanol. Store in 10 µL aliquots at –20°C; working solution of Rh-phalloidin (16 µg/mL)—dilute 10 µL of stock solution in 250 µL PME buffer; store in the dark at 4°C up to 2 wk.
5. Mounting medium: *See* Section 2.2., step 17.

3. Methods

3.1. Fluorochrome Staining Different Cell Components

3.1.1. Fluorochroming Nuclei with Acridine Orange

1. Air dry cells on slides.
2. Rinse the slides in Carnoy's fixative for 30 min.
3. Wash briefly in ethanol and air dry.
4. Immerse in 0.2M acetate buffer, pH 4.5.
5. Stain with buffered acridine orange solution for 20–120 min.
6. Rinse in acetate buffer for 2–25 min.
7. Mount a coverslip and observe.

Acridine orange does not stain DNA quantitatively and requires a careful control of pH. The procedure produces a green or yellow fluorescence of the nucleus and a brilliant orange fluorescence of the cytoplasm.

3.1.2. Fluorochroming Nuclei with Mithramycin

1. Fix cells in Carnoy's fixative for 1 h, then make slides by allowing a centrifuged sample of the fixed material to dry on a microscopic slide.
2. Stain with 2 vols of mithramycin solution for 20 min.
3. Counterstain briefly with acridine orange (1 µg/mL—see Section 2.1., step 2) if desired.
4. Apply a coverslip without rinsing and observe immediately or within 24 h if stored in dark.

The mithramycin procedure produces a bright yellow-green fluorescence of the nucleolus; the unstained area coincide with the nucleolus; discrimination of division figures is mostly obscured by the condensed chromatin.

3.1.3. Fluorochroming Nuclei with DAPI in Living Cells

1. Mix the yeast suspension with 3 vol of ethanol or resuspend pelleted cells in 70% (v/v) ethanol for 30 min.
2. Resuspend in a solution of 0.1–0.5 µg DAPI/mL.
3. Mount the stained specimen on a microscope slide, apply a coverslip, and observe immediately.

With high magnification epifluorescence microscopy, DAPI techniques overcome the difficulty of cytological resolution of the yeast nucleus. Chromosomes display a strong blue-white fluorescence; spots of weaker fluorescence localized in the cytoplasm are mitochondrial nucleoids. DAPI can also be used in combination with ethidium bromide, emitting a pronounced orange color of the nucleus and a pale orange fluorescence of the cytoplasm. Remarkable views of well-spread chromosomes during meiosis have been published (11).

3.1.4. Fluorochroming Meiotic Nuclei in Fixed Protoplasts

1. Prepare protoplast according to Chapter 5, by Curran and Bugeja.
2. Fix in Carnoy's fixative for at least 5 min.
3. Centrifuge and resuspend in Carnoy's fixative; mount a drop of the suspension on a slide.
4. Cool the slide to 0°C and then heat in flame for 3 s; allow to air dry.
5. Immerse the specimen in the buffered DAPI solution.
6. Mount using a coverslip and observe immediately or within 24 h if stored in the dark at 4°C.

3.1.5. Fluorochroming Mitochondria in Fixed Cells (see Note 1)

1. Centrifuge (e.g., 500g, 5 min) cells free from nutrient medium.
2. Fix with 70% (v/v) ethanol for 30 min.
3. Centrifuge, wash in H_2O, and centrifuge again.
4. Mix the suspension with a drop of aqueous DAPI staining solution at a concentration ranging from 0.1–0.5 µg/mL.
5. Apply a coverslip and observe immediately.

3.1.6. Fluorochroming the Cell Wall

1. Centrifuge (e.g., 500g, 5 min) the yeast sample free of the nutrient medium.
2. Mix a small amount of the pellet with a drop of Calcofluor or primulin staining solution in a concentration ranging from 0.1–0.5 µg/mL.
3. Apply a coverslip and observe (Fig. 1).

Application of Calcofluor or primulin includes determination of cell viability, characterization of the porous structure of the cell wall, quantification of bud scars and identification of regional cell wall growth (*see* Note 2 and ref. *12* for reviews).

3.2. Staining with Fluorochrome-Labeled Antibodies

3.2.1. Fixation

The fixation protocol detailed here provides reproducible staining of the microtubular system (e.g., with antibody DM 1A and/or TU-01) and proteolysis-sensitive microtubule-interacting epitopes recognized by the monoclonal antibody MA-01 *(13)* in various yeast cell and protoplasts (*see* Note 3).

1. Add 1 vol of 150 mM EGTA, pH 7.0 (KOH) to 4 vols of a suspension of cells (grown to a density of approx 10^8 cells/mL) in culture medium (YPG). For protoplasts, the medium is stabilized with 1M sorbitol.
2. Add inhibitors PMSF, leupeptin, pepstatin, and aprotinin.
3. After 5 min incubation apply 5 vols of 7.4% (w/v) formaldehyde in twice concentrated PEM buffer to the cell suspension.
4. Fix for 2 h with shaking on a reciprocal shaker.
5. Collect the fixed material by centrifugation (e.g., 1000g, 4 min).
6. Wash three times with PEM buffer.

3.2.2. Permeabilization

The procedure generally used in our laboratory to permeabilize yeast cells is as follows:

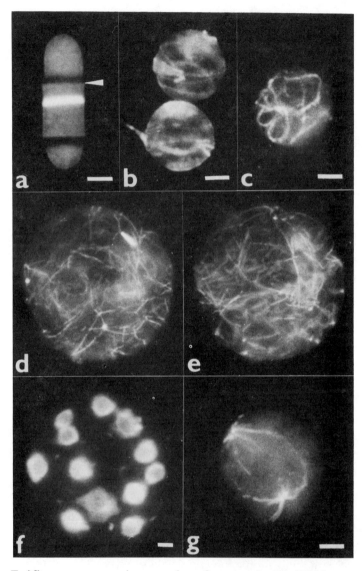

Fig.1. Epifluorescence micrographs of yeasts I. (a) Wild-type cell of *Schizosaccharomyces pombe* stained with Calcofluor (*see* Section 3.1.6.), arrow indicates one division scar; (b) The ras 1⁻ terminal phenotype of *S. pombe* stained with Calcofluor (*see* Section 3.1.6.), note the stripping off of wall rings; (c) *ras 1⁻* terminal phenotype of *S. pombe* visualized with Rh-phalloidin (*see* Section 3.3.1.); (d,e) Protoplast of *Endomyces magnusii* prepared according to Section 3.2.2., steps 1a–3 and stained with the monoclonal antibody DM 1A (*see* Section 3.2.3.), two planes of focusing throughout the protoplast are visu-

1. Centrifugally wash the fixed cells in Eppendorf tubes (500g × 5 min) three times with 0.1M potassium phosphate, pH 6.5. Perform **either** a or b.
   a. For enzymatic cell wall removal resuspend the pellet in 200 µL of the digestive mixture. Incubate the fixed cells with the digestive mixture at room temperature with gentle agitation for 15–30 min. Check the digestion of the cell wall with phase-contrast microscope. Centrifuge wall-less cells free from the digestive mixture and wash carefully twice with 0.1M potassium phosphate buffer, pH 6.5, and resuspend in PEM buffer.
   b. Mix the suspension of cells in the fixative with glass beads 0.5 mm in diameter (Sigma G-9268) to form a thick slurry. Rupture the cell wall by shaking using a vibration homogenizer for 30 s. The treatment results in stripping off cell walls from a substantial part of the cells that retain their original shape. Most of the success depends on the ratio of cells to vol of glass beads (e.g., 1:3 v/v) and on the duration of the disintegration step, which should be estimated empirically. Wash the cells from the glass beads, centrifuge, and wash two times with PEM buffer.
2. Permeabilize the specimen with 200 µL of 1% (v/v) Triton X-100 in PEM for 2 min (*see* Note 4).
3. Wash carefully (two times) with PEM buffer by centrifugation at low rpm.

3.2.3. Incubation with Antibodies

We prefer to work with a suspension, since the material can be better washed and the number of specimens prepared for observation is almost unlimited (*see* Note 5). The following protocol works well using various polyclonal and/or monoclonal antibodies.

1. Resuspend the pellet in 1 vol 2% (w/v) BSA in PEM buffer (*see* Note 6).
2. Incubate for 20 min at room temperature.
3. Add an equal vol of PEM buffer-diluted primary antibody (e.g., TU-01, DM 1A, or MA-01 diluted with PEM buffer as indicated in Section 2.2.). Generally, the dilution depends on the quality of the antibody.

---

alized; **(f)** Complementary staining of DNA (*see* Section 3.1.3.) of the same protoplast as (d) and (e); **(g)** Terminal phenotype of ts *cdc 24* mutant of *Saccharomyces cerevisiae* labeled with the monoclonal antitubulin antibody TU-01 (*see* Section 3.2.3.), the cell displays two duplicated SPBs connected with microtubules. The cell wall is removed mechanically with glass beads (*see* Section 3.2.2., steps 1b–3).

4. Incubate 60 min at room temperature.
5. Wash the cells with PEM buffer by centrifugation.
6. Apply a secondary antibody diluted in 2% (w/v) BSA in PEM buffer. Dilution depends on the quality of the antibody. Follow recommendations described in company notes. For SwAM/FITC conjugate (SEVAC, Prague, Czech Republic), add 100 µL of 10× diluted antibody.
7. Incubate 45 min at room temperature.
8. After removal of the antibody by centrifugation, wash the cells twice with PEM buffer.
9. Mix the cells with an equal volume of the mounting medium, cover with a coverslip, drain the excess medium, and press well (see Note 7).
10. When only a tiny amount of the primary antibody is available, see Note 8.
11. Examine the specimen by epifluorescence using standard FITC filter sets (excitation 490–495 nm, emission 523 nm) or standard TRIC sets (excitation 540–552 nm, emission 570 nm) (Fig. 2).

3.2.4. Double Immunolabeling

To obtain complementary pictures of more cellular components simultaneously in the same cell, multiple-label experiments are advisable. Usually, two primary antibodies with different specificities are used in two subsequent staining procedures. These antibodies should be raised in different animal species. Frequently, a combination of rabbit and mouse antibodies is used in these multiple-labeling experiments. The antigens of interest are rendered visible by appropriate secondary antibodies tagged with fluorochromes having different spectral properties (typically FITC and TRITC).

1. Apply primary and secondary antibodies according to the protocol detailed above (e.g., MA-01 mouse monoclonal antibody followed by FITC-conjugated swine antimouse antibody).
2. Apply the second primary antibody (e.g., rabbit polyclonal antibody) to the washed and pelleted, or coverslip-mounted cells and incubate for 60 min at room temperature.
3. Wash the cells carefully with PEM buffer or PBS, apply the conjugate (in this case, e.g., TRITC-labeled goat antirabbit antibody) and incubate for 45 min at room temperature.
4. After washing, mount the cells in the mounting medium and examine by epifluorescence. Complementary pictures of intracellular structures of interest are obtained using standard fluorescein or rhodamine filter sets.

3.3. Staining of Actin with Fluorochrome-Labeled Phalloidin

3.3.1. Simple Labeling

1. Collect cells or protoplasts by filtration (0.45 µm membrane) or centrifugation (500g, 10 min).
2. Resuspend the sample in PRE buffer (for protoplasts the buffer is osmotically stabilized with $1M$ sorbitol).
3. Incubate for 5 min at room temperature with gentle agitation.
4. Add an equal vol of FIX buffer (protoplasts require slow addition drop by drop).
5. Fix 5–30 min at room temperature with gentle agitation (see Note 9).
6. Wash with PME buffer by filtration or centrifugation.
7. Resuspend the cells in 20 µL of PME buffer.
8. Add 5 µL of working solution of Rh-phalloidin.
9. Stain for 5–10 min in the dark.
10. Wash with PME buffer.
11. Mount the cells in the mounting medium and examine by epifluorescence using standard rhodamine filter set (see Note 10).

3.3.2. Double Labeling

Rhodamine-tagged phalloidin can be used in conjunction with immunofluorescence to reveal colocalization of F-actin with other cellular components. The procedure of Rh-phalloidin staining of washed cells after immunolabeling with a fluorescein-conjugated secondary antibody follows the protocol (see Section 3.3.1.) mentioned earlier.

4. Notes

1. Mitochondria can be localized in vivo using rhodamine 123, berberine sulfate, or coriphospin; the classic nonfluorescence method uses Janus green (see Chapter 38 by Streiblova and Hasek).
2. Calcofluor and primulin are substances that display an intrinsic absorption anisotropy, on the basis of which one can characterize the spatial disposition of microfibrils in the native cell wall. The affinity to the microfibrillar cell wall polymers is nonspecific, although they have often been used incorrectly as a specific cytochemical test for chitin. In addition, Calcofluor and primulin are used routinely for distinguishing between living and nonliving yeast cells, since the plasma membrane of dead cells is permeable to the colloidal solution of the dyes. Although the staining procedures appear very simple, a number of factors (e.g., cell damage by irradiation during long photographic exposures or by mechanical pressure on the coverslip during preparation) must be taken into account when interpreting the fluorescence image (see Section 3.1.6.).

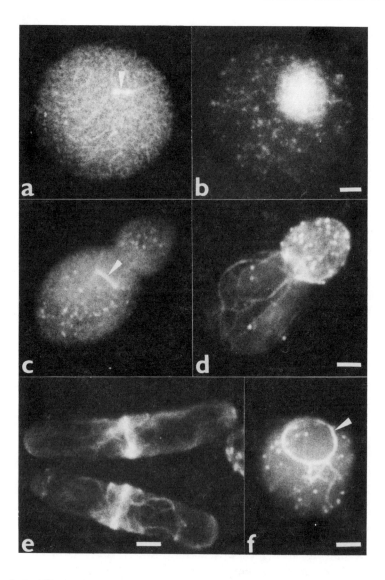

Fig. 2. Epifluorescence micrographs of yeasts II. (a) Protoplast of *Saccharomyces cerevisiae* labeled with the MA-01 antibody (*see* Section 3.2.3.), arrow indicates colocalization with the mitotic spindle; (b) DAPI staining (*see* Section 3.1.3.) of the same protoplast, note staining of both the nuclear and mitochondrial DNA; (c) Wild-type *S. cerevisiae* cell stained with the MA-01 monoclonal antibody recognizing porcine brain *MAP 2* (*see* Section 3.2.3.). Arrow indicates colocalization of the protein with the mitotic spindle; (d) The same cell as in (c) complementarily stained with Rh-phalloidin (*see* Section

3. Prior to formaldehyde fixation, we incorporate a short preincubation of cells with EGTA into the protocol. We assume that this treatment reduces the effect of the uncontrolled release of free calcium ions in response to formaldehyde permeabilization of the cytoplasm. This may help to diminish the alteration of intracellular structures (*see* Section 3.2.1.).
4. If there is a possibility that the cytoplasmic antigen could be washed out from the structures with Triton X-100, we prefer to permeabilize coverslip-dried fixed cells with cold (–20°C) methanol for 20 min (*see* Section 3.2.2.).
5. It is indispensable to have various controls concerning the incubation with antibodies. We recommend the following:
   a. To examine the reaction of the preimmune serum;
   b. To check the cells after a direct labeling with the conjugates alone;
   c. To compare the reaction of several primary antibodies of the same animal origin; and
   d. If possible, check the reaction of the primary antibody preabsorbed with the antigen (*see* Section 3.2.3.).
6. According to our experience, the preincubation of cells with BSA before the application of the primary antibody reduces substantially nonspecific interaction of the antibody in the cytoplasm. Additionally, the preincubation solution can be supplemented with 2% (w/v) lysine. Sometimes, the background fluorescence caused by nonspecifically bound antibodies may be washed out from the cell with 0.1% (v/v) Triton X-100 in PEM buffer (*see* Section 3.2.3.).
7. To diminish the fading of fluorescein and/or rhodamine fluorescence in response to the effect of the excitation light, *p*-phenylenediamine is used in the mounting medium. Similar results can be achieved with the use of propylgallate (*see* Section 3.2.3.).
8. When only a tiny amount of the primary antibody is available, we recommend working with cells or protoplasts simply dried on well-washed coverslips. After preincubation with 2% (w/v) BSA in PEM buffer, a small volume (1 µL) of a sufficiently diluted antibody is applied to the cell side of the coverslip. All following steps are performed in PBS. For incubations of coverslip-mounted cells with antibodies, a moist chamber is highly recommended (*see* Section 3.2.3.).

3.3.2.); **(e)** Cell of *Schizosaccharomyces japonicus* var. *versatilis* labeled with Rh-phalloidin (*see* Section 3.3.1.), F-actin accumulates at the cell equator prior to septum formation; **(f)** protoplast of *S. japonicus* var *versatilis* cultivated for 30 min in osmotically stabilized nutrient medium, arrow indicates the presence of the F-actin contractile ring (*see* Section 3.3.1.).

9. When the cells are exposed to the fixative for more than 30 min, cell wall removal and permeabilization with 1% (v/v) Triton X-100 is recommended in order to reduce the background fluorescence (see Section 3.3.1.).
10. F-actin patterns visualized in yeast consist of cortical patches and cytoplasmic fibers. The patches are mainly localized in regions of the cell wall growth (see Section 3.3.1.).
11. Fluorescent staining has recently been exploited in a rapid assay for killer toxin activity *(14)*.

Acknowledgments

We are indebted to J. Jochová and A. Pichová from the Laboratory of Cell Reproduction for providing us with their epifluorescence micrographs. Photographs in Fig. 1 (a–c) were taken by A. Pichová, Fig. 1 (d–g), and Fig. 2 (a–d) were taken by J. Hašek, and Fig. 2 (e,f) by J. Jochová. Bars in all micrographs, 3 µm.

References

1. Wang, Y. and Taylor, D. L., eds. (1989) *Fluorescence Microscopy in Living Cells in Culture A,B*. Academic, San Diego, CA.
2. Haughland, R. P. (1989) *Handbook of Fluorescent Probes and Research Chemicals*, Catalogue of Molecular Probes, Eugene, OR.
3. Pringle, J. R., Adams, A. E. M., Drubin, D. G., and Haarer, B. K. (1991) Immunofluorescence methods for yeast. *Meth. Enzymol.* **194**, 565–602.
4. Hašek, J., Svobodová, J., and Streiblová, E. (1986) Immunofluorescence of the microtubular skeleton in growing and drug-treated yeast protoplasts. *Eur. J. Cell. Biol.* **41**, 150–156.
5. Jochová, J., Rupes, I., Hašek, J., and Streiblová, E. (1989) Reassembly of microtubules in protoplasts of *Saccharomyces cerevisiae* after nocodazole-induced depolymerization. *Protoplasma* **151**, 98–105.
6. Wieland, T. (1986) *Peptides in Poisonous Amanita Mushrooms*. Springer Verlag, Heidelberg, Germany.
7. Kilmartin, J. V. and Adams, A. E. M. (1984) Structural rearrangements of tubulin and actin during the cell cycle of the yeast *Saccharomyces*. *J. Cell Biol.* **98**, 922–933.
8. Heath, I. B. (1987) Preservation of a labile cortical array of actin filaments in growing hyphal tips of the fungus *Saprolegnia ferax*. *Eur. J. Cell Biol.* **44**, 10–16.
9. Viklický, V., Dráber, Pa., Hašek, J., and Bártek, J. (1982) Production and characterization of a monoclonal antibutulin antibody. *Cell Biol. Internat. Rep.* **6**, 726–731.
10. Dráberová, E., Dráber, P., and Viklický, V. (1986) Cellar distribution of a protein related to neuronal microtubule-associated protein MAP2 in Leydig cells. *Cell Biol. Internat. Rep.* **10**, 881–890.
11. Kuroiwa, T., Kojima, H., Hiyakawa, I., and Sando, N. (1984) Meiotic karyotype of the yeast *Saccharomyces cerevisiae*. *Exp. Cell Res.* **153**, 259–265.
12. Streiblová, E. (1984) The yeast cell wall—a marker system for cell cycle controls, in *The Microbial Cell Cycle* (Nurse, P. and Streiblová, E., eds.), CRC, Boca Raton, FL.

13. Hašek, J., Jochová, J., Dráber, Pa., Streiblová, E., and Viklický, V. (1992) Localization of a 210 kDa microtubule-interacting protein in yeast *Saccharomyces cerevisiae. Can. J. Microbiol.* **38,** 149–152.
14. Kurzweila, H. and Sigler, K. (1993) Fluorescent staining with bromocresol purple: A rapid method for determining yeast cell dead count developed as an assay of killer toxin activity. *Yeast* **9,** 1207–1212.

CHAPTER 40

Immunoelectron Microscopy

Evert-Jan van Tuinen

1. Introduction

The immunolocalization of proteins on sections of biological materials that have been prepared for electron microscopy, requires the preservation of both antigenic and morphological structures. In this chapter, two techniques that fulfill these requirements, often difficult to reconcile, but that have been used extensively for mammalian cells, will be described in their adaptation for yeast cells: the progressive low temperature embedding technique in Lowicryl introduced by Carlemalm et al. *(1)* and cryo-ultramicrotomy as described by Tokuyasu *(2)*. The immunolocalization technique used in this chapter is based on the protein-A gold technique described by Roth et al. *(3)* and Geuze et al. *(23)*.

Embedding of aldehyde-fixed yeast cells, with an intact cell wall or sectioning of these cells with the procedures established for cryo-ultramicrotomy, has always been cumbersome because of the low permeability of the cell wall toward the media commonly used for electron microscopy. A few years ago, it was found out empirically by van Tuinen and Riezman that this cell wall effect could be eliminated by treating the cells with sodium meta-periodate after fixation *(5)*. This adaptation has made it possible to carry out immunoelectron microscopy on whole yeast cells on an almost routine basis and to displace the more delicate aspects of the technique to the actual interactions of the embedded material with the antibodies.

Immunoelectron microscopy can be carried out in a laboratory with basic electron microscopy facilities and experience. For the Lowicryl

From: *Methods in Molecular Biology, Vol. 53: Yeast Protocols*
Edited by: I. Evans Humana Press Inc., Totowa, NJ

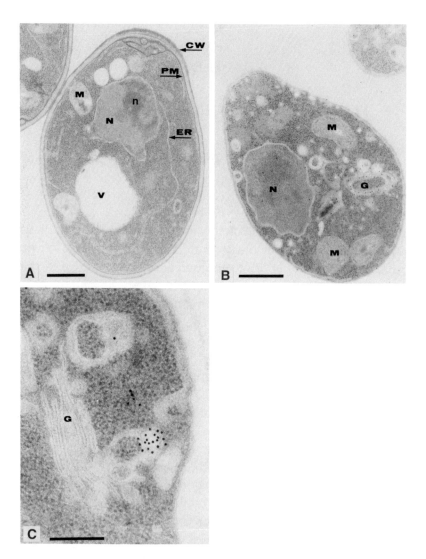

Fig. 1. (A) Morphology of an X2180 1-A yeast cell grown in medium containing 1% yeast extract, 2% peptone, and 2% glucose, fixed and embedded in Lowicryl HM-20 as described in Section 3.1.1. and 3.1.2. and stained with 3% uranyl acetate and lead acetate. Bar, 0.5 µm. (B) Morphology of AFM 67 11-B, sec-7, grown at 37°C for 1½ h in medium containing 1% yeast extract, 2% peptone, and 0.2% glucose. Invertase can be seen in the Golgi apparatus and is detected with 8-nm Protein-A-gold particles. Golgi structures accumulate specifically in this mutant. Counterstaining: 3% uranyl acetate. Bar, 0.5 µm. (C)

technique, the most important tools are a household freezer, able to achieve −35°C, a UV lamp producing an emission of 360 nm, and an aluminum block with holes drilled to accommodate the tubes used during dehydration. The Lowicryl procedure has the advantage that the sectioning of the embedded material is straightforward and very similar to the routinely used epoxy-based resins. With a possible exception made for mitochondria, the ultrastructural preservation of cell organelles and membranes and the immunoreactivity of the embedded materials is excellent (Fig. 1). Cryo-ultramicrotomy is very attractive when an ultramicrotome equipped with a freeze-sectioning device is available. The technique and the subsequent immunolabeling of ultrathin frozen sections of yeast cells, as described in this section, is based to a large extent on the method introduced for mammalian cells by Tokuyasu *(2,6)* and adapted by Griffith et al. *(7,8)*.

Immunolabeling of frozen sections is thought to be more sensitive than of resin-embedded material and can be an advantage when the results of a labeling experiment need to be known rapidly (the whole procedure takes less than 2 d). In comparison to the Lowicryl embedding, the preservation of mitochondrial ultrastructure, in particular, is superior (Fig. 2).

2. Materials

Note: All aqueous solutions are made using the cleanest possible double-distilled water.

2.1. Fixation

1. Fixative: 3% w/v formaldehyde-0.5% w/v glutaraldehyde in 0.1M phosphate buffer, pH 7.2. The formaldehyde is prepared by dissolving 3 g of paraformaldehyde in 50 mL of water at 60°C. The addition of a few µL of 1M NaOH helps to clear the solution, after which 50 mL of 0.2M phosphate buffer, pH 7.2, are added.

   After cooling down to room temperature, 2 mL of 25% w/v glutaraldehyde are finally dissolved in the formaldehyde solution.

Higher magnification of a Golgi apparatus as can be seen in AFM 67 11-B cells, grown as described under (B). The gray irregularly shaped granules that can be seen in the background, represent ribosomes. Bar, 0.2. µm. CW, cell wall; PM, plasma membrane; V, vacuole; N, nucleus; n, nucleolus; ER, endoplasmic reticulum; M, mitochondrion; G, Golgi apparatus.

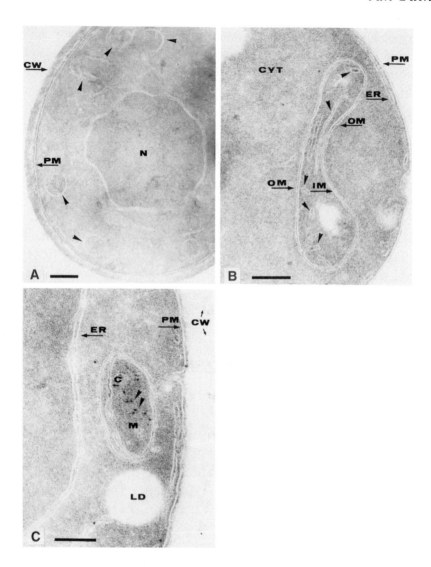

Fig. 2. (A) Morphology of Whole yeast cells, grown as described in Fig. 1A, fixed and processed for cryo-ultramicrotomy as described in Sections 3.1.3. and 3.2.3. In contrast to material embedded in Lowicryl, the granular, ribosomal staining pattern is missing but mitochondrial membranes in particular are well preserved. The arrowheads point at mitochondrial membranes. (B) High magnification of a mitochondrion. The outer and inner membrane as well as mitochondrial cristae can be discriminated. The arrowheads point at mitochondrial cristae. (C) Immunolabeling using an antibody against Mn-superoxide

The formaldehyde–glutaraldehyde fixative is prepared freshly for each experiment.
2. PBS: $0.01M$ phosphate buffer, pH 7.2 (see item 1), $0.15M$ NaCl.
3. Plastic, cylindrical Eppendorf tubes of 2.2 mL. These tubes are used throughout the whole dehydration, embedding procedure.
4. 1% w/v sodium meta-periodate is freshly prepared in water.
5. 1% w/v ammonium chloride is dissolved in PBS, and does not need to be prepared freshly.

2.2. Dehydration, Embedding, Polymerization, and Sectioning in Lowicryl

1. Agarose that is used to solidify the cell pellet, has a gelling point below 30°C (e.g., Sigma [St. Louis, MO] cat. no. A 4018).
2. For the dehydration at progressively lower temperatures, it is useful to have an aluminum block of approx $6 \times 6 \times 12$ cm with holes drilled to fit the Eppendorf tubes. The block can be insulated with a layer of plastic foam (e.g., Styrofoam) and is placed in a freezer set to the required temperature.
3. Lowicryl HM 20, produced by Chemische Werke Lowi (Waldkraiburg, Germany) can be purchased from suppliers of electron microscope material. The Lowicryl HM 20 mixture, which gives good results, has the following components: 3 g crosslinker D, 17 g monomer E, 0.1 g initiator C.
4. EMBRA nickel grids of 400 mesh (Brandsma, Hilversum, The Netherlands) (see Note 3).

Note: Lowicryl mixtures may cause eczema. Inhalation and skin contact should be avoided (see precautions taken in Methods). Additional information about embedding in Lowicryl can be found in Roth (9) and in the instruction booklet supplied with the resin.

5. Gelatin capsules for polymerization are kept in a dessicator until use.
6. An aluminum polymerization chamber of approx 20 cm wide, 40 cm high, and 50 cm long, with the two small sides left open and that is illustrated in the instruction booklet. Two UV lights of 360 nm wavelength radiation, are installed against the ceiling inside the chamber, with a right-angle reflector in order to provide indirect UV light. The chamber is installed

---

dismutase. A specific signal is seen in the mitochondrial matrix (arrowheads). CW, cell wall; PM, plasma membrane; N, nucleus; LD, lipid droplet; ER, endoplasmic reticulum; M, mitochondrion; OM, outer mitochondrial membrane; IM, inner mitochondrial membrane; C, mitochondrial cristae; M, mitochondrial matrix. Bar, 0.2 μm.

inside the freezer, which is set at –35°C, during the 24–48-h polymerization period. It is very convenient to have the coil, which triggers the UV light and produces a considerable amount of heat, mounted in a separate box and installed outside the freezer.
7. Sectioning is carried out with a glass or a diamond knife.

2.3. Cryo-Ultramicrotomy

1. Ultramicrotome equipped with a freezing device is necessary, such as that supplied by Reichert-Jung or LKB.
2. 10% w/v melted gelatin (microbiological quality).
3. 2.3M sucrose.
4. Liquid nitrogen.
5. Glass knives for sectioning (which are prepared as described by Griffith et al. *[10]*).
6. Fine hair or eyelash, and platinum loop (ca. 2-mm diameter).
7. 150 mesh EMBRA nickel grids coated with a film of 1% w/v formvar.
8. Petri dish containing a layer of 2% w/v gelatin.

2.4. Immunodecoration and Contrasting

1. PBS (*see* Section 2.1., step 2) BSA: PBS with 1% w/v bovine serum albumin (e.g., Sigma cat. A 2153) and 0.02% w/v sodium azide. This buffer is made freshly every 2–3 d and is filtered through a filter of 0.2-μm pore size, before use.
2. PBS-glycine: PBS (*see* Section 2.1., step 2) containing 0.15% w/v glycine.
3. Protein-A colloidal gold can be purchased commercially from many sources (e.g., Janssen Life Sciences, Beerse, Belgium; Amersham, Braunschweig, Germany; Bio-Rad, Hercules, CA; and Sigma). It is also possible to prepare monodisperse colloidal gold particles of various sizes and to stabilize them with protein-A in the laboratory, according to procedures described by Slot and colleagues *(11,12)*.
4. 3% w/v uranyl acetate dissolved in water (for contrasting Lowicryl sections). **Caution:** This solution is toxic and radioactive and should be handled and discarded accordingly.
5. Methyl-cellulose uranyl acetate (for contrasting/embedding frozen sections): Add 2% w/v of methylcellulose (25 cp, e.g., Sigma cat. no. M6385) to water at 95°C, then cool the mixture on ice, and stir for 24 h at 4°C. Keep the solution for several days at this temperature, centrifuge at 55,000 rpm in Beckman 60 Ti rotor for 1 h, then store the supernatant in a refrigerator.

Before each experiment, carefully remove an aliquot from the surface and mix with 3% w/v uranyl acetate (pH 4.0) in the following proportions: 1 vol of uranyl acetate to 9 vol of 2% w/v methylcellulose (25 cp). Keep at 4°C.

3. Method

3.1. Sample Preparation

3.1.1. Growth and Fixation (see Notes 1, 2, and 3)

1. Grow the cell culture overnight and centrifuge at 3000g for 5 min at room temperature (30 mL of a culture of $A_{600} = 1$ is enough material to handle with ease during the whole embedding procedure).
2. Remove the culture supernatant, add 10 mL of fixative to the cell pellet, and carefully resuspend it.
3. After 30 min, centrifuge the cells again at 3000g for 5 min, discard the fixative in a receptacle for organic waste, and add fresh fixative to the pellet. Resuspend the cells, and allow fixation to proceed for a least 3 h at room temperature or overnight at 4°C.
4. Wash the cells three times centrifugally (3000g) in PBS. Carry out the last washing step in an Eppendorf (micro) centrifuge tube. A brief centrifugation of maximum speed (ca. 16,000g) should be sufficient in a microfuge not adjustable for speed, otherwise 3000g × 5 min. After pelleting, the volume of the pellet should be approx 30 µL.
5. Resuspend the cells in 1 mL of freshly prepared 1% w/v sodium metaperiodate in water and incubate for 15 min (*see* Note 4).
6. Wash the cell three times with PBS, in the microfuge and incubate for 15 min in 1% w/v ammonium chloride in PBS.
7. Finally, wash once, in the microfuge, with PBS.

3.1.2. Dehydration and Embedding in Lowicryl HM-20

1. Pellet the cell suspension in an Eppendorf centrifuge tube and to a 30-µL pellet of cells as prepared in Section 3.1.1. Add an equal volume of 1% low melting agarose and mix with the cells.
2. Centrifuge the cells a few seconds (16,000g) and allow the agarose to solidify by putting the tube on ice for a few minutes.
3. After solidification, remove the agarose-embedded pellet from the tube. This can be best achieved by cutting the tube with a sharp razor blade just above the pellet and lifting it out with a small spatula.
4. Section the lower part of the agarose pellet with a razor blade into cubes of 1 mm^3 and transfer two to three of these to a cylindrical Eppendorf tube on ice containing 1 mL of 30% v/v ethanol. It is good practice to shake each individual tube very gently, end over end from time to time during the whole dehydration-embedding procedure, so ensuring that the agarose blocks remain immersed in the medium.

5. After 15 min, remove the 30% v/v ethanol solution and replace by 50% v/v ethanol that has been precooled in a freezer to –20°C. Keep the cells in this mixture for *30 min* on ice: *All subsequent steps are also 30 min long,* unless otherwise indicated.
6. Replace the solution of 50% v/v ethanol with 70% v/v ethanol that has been precooled to –35°C. Carry out this and all subsequent steps at –35°C using the aluminum block (described in Section 2.2., step 2, and kept in an appropriate freezer, unless stated otherwise); during the changes of ethanol and resin mixtures, briefly install the block under a hood at room temperature, in order to avoid inhalation of the resin vapors, subsequently returning the block to the –35°C freezer.
7. Replace the 70% v/v ethanol solution by 100% ethanol, twice.
8. Replace the ethanol solution by a mixture of pure ethanol and Lowicryl resin (one part by volume ethanol to one part by volume resin). (Lowicryl can cause eczema on sensitive persons. All changes of medium should be carried out under a hood and Lowicryl-resistant [latex] gloves should be used.)
9. Replace this mixture by a solution of 1 vol ethanol to 2 vol resin.
10. Now exchange the mixture twice with pure resin, and allow the resin to infiltrate overnight, after the second change.
11. The next day, fill the 0.5-mL gelatin capsules almost to the top, with precooled pure resin. This is best achieved if the capsules are installed in the same aluminum block used for the previous manipulations.
12. Transfer the samples to the capsules using a toothpick. Close the capsules and install them in the freezer, under the UV lamp.
13. Allow polymerization to take place under indirect UV light, at 360 nm, –35°C, for 24–48 h.
14. Remove the capsules from the freezer and allow polymerization to continue for an additional 2–3 d at room temperature.
15. The embedded material is now ready for sectioning: Lowicryl HM-20 resins can be sectioned with glass or diamond knives at a speed of 2–5 mm/s. The resin is hydrophobic and its sectioning properties closely resemble epoxy embedding media such as Epon or Araldite. Collect the sections on clean (*see* Note 5) 400-mesh EMBRA nickel grids.

3.1.3. Cryo-Ultramicrotomy

1. Pellet the cell suspension, as prepared in Section 3.1.1. in an Eppendorf centrifuge, and resuspend in an equal volume of 10% w/v melted gelatin at 37°C.
2. Allow to solidify, then remove the pellet from the tube and section a few small pyramids, with a base of 1 mm² using a razor blade.
3. Soak the pyramids in a solution of $2.3 M$ sucrose for 30 min and place on the specimen holder of the freezing apparatus.

Immunoelectron Microscopy

4. Take the holder with forceps and drop from a height of approx 30 cm into liquid nitrogen.
5. When the bubbling of the gas stops, the specimen is ready for transfer to the cryochamber of the freeze microtome, previously adjusted to a temperature of $-120°C$.
6. Section using a glass knife at a temperature of $-100°C$.
7. Assemble sections with an eyelash and retrieve by touching them with a drop of $2.3M$ sucrose that is contained in a platinum loop of 2-mm diameter. This action needs some practice and should be carried out calmly but rapidly, before the sucrose solution freezes.
8. Transfer the sections, floating at the lower surface of the sucrose drop, to a film-coated nickel grid of 150 mesh, by touching the grid with the drop.
9. Using nonmagnetic forceps, place the grid, with the drop of sucrose attached to it, upside down, on a Petri dish containing a 2% w/v gelatin gel at room temperature.

3.2. Immunodecoration

3.2.1. Immunolabeling of Lowicryl Sections (see Notes 6 and 7)

1. Incubate the grids for 5 min on a drop of PBS-BSA without antibody.
2. Carefully transfer the grids for 2–3 h to the surface of a drop of PBS-BSA containing the antibody (*see* Note 8).
3. Transfer the grids to a drop of PBS-BSA and incubate for at least 1 min. Repeat this step five times.
4. Incubate the section for 1 h in PBS-BSA in which the protein-A gold has been diluted. (*see* Note 9).
5. Repeat step 3; when grids have been incubated on the last drop of washing buffer, "jet wash" the grid by a mild spray of distilled water from a spray bottle (*see* Note 10).
6. Allow the grids to dry for 30 min and then transfer them to a drop of 3% w/v aqueous uranyl acetate *(care)* for 5 min.
7. Again "jet wash" the grids for several seconds and then air dry them.

The sections are now ready to be examined under the electron microscope (*see* Note 11).

3.2.3. Immunolabeling of Frozen Sections

1. Warm the Petri dish with 2% w/v gelatin, containing the grids floating on drops of sucrose (from Section 3.1.3., step 9), to 35–40°C under a lamp—allowing the gelatin to melt. Collect the grids in the center of the dish and incubate on the melted gelatin to 10–30 min.

2. Transfer a grid to a drop of PBS-glycine for 3 min; repeat this step twice.
3. Incubate the grid for 1 h on a drop of PBS-BSA that contains the relevant antibody at an appropriate dilution: As a guideline, the dilution of monoclonal or polyclonal antibodies is approx 10–50 μg of specific antibody per mL.
4. Transfer the grid to a drop of PBS-glycine for 1 min; repeat this step 5 times.

    After stage 4, if the primary antibody is a mouse monoclonal antibody, an additional incubation should be done with an affinity-purified rabbit anti-mouse Ig antiserum. The dilution of this antiserum is approx 10 μg/mL of specific antibody, the incubation time is 1 h and is followed by a washing step as described herein.
5. Transfer the grid to a drop of PBS-BSA in which the protein-A-gold has been diluted (see Note 12): The incubation time is 20–30 min.
6. This step is identical to step 4; follow by three incubations of 1 min each on 3 drops of distilled water.
7. The contrasting step is combined with embedding in methylcellulose (8): Incubate the grids at 4°C for 10 min in methylcellulose/uranylocetate, then remove using a platinum loop that is slightly larger than the grid. Blot the excess of methylcellulose by gently touching the surface of a filter paper. Finally, dry the grid, still in the loop, in the air for a few minutes.

    For a discussion of a number of general issues in immunoelectron microscopy, see Note 13.

4. Notes

1. The electron microscopy procedure described in this chapter concerns the embedding of whole yeast cells. This is made possible by treating the cells with sodium metaperiodate after fixation (see Section 3.1.1., step 5). This treatment allows optimal infiltration of the resin in the cell, probably because of the oxidation of the mannose residues of the cell wall by the periodate. The fixation-periodate treatment is independent of the culture conditions of the cells. It is therefore equally possible to use cells grown to early log phase or stationary cultures, according to the interest of the investigator.
2. The low permeability of the yeast wall toward media used in electron microscopy has been a major drawback in yeast ultramicrotomy. The infiltration of the cell with the resin, and the sectioning properties in particular, are impaired when the cell wall is left intact. This difficulty has been overcome in the past by either (post)-fixing the cells in potassium permanganate, highlighting membrane structures especially *(14,15)*, or by removing the cell wall enzymatically, after fixation with aldehydes *(16,17)*. These two approaches, in combination with embedding in epoxy resins,

have been the standard embedding techniques for yeast cells for many years. Fixation with potassium permanganate prevents immunolocalization studies from being carried out, but can be useful when only morphological information is needed.
3. Permeabilization of the cell wall with periodate as described in Section 3.1.1., can also be applied to embedding in epoxy resins, avoiding the need to remove the cell wall. A short protocol for embedding whole yeast cells in Araldite is therefore included here; it should be emphasized that although immunolocalization on epoxy-embedded material is possible, it is not as efficient as on Lowicryl sections *(3)* or on cryosections *(16)*.

The cells are prepared as described in Section 3.1.1. They can be optionally postfixed in 2% w/v osmium tetroxide. The dehydration and infiltration steps are then carried out, at room temperature, as follows:
  a. 30% v/v ethanol, 15 min.
  b. 50% v/v ethanol, 15 min.
  c. 70% v/v ethanol, 15 min.
  d. 95% v/v ethanol, 15 min.
  e. 100% ethanol, 15 min (two changes).
  f. Propylene oxide, 15 min (two changes).
  g. 1 v/1 v propylene oxide/resin mixture, 30 min to overnight.
     The vial should not be covered in order to allow the propylene oxide to evaporate and, as a consequence, to concentrate the resin in the medium.
  h. Pure resin, 30 min.
  i. Pure resin, 30 min.
  j. The agarose blocks containing the cells can now be transferred to gelatin capsules with pure resin, after which polymerization is carried out at 60°C for 3 d.
  k. Sectioning and contrasting with uranyl acetate are done according to standard procedures.
4. The reaction of carbohydrate residues with periodate creates aldehyde groups that are very reactive and could at a later stage interfere with the antibody reaction. Sections 3.1.1., step 6 and Section 3.1.1., step 7 prevent this.
5. Grids without film have to be cleaned periodically in order to keep the grid surface hydrophilic. Sections from resin-embedded material are collected from the water reservoir of the knife by lowering the grid into it. Together with the sections, a drop of water will adhere to the grid and should spread evenly over its entire surface. If this is not the case and the drop collects in the center, it is time to clean the grids in the following way:
  a. Place the grids in a small beaker with a sufficient amount of acetic acid to cover them. **Caution:** This manipulation should be done carefully in a fume hood.

b. Remove the acid and rinse the grids several times with distilled water.
c. Rinse the grids twice with 95% v/v ethanol.
d. Decant the alcohol and transfer the grids with forceps to a dry piece of filter paper in a Petri dish, and allow to dry.
6. Incubation with antibody and washing steps are carried out on drops of PBS-BSA of approx 20 µL on which the grids are floated, with the sections on the underside. Care should be taken not to wet the upperside of the grid. The transfer of a grid from one drop to the next is easily carried out using nonmagnetic forceps and holding the grid by the outer rim.

It is good practice to shake off the drop of buffer that is attached to the grid, by gently bouncing the hand holding the forceps with the grid on the table in order to release the drop. During transfer, the sections should not be allowed to dry, not even for seconds. A convenient incubation chamber consists of a Petri dish of approx 10 cm containing a piece of Parafilm firmly attached to the bottom. It is important when attaching the parafilm not to touch the Parafilm directly with fingers but to use the protective paper to press the film against the bottom of the dish. For each experiment a new piece of Parafilm should be used. Drops of PBS-BSA are laid on the Parafilm. To avoid any possible evaporation during incubation time, a wet piece of blotting paper soaked with water can be positioned beside the Parafilm in the dish, which is then closed.
7. For the correct interpretation of the results, it is important to include in each labeling experiment, sections of cells that do not express the protein of interest and/or to use a preimmune antiserum as a control for specificity.
8. The optimal dilution has to be found out empirically and depends on the affinity of the antibody; it can vary from 1120 to 1/2000. In principle it corresponds to a concentration of 10–50 µg/mL of specific antibody.
9. Since the maximal light absorption of colloidal gold is at 525 nm wavelength, the OD_{525} of a gold dilution giving the strongest signal and the lowest background can be measured as a reference. For example, the best dilution for 15 nm gold or 8 nm gold, on Lowicryl HM 20 sections is: A_{525} = 0.4 or 0.05, respectively.
10. The "jet wash" is important because it eliminates phosphate ions that could interfere with the subsequent contrasting steps.
11. If increased staining contrast is needed, the sections can be stained, in addition, with lead acetate according to Millonig *(13)*, under an atmosphere of nitrogen, for 1 min.
12. The dilution depends on the size of the gold particles and is approx OD_{525} = 0.4 or 0.1 for 15 nm or 8 nm gold sizes, respectively.
13. The fixation-periodate treatment of whole yeast cells as described in this chapter, allow embedding in Lowicryl, Araldite, and cryo-ultramicrotomy

to be carried out as a routine procedure in an electron microscopy laboratory. Recently, the resin LR White has been proposed by Wright and Rine for whole yeast cells *(18)*. The advantages of this resin are its low toxicity and its ease in handling: The embedding and polymerization are carried out at room temperature and the immunoreactivity is good. The localization of proteins with the protein-A gold technique is not always straightforward, the major source of difficulties being related to the sensitivity and the reproducibility of the method. The sensitivity of the immunolabeling procedure itself is a function of several parameters, such as the preservation of the protein during fixation, the level of expression of the protein in vivo and the quality of the antiserum. The absence of a signal can be owing to the destruction of the antigenic sites during the fixation step. The crosslinking by glutaraldehyde can be a reason for such an absence and, in this case, reducing its concentration in the fixation mixture should be considered. More detailed information about fixation can be found in Glauert *(19)*.

Immunolocalization on frozen sections is thought to give a higher signal than on Lowicryl sections *(20,21)*. In order to estimate the sensitivity of the Lowicryl technique, in relation to the level of expression of a given protein, the labeling efficiency of the vacuolar enzyme carboxypeptidase Y (CPY) was calculated *(5)*: This enzyme, which represents 0.3% of total yeast protein, gave a signal of approx five gold particles of $15 \text{ nm}/\mu m^2$ in the vacuole. This signal was more than tenfold the signal found in the cytoplasm of the same cells, or in the vacuole of a strain not producing CPY. If the abundance of a given protein is low, expression of proteins that are encoded by cloned genes can be manipulated to higher levels than found in the wild-type situation, by choosing the right kind of promoter or plasmid. This can result in the localization of molecules that are undetectable in the normal context of the cell. It is, however, important to bear in mind that the overproduction of a protein can produce undesirable effects such as its aggregation or its mislocalization *(22)*. Finally, the quality of the antiserum is also critical. Affinity purification should be considered when crossreactivity is suspected. This can be carried out by affinity chromatography or by eluting the specific antibody from a Western blot *(5)*. Crossreactivity is easily detected on protein blots but also on cells that do not express the protein of interest owing to an inactivated gene, for example *(5)*. Whenever possible, "nonexpressing" cells should be included as control, in an embedding-immunolabeling experiment. It is certainly also good practice to use for each experiment a fresh aliquot of an antiserum that has been kept in a deep freezer for storage. If problems owing to nonspecific binding are encountered, the investigator must try to modify different parameters: adjustment of the dilution of the antibody or gold solutions, shortening of the incubation times, addition of more BSA or Tween-20 to the blocking/washing buffers, and so on.

The immunolabeling protocols described in this chapter use protein-A as an immunomarker. Protein-A binds well to rabbit immunoglobulins but has a lower affinity for certain classes of mouse 1 gs. When mouse monoclonal antibodies are used in the primary incubation step, better results will be obtained by using either a rabbit or goat anti-mouse IgG antibody, conjugated to gold, which can be obtained commercially or prepared in the laboratory *(23)*. It is also possible to include an intermediate incubation step using an affinity-purified rabbit anti-mouse 1 g, as indicated in paragraph 4 or Section 3.2.3.

Another aspect that can represent a problem is the reproducibility of a given signal from one experiment to the next. In this respect it is important to repeat a faithfully as possible the growth conditions in every experiment. A fixation protocol described by Wright and Rine *(18)*, who prefix their cells directly in culture in 1% formaldehyde-1% glutaraldehyde, using a tenfold concentrated fixative, could become a worthwhile alternative to the protocol described in this chapter, which will certainly favor reproducibility.

Finally, it is to be hoped that this chapter will encourage ultrastructural studies of yeast, as these continue to illuminate the molecular biology of these important and interesting organisms *(24)*.

Acknowledgments

I would like to thank K. M. van Tuinen-Whitburn, L. A. Grivell, C. L. Woldring, and F. M. Klis for carefully reading the manuscript. I am grateful to E. C. M. Hoefsmit, in whose laboratory the cryo-ultramicrotomy was carried out, and B. Roberts and G. Scholten for making their E. M. facilities available. I also thank A. F. Franzusoff for sending the AEM 67 ll-B *sec*-7 strain.

References

1. Carlemalm, E., Garavito, M., and Villiger, W. (1982) Resin development for electron microscopy and an analysis of embedding at low temperature. *J. Microsc.* **126,** 123–143.
2. Tokuyasu, K. T. (1973) A technique for ultracryotomy of cell suspensions and tissues. *J. Cell Biol.* **57,** 551–565.
3. Roth, J., Bendayan, M., Carlemalm, E., Villiger, E., and Garavito, M. (1981) Enhancement of structural presentation and immunocytochemical staining in low temperature embedded pancreatic tissue. *J. Histochem. Cytochem.* **29,** 663–671.
4. Geuze, H. J., Slot, J. W., van der Ley, P. A., and Scheffer, R. C. T. (1981) Use of colloidal gold particles in double-labelling immunoelectron microscopy of frozen tissue sections. *J. Cell Biol.* **89,** 653–665.

5. van Tuinen, E. and Riezman, H. (1987) Immunolocalization of glyceraldehyde-3-phosphate dehydrogenase, hexokinase and carboxypeptidase Y in yeast cells at the ultrastructural level. *J. Histochem Cytochem.* **35**, 327–333.
6. Tokuyasu, K. T. and Singer, S. J. (1976) Improved procedures for immunoferritin labelling of ultrathin frozen sections. *J. Cell Biol.* **71**, 894–906.
7. Griffith, G., Brands, R., Burke, B., Louvard, D., and Warren, G. (1982) Viral membrane proteins acquire galactose in trans golgi cisternae during intracellular transport. *J. Cell Biol.* **95**, 781–792.
8. Griffiths, G., McDowall, A., Back, R., and Dulbochet, J. (1984) On the preparation of cryosections of immunocytochemistry. *J. Ultrastruct. Res.* **89**, 65–78.
9. Roth, J. (1989) Postembedding labelling on Lowicryl K4M tissue sections: detection and modification of cellular components, in *Meth. Cell Biol.* **31**, 513–551.
10. Griffith, G., Simons, K., Warren, G., and Tokuyasu, K. T. (1983) Immunoelectron microscopy using thin, frozen sections: application to studies of the intracellular transport of semliki forest virus spike glyroproteins. *Methods Enzymol.* **96**, 466–483.
11. Slot, J. W. and Geuze, H. J. (1985) A new method of preparing gold probes for multiple-labelling cytochemistry. *Eur. J. Cell Biol.* **38**, 87–93.
12. Slot, J. W., Geuze, H. J., and Weerkamp, A. H. (1988) Localization of macromolecular components by application of the immunogold technique on cryosectioned bacteria. *Methods Microbiol.* **20**, 211–236.
13. Millonig, G. (1961) A modified procedure for lead staining of thin sections. *J. Biophys. Biochem.* **11**, 736–739.
14. Stevens, B. J. (1977) Variation in number and volume of the mitochondria in yeast according to growth conditions. *Biol. Cellulair* **28**, 37–56.
15. Stevens, B. (1981) The molecular biology of the yeast *Saccharomyces*: life cycle and inheritance (Strathern, J. N., Jones, E. W., and Broach, J. R., eds.), Cold Spring Harbor Laboratory, Cold Spring Harbor, NY, pp. 471–504.
16. Byers, B. and Goetsch, L. (1975) Behaviour of spindles and spindle plaques in the cell cycle and conjugation of *Saccharomyces cerevisiae*. *J. Bacteriol.* **24**, 511–523.
17. Byers, B. (1982) Cytology of the yeast life cycle, in *The Molecular Biology of the Yeast* Saccharomyces: *Life Cycle and Inheritance* (Strathern, J. N., Jones, E. W., and Broach, J. R., eds.), Cold Spring Harbor Laboratory, Cold Spring Harbor, NY, pp. 59–96.
18. Wright, R. and Rine, J. (1989) Transmission Electron Microscopy and immunocytochemical Studies of Yeast, in *Methods in Cell Biology*, Academic, San Diego, CA, vol. 31, pp. 473–512.
19. Glauert, M. (1974) *Practical Methods in Electron Microscopy*, vol. 3, North Holland, Amsterdam, pp. 1–110.
20. Griffith, G. and Hoppeler, H. (1986) Quantitation in immunocyto-chemistry: correlation of immunogold labelling to absolute number of membrane antigens. *J. Histochem. Cytochem.* **34**, 1389–1398.
21. Kellenberger, E., Durrenberger, M., Villiger, W., Carlemalm, E., and Wurtz, M. (1987) The efficiency of immunolabel on Lowicryl sections compared to theoretical predictions. *J. Histochem Cytochem.* **35**, 959–969.

22. Stevens, T. H., Rotheman, J. H., Payne, G. S., and Schekman, R. (1986) Gene dosage-dependent secretion of yeast vacuolar carboxypeptidase Y. *J. Cell Biol.* **102,** 1551–1557.
23. Slot, J. W. and Geuze, H. (1984) Gold markers of single and double immunolabelling of ultrathin cryosections, in, *Immunolabelling for electron microscopy* (Polak, J. M. and Varndell, I. M., eds.), Elsevier, Amsterdam, pp. 129–142.
24. Webster, D. L and Watson, K. (1993) Ultrastructural changes in yeast following heat shock and recovery. *Yeast* **9,** 1165–1175.

Index

A

α amylase in genomic DNA isolation, 165
(A + T)-rich DNA, 109
Adenine auxotrophs, in Schizosaccharomyces pombe, 344
Agarose gel electrophoresis, 166, 178, 179
  buffers for electrophoresis, 158, 179, 180
  DNA fractionation ranges for different gel strengths, 178
  DNA size markers, 179
  ethidium bromide, for staining DNA in gels, 158, 166 and disposal of, 180
  gel preparation and running, 178, 179
  loading of gels and gel loading "buffers," 158, 166, 179
  photography of agarose gels, 180, 181
  transilluminator, use in viewing ethidium bromide-stained DNA gels, 158, 180
Agarose plugs (containing yeast chromosomal DNA), 72–74
Alkaline phosphatase, 182
Aneuploidy, *see* Chromosome ploidy
Ascus dissection, 59–67, *see also* Meiotic analysis, 51–58
Autofluorescence, 391
Autolysis, *see* Lysis of yeast cell walls
Automated microdissection, *see* Ascus dissection
Autoradiography of labeled DNA, 84, 85, 200–202, 210, 211
Auxotrophic marker agar plates, 141, *see also* Transformation of yeast cells

B

Batch-type calorimetry, *see* Calorimetry
Biolistic transformation of yeasts, 147–153, *see also* Transformation of yeast cells
  microprojectiles/microparticles, 148–152
  in mitochondrial transformation, 148
  operation of the biolistic device, 151, 152
  preparation of the microprojectiles, 150, 151
  in transformation with linear and circular DNA, 147
  in transformation of stationary phase cells, 147
Bioruptor, 115
bis-benzimide, 109–113
Bradford protein assay, *see* Protein estimation
Braun bead mill, for disrupting yeast cells, 363–382
Braun homogenizer, *see* Lysis of yeast cell walls
Brighteners, *see* Fluorochromes

C

Calorimetry,
  calorimeter set-up, 373, 374
  controlled organisms, use of, 372, 373, 376, 377

423

drug bioassay, 370, 375, 376
  growth curves, drug modified, 376
  media identification, 370, 374, 375
  strain identification, 370, 374, 375
Carcinogenicity testing, see
  Genotoxicity testing
Cascade fusion of protoplasts, 337
Cationic silica beads, see Plasma
  membranes, isolation of
Cell extracts, for soluble protein, 243
Cell-free translation lysate, preparation
  and use, 249–257
Cell-free translation,
  cell disruption and lysate preparation,
    251–253
  Dounce homogenizer use of, 252
  fractionation of lysate, 252, 253
  polysomes, isolation of, 252, 253
  in vitro translation assays, 253–255
  elongation assays with synthetic
    templates, 256
Cesium chloride density gradients,
  109, 173, 174, 186
Carcinogenicity testing, see
  Genotoxicity testing
Chemostat, for growth of
  Schizosaccharomyces pombe,
  346, 347, 351, 252
Chloramphenicol-resistance, use in
  ploidy determination, see
  Chromosome ploidy
Chromatographic resins, in purification
  of Cytochrome P 450, 356–360
Chromosomal integration of probes/
  plasmids, 206, 213
Chromosome engineering, using
  site-specific recombination,
  217–225
Chromosome-length polymorphism
  (CLP), 73
Chromosomes, see Karyotyping and
  Meiotic analysis
Chromosome ploidy, determination
  of, 205–216

Classification, see Isolation and
  identification of yeasts
Clumping of yeast cells, 143
Competent yeast cells, production of,
  see Transformation of yeast cells
Continuous cultures, of
  Schizosaccharomyces pombe, 346
Contour clamped homogenous
  electric fields (CHEF), 70
Cryo-ultramicrotomy, see
  Immunoelectron microscopy
Culture of yeasts, see Maintenance
  and culture of yeasts, 5–15
Curing of killer strains, 336, 337
Cytochrome P 450, purification and
  quantification, 355–366
  microsome preparation, 358, 359
    Braun bead mill, use of, 363,
    364
  purification, 359, 360
    ammonium sulfate
      fractionation, 359
    chromatography, 359, 360, 361,
    365
    PMSF, use of, 358, 361
    DTT, use of, 360
  quantification, 357, 358
    CO-reduced difference spectrum,
    358
Cytoduction, 39–44

D

Deletion, of chromosomal segments,
  see Chromosome engineering
DAPI (4',6'-diamidino-2-
  phenylindole), 109
Denaturing gels, in RNA analysis,
  270, 279, 280
Densitometric scanning of
  autoradiographs, 203, 210–212
Diethylpyrocarbonate, use in
  inactivating RNase, 275
Disruption of yeast cells, see Lysis of
  yeast cell walls

Index

DNA (radio)labeling, by the random primer technique, 83, 84, 198–200
DNA base composition, 89, 90
DNA mobility determination, 186
DNA preparation/isolation from yeast
   for Karyotyping, 72–74
   for purposes other than Karyotyping, 103–107
     small-scale isolation of plasmid DNA, 105, 106, 175, 176
     large-scale isolation of high MWt DNA, 105, 106, 157, 158
   for taxonomy, 89–102
DNA reassociation, 89, 90
   optical reassociation to determine DNA sequence homology, 98
DNA shearing, 97, 98
   using French Pressure Cell, 97, 98
   using sonication, 98
   using a syringe, 98
DNA size markers (λ-*Hind*III), 86
Dominant marker selective plates, 141, *see also* Transformation of yeast cells
Dounce homogenizer in shearing spheroplasts/protoplast, 135
Drug bioassay, *see* Calorimetry

E

E. Coli,
   in DNA clone bank construction, 155
   isolation of plasmid from, 171–174, 184
   transformation of, 170, 183
EDTA stock solution, preparation of, 186
Electrophoretic analysis of yeast proteins, 259–267
   isoelectric focusing gel electrophoresis (IEF), 264–266
   nonequilibrium pH gradient gel electrophoretics (NEPHGE), 259, 266

   radiolabeling and extraction of proteins, 263
   radiolabeled amino acid mixtures, use of, 261
   radiolabeled leucine, use of, 261
   radiolabeled methionine, use of, 261
   SDS polyacrylamide gel electrophoresis (PAGE), 264
Endoplasmic reticulum, isolation of, *see* Microsomes, isolation of
Epifluorescence, *see* Fluorescence microscopy methods
Escherichia coli, see E. coli
Expression vectors for yeast gene cloning, 157

F

Field inversion gel electrophoresis (FIGE), 69, 70
Flocs, dispersion of, 75
Flow calorimetry, *see* Calorimetry
Fluorescence microscopy methods, 391–404
   fluorochrome staining of different cell components, 395, 396
   cell wall, 387
   meiotic nuclei, 396
   mitochondria, 397
   nuclei with acridine orange, 395
   nuclei with mithramycin, 396
   nuclei with DAP1, 396
   staining actin with fluorochrome-labeled phalloidin, 401
   simple labeling, 401
   double labeling, 401
   using fluorochrome-labeled antibodies, 397–400
   double immunolabeling, 400
   fixation, 397
   incubation with antibodies, 399, 400
   permeabilization, 397–399

Fluorescence microscopy methods, 391–404
Fluorochromes, 391–393

G

Galactose induction, 230
Gel electrophoresis, *see* Agarose gel electrophoresis and Karyotyping
Gene library/gene banks, *see* Genomic yeast DNA clone banks
Gene localization on chromosomes, by PFGE, 79–87, *see also* Meiotic analysis
Genetic mapping, *see* Meiotic analysis and Chromosomal localization of genes
Genomic yeast DNA clone banks, 155–186
  assessment of the *E. coli* clone bank, 170, 171, 183
  number of clones needed, 183
  storage of clone bank, 184
  end-filling of vector and genomic DNA, 168, 169, 182
  ligation of vector and genomic DNA, 169, 170
  isolation of desired yeast transformants, 175, 184, 185
  isolation of genomic DNA, suitable for gene library construction, 106, 163–165
  isolation of DNA from lysate, 164, 165
  lysis of protoplasts, 164
  preparation of protoplasts, 163, 164
  isolation of plasmid (vector) DNA from *E. coli*, 171–174
  small scale, by alkaline lysis, 171
  large scale, by alkaline lysis, 172
  purification of plasmid DNA by isopycnic centrifugation using CsCl gradients, 173, 174

  isolation of plasmid DNA from *Saccharomyces cerevisiae* 175, 176
  partial restriction enzyme digestion of genomic DNA, 165–167
  pilot digestion, 165, 166
  scaled-up digestion, 167
  plasmid vectors for cloning yeast genes, 155–157
  pRS, 156
  YAc, 156
  YCp, 156
  YEp, 156
  YIp, 217
  YRp, 156
  shuttle vectors, 155, 156
  size fractionation of DNA, 167
  transformation of *E. coli*, with ligated DNA, 170, 183
  transformation of yeast with library DNA, 174, 175, *see also* Transformation of yeast cells and Biolistic transformation of yeasts
  treatment of purified vector DNA, prior to end-filling, 168
Genotoxicity testing, in *Schizosaccharomyces pombe*, 343–353
  in batch culture, 346–348
  in continuous cultures, 348–350
Germination media, for teliospores of basidiomycetous yeasts, 8, 9
Growth media, 5–7
  acidified malt extract, 1
  anisomycin medium, 345
  cycloheximide medium, 345
  defined growth medium, 371, 372
  Edinburgh minimal medium number 2, 344, 345
  Gieses' salts vitamins buffer, 345
  hay infusion agar, 8
  heat-inducible genes, *see* Induction of heat-shock proteins

minimal, 18
minimal medium (defined), 333
SD (minimal), 241
sucrose-yeast extract agar, 8
supplemented minimal media,
 trichodermin medium, 345
Wickerham's vitamin solution, 333
yeast nitrogen base, 1

H

Heat-inducible promoters, 317
Heat-shock element (HSE), see
 Induction of heat-shock proteins
Hoechst No. 332258, see bis-
 benzimide
Hemocytometer, use of, in cell
 concentration determination, 143
Hemoprotein enzymes, see
 Cytochrome P 450
Hybridization of DNA, 84, 85, 200
Hydroxylapatite, in DNA isolation,
 96, 97
Hybridization of industrial strains,
 see Killer toxins in selection of
 hybrids and cybrids

I

Identification of yeasts, see Isolation
 and identification of yeasts, 1–4
Immunoelectron microscopy, 407–422
 embedding in Araldite, 417
 immunodecoration, 415, 416
 immunolabeling of Lowicryl
  sections, 415
 immunolabeling of frozen
  sections, 415, 416
 permeabilization of the cell wall,
  using periodate, 407, 413–
  416
 sample preparation, 413, 414
 growth and fixation, 413
 dehydration and embedding in
  Lowicryl, 413, 414
 cryo-ultramicrotomy, 414, 415
use of resin LR White, 419

Improved Neubauer hemocytometer,
 see Hemocytometer
Indirect immunofluorescence, see
 Fluorescence microscopy methods
Induction of heat-shock proteins,
 313–317
 induction of thermotolerance, 315
 optimum induction, 314, 315
 induction by temperature, 314
 induction using chemicals, 314,
  315
 temperature-regulated heterologous
  gene expression, 315, 316
Industrial strains, ploidy determination,
 see Chromosome ploidy
Insertional mutagenesis, by Ty
 elements, 227–237
 introduction of pGTg plasmids
  into yeast, 228
 retrotransposition, 230
 ROAM mutations, 233
 spt 3 gene, role in Ty mutagenesis,
  232
 transposition libraries, 232
 Ty mutagenesis (e.g., isolation of
  lys2 and *lys5* mutants), 229
 Ty transposition assay, 229
Inversion, of chromosomal segments,
 see Chromosome engineering
Isolation and identification of
 yeasts, 1–4

K

Kapton disks, in biolistic
 transformation, 148, 149
Karyotyping, 69–78, see also
 Chromosomal localization of
 genes by PFGE, 79–87
Killer plaque technique, in selection
 of hybrids and cybrids, 339–342
 killer plaque technique, 342
 protoplasting and induced
  protoplast fusion, 341

Killer toxin, nystatin-Rhodamine B
    assay, 325–329
Killer toxins in selection of hybrids
    and cybrids, 331–338
    induced fusion of protoplasts, 334,
        335
    protoplasting of parental cells,
        334
    stepwise selection technique, 335
Killer toxins, Rhodamine B assay,
    319–324
    superkiller toxin K1 from *Saccharomyces cerevisiae*, 320
    use of flow cytometer counting, 323
Killer virus-like particles, 39, 42
Klenow enzyme, 182, 186
Kluyveromyces lactis killer toxin,
    see Killer toxin, nystatin-
    Rhodamine B assay, and killer
    toxins in hybrid and cybrid
    selection and Killer plaque
    technique

L

Labeling of DNA, see DNA
    (radiolabeling)
Light microscopy, methods, 383–390
    immobilization of cells, 385
    staining nuclei in fixed cells, 388,
        389
    aceto-orcein staining, 389
    acid fuchsin staining, 388
    fixation, 388
    HCl-Giemsa staining, 389
    vital staining, of unfixed cells,
        386–388
    glycogen, 387
    lipids, 387, 388
    mitochondria, 386
    nucleus, 386
    vacuoles, 386
    volutin, 387
Lithium acetate, in transformation,
    see Transformation of yeast cells

Low-gelling-temperature (LGT)
    agarose, 80, 83, 84
Lowicryl, use of, see Immunoelectron
    microscopy
Lysis of yeast cell walls, 93
    by autolysis, 93
    by dry cell disintegration, 94
    by enzymes, 46, 48, 71–75, 135
        lyticase, 71, 72
        Novozym(e), 157
        zymolyase, 134, 136
    by manual disruption with glass
        beads, 225–256, 263
    by wet cell disintegration using a
        Braun homogenizer, 93

M

Maintenance and culture of years, 5–15
    fed-batch, continuous and high
        density culture, 14
    liquid medium, 13, 14
        fermenters, use of, 13, 14
    solid medium, 13
Medium identification, see
    Calorimetry
Meiotic analysis, 51–58, see also
    Ascus dissection, 59–67
    micromanipulation/
        micromanipulator, 51, 53,
        54, 59–67
    mitotic recombination, 44
Microcalorimetry, see Calorimetry
Microdissection, see Meiotic analysis
    and Ascus dissection
Microsomes, isolation of, 358, 359
Mitochondria, 39, 42, 43
    mitochondrial transfer between
        strains, 42, 43
Mitochondria, isolation of, 112–116
Mitochondrial(mt) DNA, isolation,
    109–116
    using CsCl/*bis*-benzimide, 111, 112
    via a mitochondrial preparation,
        112, 113, 115

Index

Mol % G + C determination, 97
   by buoyant density determination in cesium chloride, 101
   by HPLC, 101
   by optical reassociation, 98
   by thermal denaturation, 97
Molecular taxonomy, 89
mRNA (poly [A] + RNA),
   purification of, 270
mRNA, abundance and half-life measurements, 277–295
   dot-blotting, 279–282
      dot-blotting of RNA samples, 280–287
      filter-hybridization, 281, 282
      gel analysis of RNA samples, 279–290
   half-life measurement, using 1,10-phenanthroline, 287–291
      cell growth and RNA preparation, 289, 290
      mRNA half-life calculation, 291
   half-life measurement, using the *rpb1-1* mutation, 291, 292
   Northern analysis, 282–287
      filter hybridization, 285
      Northern blotting, 283–285
      quantitation, 286, 287
      reprobing filters, 287
mtDNA, *see* Mitochondrial (mt) DNA
Mucor racemosus, mtDNA of, 114
Mutagenesis of yeasts, 17–44, *see also* Insertional mutagenesis
   mutants, 17–38
      antibiotic, 19, 25, 35, 344
      antibiotic-sensitive, 28, 36
      auxotrophic, 23
      auxotrophic for dTMP, 30, 31, 36
      cell wall, 27, 28, 36
      glycolytic, 31, 32, 37
      kar 1-1, 27, 35, 39, 40
      lys2 and *lys5,* 229
      membrane, 27, 36
      mit, 19, 24, 25
      mutator, 19, 25, 26
      pep4, 28, 29
      petite, 17, 19, 23, 34, 35
      secretory, 33, 37
      sulfite-resistant, 37
      temperature-sensitive, 26, 27, 35
      transport, 32, 33
   mutations, induction of, 17–44
      induction by radiation, 18, 34
      induction by chemical mutagens, 18, 34
      acriflavin, 19, 40, 41
      ethidium bromide, 19, 40, 41
      ethyl methane sulfonate (EMS), 17, 18
      manganese chloride, 19
      methyl methane sulfonate (MMS), 18
      nitrosoguanidine ((M)NNG), 18
      nitrous acid (NA), 18
Mutagenicity testing, *see* Genotoxicity testing
Mutations, in relation to taxonomy, 89

N

Nitrocellulose membranes, for DNA hybridization, 86, 189–191
Nuclear DNA isolation and use in yeast taxonomy, 89–102
Nylon membranes, for DNA hybridization, 86, 189–191

O

Orthogonal field alternation gel electrophoresis (OFAGE), 69

P

Pancreatic RNase in genomic DNA isolation, 157
Permeabilization of cytoplasmic membranes, *see* Killer toxins, Rhodamine B assay
Peroxisomes, isolation, 133–138

induction by growth on a fatty
acid, 133–135
stability at low pH, 133–136
Phalloidin, 393
Phenol, use in DNA isolation, 185
Phosphoimager technology, use of,
203, 209, 210
Plasma membranes isolation of,
117–121
ATPase activity, 120, 121
avoiding use of spheroplasts, 120
from spheroplasts, using cationic
silica beads, 119
Plasmid copy number in yeast,
determination of, 193–203
Plasmid DNA, 217–219, *see also*
Genomic yeast DNA clone banks
E. coli miniplasmid πN, 230
pGTy plasmid DNA, 227
pSRI DNA, 218
2 µm DNA, 217
Ploidy, chromosomal, *see*
Chromosome ploidy
Ploidy, plasmid, *see* Plasmid copy
number in yeast
PMSF (phenylmethylsulfonyl
fluoride), 118, 134, 135
Polysome analysis, 297–311
in vivo translatability of mRNAs,
298
mRNA analysis across polysome
gradients, 305–307
fractionation across polysome
gradients for dot blotting,
305
fractionation of polysome
gradients for Northern
analysis, 305–307
polysome isolation on gradients,
from sheared cells, 302–304
polysome isolation on gradients, from
spheroplasted cells, 300–302
polysome preparation, using
sucrose shelves, 307, 308

polysome profiles, related to cell
growth, 297–310
Probe preparation (radiolabeling of
DNA probes), 83, 84
Probes for ploidy determination, *see*
Chromosome ploidy
Programmable autonomously
controlled electrode (PACE), 70
Protoplast fusion, 45–49
G1, arresting cells in, 48
heterokaryon, 45
polyethylene glycol (PEG), 45, 48
protoplast formation, 46, 47
Protein A-gold labeling, *see*
Immunoelectron microscopy
Protein estimation, by Bio-Rad
microassay (Bradford),
243–245
Pulsed field gel electrophoresis
(PFGE), *see* Karyotyping

Q
Quick-seal centrifuge tubes, 134

R
Radiolabeled amino acids, in
analyzing yeast proteins/protein
synthesis, *see* Electrophoretic
analysis of yeast proteins
Radiolabeling of DNA, *see* DNA
(radiolabeling)
Random spore isolation, 52–57
Rare-mating, 39–44
Reaction enthalpies of biological
systems, *see* Calorimetry
Reaction kinetics of Biological
systems, *see* Calorimetry
Red pigmentation of *ade* strains, 351
Reporter gene systems, 239–248
aminoglycoside
phosphetransferase (APT),
239–241
APT enzyme assay, 245, 246
G418 selective medium, 241

Index

β-galactosidase, 239, 240
  β-galactosidase enzyme assay, 244
  β-lactamase, 248
  ONPG, 239
  X-gal, 240
  chloramphenicol acetyl transferase (CAT), 239, 240
  CAT enzyme assay, 244, 245
  chloramphenicol selective medium, 241
  DTNB, 240
  transcriptional fusions, 239
  translational fusions, 239
Restriction enzymes use in constructing and analyzing recombinants, 177, 178
  buffers for restriction enzymes, 178
Restriction enzymes, use in cleaving genomic DNA, 165–167, 197, 198, 209, 210
Rhodamine B, *see* Killer toxins, Rhodamine B assay
Rhodamine B assay, and Killer toxins in hybrid and cybrid selection, and Killer plaque technique
Ribosome, role in polysomes, 297
RNA, preparation of, *see* Total RNA, preparation of
RNase, removal of contamination, 269, 274, 275, 292–294, 308, 309
RNase A, *see* Pancreatic RNase

S

Saponification, 123
Scintillation counting, errors in, 215
Sectored colonies, of *Schizosaccharomyces pombe,* 351
Selection of hybrids and cybrids, *see* Killer toxins in selection of hybrids and cybrids

Selection of transformants, 139–145
  using antibiotic resistance markers, 140–143
  chloramphenicol, 241
  endogenous resistance to G418, 231
  G418, 241
  heavy metals, 140
  hygromycin B, 140
  methotrexate, 140
  sulfometuron methyl, 140
  using auxotrophic markers, 139, 142
Self-ligation of vector, prevention of
  by dephosphorylation, 182
  by end-filling, 182
Shuttle plasmids, 104, 106, 193, 194
Site-specific recombination, *see* Chromosome engineering
Sodium iodide gradients, 115
Southern blotting, 82, 83, 194–196
Spheroplasts, 72–74
Sporulation media, 7, 8
  cornmeal agar, 8
  dilute V8 agar, 7
  Gorodkowa agar, 7
  gypsum blocks, 8
  McClary's medium, 7
  potato agar, 8
  raffinose-acetate medium, 7
  sodium acetate medium, 7
  vegetable juice agar, 7
  vegetable wedges, 8
Sporulation, 7–9, 52, 53, 56, 57
  ballistospore formation, 15
Stationary phase cells, thermotolerance of, 316
Sterols, extraction and analysis of, 123–131
  biological roles, 123
  episterol, 123
  ergosterol, 123
  fecosterol, 123
  fluconazole, effects on sterols, 126–130

gas chromatography mass spectrometry (GCMS), in analysis of, 124–130
HPLC, in identification of sterols, 124
mass fragmentation patterns of common fungal sterols, derivatized, 127–129
UV spectrophotometry, in analysis of, 123–125
Storage of yeasts, 10–12
  long term, 11–13
    freezing in glycerol solution, 12
    freezing in liquid nitrogen, 12
    lyophilization, 11, 12
    of *Schizosaccharomyces pombe*, 350
    silica gel, 14
    viability in storage, 14
  medium term, 10, 11
    drying on filter paper, 10, 11
    under mineral oil, 10
  short term of *Schizosaccharomyces pombe*, 350
Strain identification, *see* Calorimetry
Stringency of hybridization, 87
Sucrose density gradients
  continuous, for DNA fractionation, 167, 181, 182
    construction of sucrose density gradients, 181
    fractionation of sucrose density gradients, 181, 182
  continuous, for polysome analysis, 299
    fraction of polysome gradients, 302
  discontinuous, 118, 120
    in peroxisome isolation, 134–138
    in plasma membrane isolation, 118, 120
    in polysome isolation, 307, 308

T
T1 RNase in genomic DNA isolation, 158
Taxonomy, *see* Isolation and identification of yeasts
Tetrad analysis, *see* Meiotic analysis and Ascus dissection
Thermotolerance, *see* Induction of heat-shock proteins
Total RNA, preparation of, 269–276
  cell storage, 271
  large-scale RNA preparation, 272, 273
  small-scale RNA preparation, 273, 274
Transformation of yeast cells, 139–145, *see also* Biolistic transformation of yeast
  by electroporation, 144
  by lithium treatment, 139–145
Translation, *see* Cell-free translation lysate
Transposition, *see* International mutagenesis
Transverse alternating field electrophoresis (TAFE), 70
Tris-HCL stock buffer, preparation, 186
TRPI, yeast gene encoding N-(5-phosphoribosyl) anthranilate isomerase, 157
Ty elements, *see* Insertional mutagenesis

V
Vectors, in gene cloning, 193, 194, *see also* Genomic yeast DNA clone banks
Vertical rotors, 186
  in peroxisome isolation, 135

X
Xenobiotic compounds, metabolism of, *see* Cytochrome P 450, 356–360

Index

Y
Yeast colony hybridization, 189–192
Yeast(s)
  growth media, 5–7
    5-fluoro-orotic acid medium, 228
    G418 medium, 228
    GAL-indicator medium, 227, 228
    growth factor mixture, 6
    L-α-aminoadipic acid medium, 228
    malt agar, 5
    MYGP, 5
    Saboraud's glucose broth, 6
    synthetic complete medium, 228
    trace elements mixture, 6
    Wickerham's chemically defined medium, 6
    yeast extract-peptone-dextrose agar, 6
  identification, 3, 4
  isolation, 1, 2
  maintenance and culture, 5–15
  species
    Brettanomyces spp, 3
    Bullera spp, 15
    Candida albicans, 103, 104, 377
    Candida mogii, 3
    Candida parapsilosis, 100
    Candida sonorensis, 2
    Candida spp, 2
    Candida tsukubaensis, 156
    Candida utilis, 2
    Citeromyces spp, 2
    Clavispora spp, 2
    Cryptococcus spp, 2, 99
    Cryptococcus cereanus, 2
    Cryptococcus neoformans, 148
    Cyniclomyces guttulatus, 3
    Debaryomyces hansenii, 2
    Hansenula polymorpha, 103, 104
    Hansenula spp, 2
    Kluyveromyces fragilis, 380
    Kluyveromyces lactis, 325–342
    Kluyveromyces spp, 99
    Lipomyces starkeyi, 157, 177
    Metschikowia bicuspidata, 3
    Pichia cactophila, 2
    Pichia mexicana, 2
    Pichia norvegensis, 2
    Pichia pseudocactophila, 2
    Pichia spp, 2
    Saccharomyces cerevisiae, throughout text
    Schizosaccharomyces japonicus var versatilis, 402, 403
    Schizosaccharomyces octosporus, 3
    Schizosaccharomyces pombe, 343–353
    Schizosaccharomyces spp, 99
    Schwanniomyces occidentalis, 156
    Sporobolomyces, 15
    Torulopsis spp, 4
    Yarrowia lipolytica, 156
    Zygosaccharomyces rouxii, 218
    Zygosaccharomyces spp, 3
Yeast integrative vectors (YIp), in chromosome engineering, 217

Z
Zymolyase, 119, *see also* Lysis of yeast cell walls

Methods in Molecular Biology™

Methods in Molecular Biology™ manuals are available at all medical bookstores. You may also order copies directly from Humana by filling in and mailing or faxing this form to: Humana Press, 999 Riverview Drive, Suite 208, Totowa, NJ 07512 USA, Phone: 201-256-1699/Fax: 201-256-8341.

- [ ] 60. **Protein NMR Protocols**, edited by *David G. Reid*, 1996 • 0-89603-309-0 • Comb $69.50 (T)
- [ ] 59. **Protein Purification Protocols**, edited by *Shawn Doonan*, 1996 • 0-89603-336-8 • Comb $64.50 (T)
- [ ] 58. **Basic DNA and RNA Protocols**, edited by *Adrian J. Harwood*, 1996 • 0-89603-331-7 • Comb $69.50 • 0-89603-402-X • Hardcover $99.50
- [ ] 57. **In Vitro Mutagenesis Protocols**, edited by *Michael K. Trower*, 1996 • 0-89603-332-5 • Comb $69.50
- [ ] 56. **Crystallographic Methods and Protocols**, edited by *Christopher Jones, Barbara Mulloy, and Mark Sanderson*, 1996 • 0-89603-259-0 • Comb $69.50 (T)
- [ ] 55. **Plant Cell Electroporation and Electrofusion Protocols**, edited by *Jac A. Nickoloff*, 1995 • 0-89603-328-7 • Comb $49.50
- [ ] 54. **YAC Protocols**, edited by *David Markie*, 1995 • 0-89603-313-9 • Comb $69.50
- [ ] 53. **Yeast Protocols: *Methods in Cell and Molecular Biology***, edited by *Ivor H. Evans*, 1996 • 0-89603-319-8 • Comb $74.50 (T)
- [ ] 52. **Capillary Electrophoresis: *Principles, Instrumentation, and Applications***, edited by *Kevin D. Altria*, 1996 • 0-89603-315-5 • Comb $74.50
- [ ] 51. **Antibody Engineering Protocols**, edited by *Sudhir Paul*, 1995 • 0-89603-275-2 • Comb $69.50
- [ ] 50. **Species Diagnostics Protocols: *PCR and Other Nucleic Acid Methods***, edited by *Justin P. Clapp*, 1996 • 0-89603-323-6 • Comb $69.50
- [ ] 49. **Plant Gene Transfer and Expression Protocols**, edited by *Heddwyn Jones*, 1995 • 0-89603-321-X • Comb $69.50
- [ ] 48. **Animal Cell Electroporation and Electrofusion Protocols**, edited by *Jac A. Nickoloff*, 1995 • 0-89603-304-X • Comb $64.50
- [ ] 47. **Electroporation Protocols for Microorganisms**, edited by *Jac A. Nickoloff*, 1995 • 0-89603-310-4 • Comb $64.50
- [ ] 46. **Diagnostic Bacteriology Protocols**, edited by *Jenny Howard and David M. Whitcombe*, 1995 • 0-89603-297-3 • Comb $69.50
- [ ] 45. **Monoclonal Antibody Protocols**, edited by *William C. Davis*, 1995 • 0-89603-308-2 • Comb $64.50
- [ ] 44. **Agrobacterium Protocols**, edited by *Kevan M. A. Gartland and Michael R. Davey*, 1995 • 0-89603-302-3 • Comb $64.50
- [ ] 43. **In Vitro Toxicity Testing Protocols**, edited by *Sheila O'Hare and Chris K. Atterwill*, 1995 • 0-89603-282-5 • Comb $69.50
- [ ] 42. **ELISA: *Theory and Practice***, by *John R. Crowther*, 1995 • 0-89603-279-5 • Comb $59.50
- [ ] 41. **Signal Transduction Protocols**, edited by *David A. Kendall and Stephen J. Hill*, 1995 • 0-89603-298-1 • Comb $64.50
- [ ] 40. **Protein Stability and Folding: *Theory and Practice***, edited by *Bret A. Shirley*, 1995 • 0-89603-301-5 • Comb $69.50
- [ ] 39. **Baculovirus Expression Protocols**, edited by *Christopher D. Richardson*, 1995 • 0-89603-272-8 • Comb $64.50
- [ ] 38. **Cryopreservation and Freeze-Drying Protocols**, edited by *John G. Day and Mark R. McLellan*, 1995 • 0-89603-296-5 • Comb $79.50
- [ ] 37. **In Vitro Transcription and Translation Protocols**, edited by *Martin J. Tymms*, 1995 • 0-89603-288-4 • Comb $69.50
- [ ] 36. **Peptide Analysis Protocols**, edited by *Ben M. Dunn and Michael W. Pennington*, 1994 • 0-89603-274-4 • Comb $64.50
- [ ] 35. **Peptide Synthesis Protocols**, edited by *Michael W. Pennington and Ben M. Dunn*, 1994 • 0-89603-273-6 • Comb $64.50
- [ ] 34. **Immunocytochemical Methods and Protocols**, edited by *Lorette C. Javois*, 1994 • 0-89603-285-X • Comb $64.50
- [ ] 33. **In Situ Hybridization Protocols**, edited by *K. H. Andy Choo*, 1994 • 0-89603-280-9 • Comb $69.50
- [ ] 32. **Basic Protein and Peptide Protocols**, edited by *John M. Walker*, 1994 • 0-89603-269-8 • Comb $59.50 • 0-89603-268-X • Hardcover $89.50
- [ ] 31. **Protocols for Gene Analysis**, edited by *Adrian J. Harwood*, 1994 • 0-89603-258-2 • Comb $69.50
- [ ] 30. **DNA–Protein Interactions**, edited by *G. Geoff Kneale*, 1994 • 0-89603-256-6 • Paper $64.50
- [ ] 29. **Chromosome Analysis Protocols**, edited by *John R. Gosden*, 1994 • 0-89603-243-4 • Comb $69.50 • 0-89603-289-2 • Hardcover $94.50
- [ ] 28. **Protocols for Nucleic Acid Analysis by Nonradioactive Probes**, edited by *Peter G. Isaac*, 1994 • 0-89603-254-X • Comb $59.50
- [ ] 27. **Biomembrane Protocols: *II. Architecture and Function***, edited by *John M. Graham and Joan A. Higgins*, 1994 • 0-89603-250-7 • Comb $64.50
- [ ] 26. **Protocols for Oligonucleotide Conjugates: *Synthesis and Analytical Techniques***, edited by *Sudhir Agrawal*, 1994 • 0-89603-252-3 • Comb $64.50
- [ ] 25. **Computer Analysis of Sequence Data: *Part II***, edited by *Annette M. Griffin and Hugh G. Griffin*, 1994 • 0-89603-276-0 • Comb $59.50
- [ ] 24. **Computer Analysis of Sequence Data: *Part I***, edited by *Annette M. Griffin and Hugh G. Griffin*, 1994 • 0-89603-246-9 • Comb $59.50
- [ ] 23. **DNA Sequencing Protocols**, edited by *Hugh G. Griffin and Annette M. Griffin*, 1993 • 0-89603-248-5 • Comb $59.50
- [ ] 22. **Microscopy, Optical Spectroscopy, and Macroscopic Techniques**, edited by *Christopher Jones, Barbara Mulloy, and Adrian H. Thomas*, 1993 • 0-89603-232-9 • Comb $69.50
- [ ] 21. **Protocols in Molecular Parasitology**, edited by *John E. Hyde*, 1993 • 0-89603-239-6 • Comb $69.50
- [ ] 20. **Protocols for Oligonucleotides and Analogs: *Synthesis and Properties***, edited by *Sudhir Agrawal*, 1993 • 0-89603-247-7 • Comb $69.50 • 0-89603-281-7 • Hardcover $89.50
- [ ] 19. **Biomembrane Protocols: *I. Isolation and Analysis***, edited by *John M. Graham and Joan A. Higgins*, 1993 • 0-89603-236-1 • Comb $64.50
- [ ] 18. **Transgenesis Techniques: *Principles and Protocols***, edited by *David Murphy and David A. Carter*, 1993 • 0-89603-245-0 • Comb $69.50
- [ ] 17. **Spectroscopic Methods and Analyses: *NMR, Mass Spectrometry, and Metalloprotein Techniques***, edited by *Christopher Jones, Barbara Mulloy, and Adrian H. Thomas*, 1993 • 0-89603-215-9 • Comb $69.50
- [ ] 16. **Enzymes of Molecular Biology**, edited by *Michael M. Burrell*, 1993 • 0-89603-322-8 • Paper $59.50
- [ ] 15. **PCR Protocols: *Current Methods and Applications***, edited by *Bruce A. White*, 1993 • 0-89603-244-2 • Paper $54.50
- [ ] 14. **Glycoprotein Analysis in Biomedicine**, edited by *Elizabeth F. Hounsell*, 1993 • 0-89603-226-4 • Comb $64.50
- [ ] 13. **Protocols in Molecular Neurobiology**, edited by *Alan Longstaff and Patricia Revest*, 1992 • 0-89603-199-3 • Comb $59.50
- [ ] 12. **Pulsed-Field Gel Electrophoresis: *Protocols, Methods, and Theories***, edited by *Margit Burmeister and Levy Ulanovsky*, 1992 • 0-89603-229-9 • Hardcover $69.50
- [ ] 11. **Practical Protein Chromatography**, edited by *Andrew Kenney and Susan Fowell*, 1992 • 0-89603-213-2 • Hardcover $59.50
- [ ] 10. **Immunochemical Protocols**, edited by *Margaret M. Manson*, 1992 • 0-89603-270-1 • Comb $69.50
- [ ] 9. **Protocols in Human Molecular Genetics**, edited by *Christopher G. Mathew*, 1991 • 0-89603-205-1 • Hardcover $69.50
- [ ] 8. **Practical Molecular Virology: *Viral Vectors for Gene Expression***, edited by *Mary K. L. Collins*, 1991 • 0-89603-191-8 • Paper $54.50
- [ ] 7. **Gene Transfer and Expression Protocols**, edited by *Edward J. Murray*, 1991 • 0-89603-178-0 • Hardcover $79.50
- [ ] 6. **Plant Cell and Tissue Culture**, edited by *Jeffrey W. Pollard and John M. Walker*, 1990 • 0-89603-161-6 • Comb $59.50
- [ ] 5. **Animal Cell Culture**, edited by *Jeffrey W. Pollard and John M. Walker*, 1990 • 0-89603-150-0 • Comb $69.50

Name _____

Department _____

Institution _____

Address _____

City/State/Zip _____

Country _____

Phone # _____ Fax # _____

"T" denotes a tentative price. Prices listed are Humana Press prices, current as of October 1995, do not reflect the prices at which books will be sold to you by suppliers other than Humana Press. All prices subject to change without notice.

UK, Europe, Middle East, and Africa: Order directly from Chapman & Hall by faxing to: +44-171-522-9623.

Postage & Handling: *USA Prepaid (UPS):* Add $4.00 for the first book and $1.00 for each additional book. *Outside USA (Surface):* Add $5.00 for the first book and $1.50 for each additional book.

- [ ] My check for $ _____ is enclosed
  (Drawn on US funds from a US bank).
- [ ] Visa   - [ ] MasterCard   - [ ] American Express

Card # _____

Exp. date _____

Signature _____